Applications
and Investigations
in Earth Science

Applications and Investigations in Earth Science

FIFTH EDITION

Edward J. Tarbuck

Frederick K. Lutgens

Kenneth G. Pinzke

Illustrations by Dennis Tasa

PEARSON

Prentice Hall

Upper Saddle River, NJ 07458

Library of Congress Cataloging-in-Publication Data

Tarbuck, Edward J.
    Applications & investigations in Earth science / Edward J. Tarbuck, Frederick K.
  Lutgens, Kenneth G. Pinzke. —5th ed. / illustrations by Dennis Tasa.
        p. cm.
      ISBN 0-13-149754-5
      1. Earth sciences—Laboratory manuals. I. Title: Applications and investigations in Earth
  science. II. Lutgens, Frederick K. III Pinzke, Kenneth G. IV. Title.
QE44.T372006
550'.78—dc22   2005045937

Executive Editor: *Patrick Lynch*
Executive Managing Editor: *Kathleen Schiaparelli*
Production Editor: *Edward Thomas*
Managing Editor, Science Media and Supplements:
    Nicole M. Jackson
Media Editor: *Chris Rapp*
Director of Marketing, Science: *Linda Taft-Mackinnon*
Manufacturing Buyer: *Alan Fischer*
Director of Creative Services: *Paul Belfanti*
Art Director: *Heather Scott*
Interior Designer: *Susan Anderson-Smith*
Cover Designer: *Tamara Newnam*
Copy Editor: *Barbara Booth*
Editorial Assistant: *Sean Hale*
AV Manager: *Abigail Bass*

Development Editor: *Anne Scanlan-Rohrer*
Marketing Assistant: *Larry Grodsky*
Photo Research Administrator: *Melinda Reo*
Proofreader: *Alison Lorber*
Image Permission Coordinator: *Debbie Hewitson*
Color Scanning Supervisor: *Joseph Conti*
Production Assistant: *Nancy Bauer*
Composition: Laserwords
Senior Manager, Artworks: *Patty Burns*
Production Manager, Artworks: *Ronda Whitson*
Manager, Production Technologies, Artworks: *Matt Haas*
Project Coordinator, Artworks: *Jessica Einsig*
Illustrator, Artworks: *Mark Landis*
Illustrations by: *Dennis Tasa, Tasa Graphic Arts*
Cover Photo Credit: *Tom and Susan Bean, Inc.*

© 2006, 2003, 2000, 1997, 1994 by Pearson Education, Inc.
Pearson Prentice Hall
Pearson Education, Inc.
Upper Saddle River, New Jersey 07458

Pearson Prentice Hall™ is a trademark of Pearson Education, Inc.

Printed in the United States of America

10 9 8 7 6 5 4 3

ISBN 0-13-149754-5

Pearson Education Ltd., *London*
Pearson Education Australia Pty., Limited, *Sydney*
Pearson Education *Singapore*, Pte. Ltd
Pearson Education North Asia Ltd., *Hong Kong*
Pearson Education Canada, Ltd., *Toronto*
Pearson Educación de *Mexico*, S.A. de C.V.
Pearson Education—Japan, *Tokyo*
Pearson Education Malaysia, Pte. Ltd

I hear and I forget.
I see and I remember.
I do and I understand.

—ANCIENT PROVERB

All lives are precious, but those that end suddenly and too soon are remembered forever in a special place in our minds and hearts. This manuscript is dedicated in loving memory to one of those lives. His name is Tristan. He will be missed.

Contents

Preface

Earth is a very small part of a vast universe, but it is our home. It provides the resources to carry on a modern society and the ingredients necessary to support life. Therefore, a knowledge and understanding of our planet is critical to our economic and social welfare and, indeed, vital to our survival.

In recent years, media reports have made us increasingly aware of our place in the universe and the forces at work in our physical environment, as well as how human interaction with natural systems can often upset a delicate equilibrium. News stories inform us of new discoveries in the solar system and beyond. Daily reports remind us of the destruction created by hurricanes, flooding, mudflows, and earthquakes. We have been made aware of the depletion of ozone, an increase in carbon dioxide in the atmosphere caused by human activities, and growing environmental concerns over the fate of the oceans. To comprehend, prepare for, and solve these and other concerns requires a full awareness of how science is done and the scientific principles that govern the operation of Earth.

To achieve scientific understanding requires not only a knowledge and appreciation of broad scientific theories but also the ability to gather scientific data and to solve problems creatively with critical reasoning skills. As part of the educational process, laboratory experience is essential to developing these competencies. In the laboratory, nature can be explored, principles examined, and hypotheses developed, tested, and compared with experimental data. It is often in the laboratory that understanding begins.

Applications and Investigations in Earth Science is an introductory-level laboratory manual consisting of twenty-two exercises designed to examine many of the basic principles of geology, meteorology, oceanography, and astronomy. The format of each of the exercises is the same. Each investigation includes

Statement of Purpose that outlines the major topic(s) to be investigated in the exercise.

Objectives that clearly state what should be learned by completing the exercise.

Materials other than the text, paper, and pencil that are needed and supplied by the student. Also indicated are those materials that are supplied by the instructor and should be located in the laboratory.

Terms are the key words that are shown in bold type in order of appearance in the exercise. In most cases the terms are defined within the exercise.

Introduction that briefly discusses the material to be investigated.

Questions and Problems that investigate and apply the principles examined in the exercise. The questions require various levels of understanding and build from basic comprehension to more demanding applications. When necessary, adequate space has been provided after each question so that the answer can be recorded for future reference.

Investigations using the Internet explores the vast World Wide Web (WWW) of resources that are available to conduct both historical and real-time Earth science studies. These web-based activities are found on the manual's companion website located at http://prenhall.com/earthsciencelab

Summary/Report Page located at the end of each exercise to be filled out after each exercise is completed. Most of the information requested will have been examined and answered in the exercise. Your laboratory instructor may require that a copy of the Summary/Report page be submitted for evaluation.

A Note to the Student

The purpose of *Applications and Investigations in Earth Science* is to assist you in understanding the basic concepts of Earth science that are presented in most introductory courses by providing the experience of participation. Often, simply being told that something is true is not enough. However, the opportunity to collect facts, examine information, and draw conclusions in a scientific manner frequently results in comprehension. Learning is much more than the accumulation of knowledge; it is also the understanding that comes from "doing." In the laboratory we hope you will refine the critical and creative thinking skills that are vital to your effective participation in a modern society.

Each of the manual's exercises is self-contained and will require approximately two hours to complete. However, your instructor may not follow the exercises in the order presented in the manual. Also, exercises may be assigned for completion *outside* of the regularly scheduled laboratory session.

Prior to beginning each exercise you should

1. Read the required text assignment, paying particular attention to the key terms that are listed in the exercise.

2. Examine the exercise and read the purpose and list of objectives.

3. Gather any additional materials listed at the beginning of the exercise. Bring these materials along with your laboratory manual, text, course notes, paper, and pencils to the laboratory session.

After you have completed each exercise, complete the Summary/Report page at the end of the exercise. Your instructor may require that you submit this report for evaluation by a specified date.

A Note to the Instructor

Applications and Investigations in Earth Science is designed to supplement an introductory general education Earth science course. We have attempted to include a general survey of most of the major topics in Earth science, building upon questions that require various levels of understanding to answer. The first type of question is concerned with the basic concepts and fundamental principles of Earth science. These can be answered by reflective thought and careful reading of class notes and textbook materials. Other sections involve conducting experiments, gathering data, and drawing conclusions. The third type of investigation entails the application of principles to achieve solutions.

Recognizing that different aspects of Earth science are emphasized by each of us, we have included exercises that cover a wide variety of subjects. Since each exercise is basically self-contained and covers a specific topic, individual exercises or whole units may be omitted or introduced in a different sequence without difficulty. Furthermore, with only minor modification, several exercises, or portions of exercises, may be assigned for completion outside the regularly scheduled laboratory sessions.

The last two exercises in the manual investigate the basic skills of determining location and distance on Earth, working with the metric system, and conducting a scientific inquiry. Should you decide that completing all, or any part, of these exercises would be of benefit to your students, we suggest that you introduce them near the beginning of the course. In addition, it should be noted that Exercise 17, "Astronomical Observations," will take the student several weeks to complete.

Many of the exercises require that equipment and materials be supplied and present in the laboratory. When needed, a general list has been included at the beginning of each exercise. Realizing that some laboratory settings have limited access to materials such as minerals, rocks, and fossils, we have left the specific choices of these specimens to the instructor.

Applications and Investigations in Earth Science includes an **Investigating the Internet** section with each of the exercises. By accessing the manual's online companion website at http://prenhall.com/earthsciencelab students will be presented with several challenging web-based activities that use World Wide Web resources to reinforce and extent the topics presented in each exercise. We recommend that these web-based activities be examined after the main body of the exercise has been completed. For those with limited time and/or resources, the entire section may be assigned as an out-of-class activity or omitted entirely. To facilitate grading, students may either download and submit the activity response form provided with each exercise or, at the discretion of the instructor, submit their answers electronically directly to any email address provided. Additional information and Internet links to Earth science topics can also be explored on the Internet at the *Earth Science*, 11th edition textbook site, http://www.prenhall.com/tarbuck/

In summary, we have attempted to put together a versatile and adaptable collection of laboratory experiences that investigate many of the topics in introductory Earth science. We sincerely feel that each exercise has merit and, through active student participation, the learning process is carried one step closer to complete understanding.

Acknowledgments

The authors wish to express their sincere thanks to each of the many individuals, institutions, and agencies that provided photographs and illustrations for the manual. In particular, the manual has benefited greatly from the creativity and imaginative production of Dennis Tasa and the talented staff at Tasa Graphic Arts, Inc. Further, we would like to acknowledge the aid of our many students, past and present. Their comments have helped us maintain our focus on readability and understanding.

Furthermore, we would like to acknowledge with a special thanks the many dedicated individuals at Prentice Hall. The open communication and strong support of excellence of Patrick Lynch and others have helped guide this project along. And, as always, the production team has done an outstanding job transforming our manuscript into a finished product we can all be proud of. They are each true professionals with whom we are very fortunate to be associated.

(Source: Art Wolfe, Inc.)

Geology

The Study of Minerals

The ability to identify minerals using the simplest of techniques is a necessity for the Earth scientist, especially those scientists working in the field. In this exercise you will become familiar with the common physical properties of minerals and learn how to use these properties to identify minerals (Figure 1.1). In order to understand the origin, classification, and alteration of rocks, which are for the most part aggregates (mixtures) of minerals, you must first be able to identify the minerals that comprise them.

Figure 1.1 Quartz crystals. Slender, six-sided, transparent crystals that will scratch glass. The shape of the crystals and their hardness are two physical properties used to identify this mineral. (Photo by E. J. Tarbuck)

Objectives

After you have completed this exercise, you should be able to:

1. Recognize and describe the physical properties of minerals.
2. Use a mineral identification key to name minerals.
3. Identify several minerals by sight.
4. List the uses of several minerals that are mined.

Materials

hand lens

Materials Supplied by Your Instructor

mineral samples	dilute hydrochloric acid
streak plate	set of quartz crystals
magnet	(various sizes)
glass plate	contact goniometer
binocular microscope	crystal growth solution(s)

Terms

mineral	translucent	cleavage plane
rock-forming	transparent	direction of
mineral	hardness	cleavage
luster	color	fracture
metallic luster	streak	specific gravity
nonmetallic	crystal form	magnetism
luster	contact goniometer	striations
opaque	cleavage	tenacity

Introduction

A **mineral** is a naturally occurring, inorganic solid with an orderly internal arrangement of atoms (called *crystalline structure*) and a definite, but not fixed, chemical composition. Some minerals, such as gold and diamond, are single chemical elements. However, most minerals are compounds consisting of two or more elements. For example, the mineral halite is composed

Table 1.1 Mineral Uses

MINERAL	USE
Chalcopyrite	Mined for copper
Feldspar	Ceramics and porcelain
Fluorite	Used in steel manufacturing
Galena	Mined for lead
Graphite	Pencil "lead," lubricant
Gypsum	Drywall, plaster of paris, wallboard
Halite	Table salt, road salt, source of sodium and chlorine
Hematite	Mined for iron
Magnetite	Mined for iron
Pyrite	Mined for sulfur and iron
Quartz	In the pure form, for making glass
Sphalerite	Mined for zinc
Talc	Used in ceramics, paint, talcum powder

of the elements sodium and chlorine. The distinctive crystalline structure and chemical composition of a mineral give it a unique set of physical properties such as its luster, its hardness, and how it breaks. The fact that each mineral has its own characteristic physical and chemical properties can be used to distinguish one mineral from another.

Of the nearly 4000 known minerals, only a few hundred have any current economic value. An example would be the mineral gypsum, used for making drywall and wallboard. Table 1.1 lists a few of the minerals that are mined as well as their uses. Of the remaining minerals, no more than a few dozen are abundant. Collectively, these few often occur with each other in the rocks of Earth's crust and are classified as the **rock-forming minerals**.

Physical Properties of Minerals

The physical properties of minerals are those properties that can be determined by observation or by performing some simple tests. The primary physical properties that are determined for all minerals include optical properties (in particular, luster and the ability to transmit light), hardness, color, streak, crystal form, cleavage or fracture, and specific gravity. Secondary (or "special") properties, including magnetism, taste, feel, striations, tenacity, and the reaction with dilute hydrochloric acid, are also useful in identifying certain minerals.

Optical Properties

Of the many optical properties of minerals, two—luster and the ability to transmit light—are frequently determined for hand specimens.

Luster describes the manner in which light is reflected from the surface of a mineral. Any mineral that shines with a metal-like appearance has a **metallic luster**. Those minerals that do not have a metallic luster are termed **nonmetallic** and may have one of a variety of lusters that include vitreous (glassy), pearly (like a pearl), or earthy (dull, like soil or concrete). In general, many minerals with metallic luster produce a dark gray, black, or other distinctively colored powder when they are rubbed on a hard porcelain plate (this property, called *streak*, will be investigated later in the exercise).

Observe the mineral photographs shown in Figures 1.2 through 1.13. The minerals illustrated in Figures 1.5 and 1.6 have definite metallic lusters. The minerals in Figure 1.9 have nonmetallic, vitreous (glassy) lusters. Some minerals, such as hematite (Figure 1.7), occur in both metallic and nonmetallic varieties.

The ability of a mineral to transmit light can be described as either **opaque**, when no light is transmitted (e.g. Figure 1.3); **translucent**, when light but not an image is transmitted; or, **transparent**, when an image is visible through the mineral (e.g. Figure 1.1). In general, most minerals with a metallic luster are opaque, while vitreous minerals are either translucent or transparent.

Examine the mineral specimens provided by your instructor, and answer the following questions.

1. How many of your specimens can be grouped into each of the following luster types?

   Metallic: _____   Nonmetallic-glassy: _____

2. How many of your specimens are transparent, and how many are opaque?

   Transparent: _____   Opaque: _____

Hardness

Hardness, one of the most useful diagnostic properties of a mineral, is a measure of the resistance of a mineral to abrasion or scratching. It is a relative property in that a harder substance will scratch, or cut into, a softer one.

In order to establish a common system for determining hardness, Friedrich Mohs (1773–1839), a German mineralogist, developed a reference scale of mineral hardness. The Mohs scale of hardness (Figure 1.14), widely used today by geologists and engineers, uses 10 index minerals as a reference set to determine the hardness of other minerals. The hardness value of 1 is assigned to the softest mineral in the set, talc, and 10 is assigned to the hardest mineral, diamond. Higher-numbered minerals will scratch lower-numbered minerals. For example, quartz, with a hardness of 7, will scratch calcite, which has a hardness of 3. It should be remembered that Mohs scale is a *relative ranking* and does *not* imply that mineral number 2, gypsum, is twice as hard as mineral 1, talc.

Most people do not have a set of Mohs reference minerals available. However, by knowing the hardness of some common objects, such as those listed on

Figure 1.2 Fluorite (light), halite (center), and calcite (right) exhibit smooth cleavage planes that are produced when the mineral is broken. (Photo by GeoScience Resources/American Geological Institute).

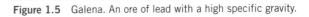

Figure 1.5 Galena. An ore of lead with a high specific gravity.

Figure 1.3 Sphalerite. An ore of zinc.

Figure 1.6 Pyrite. A brassy-yellow mineral with a metallic luster that is commonly known as "fool's gold."

Figure 1.4 Graphite. A soft silver-gray mineral.

Figure 1.7 Hematite. An ore of iron that has both a metallic (right) and nonmetallic (left) form.

Figure 1.8 Two varieties of the mineral quartz. Rose quartz (right) and smoky quartz (left).

Figure 1.11 Augite. A dark green to black, rock-forming, pyroxene mineral.

Figure 1.9 Biotite mica (black) and muscovite mica (light color) are similar in appearance, except for color.

Figure 1.12 Potassium feldspar, variety microcline.

Figure 1.10 Hornblende. A generally green to black, rock-forming, amphibole mineral.

Figure 1.13 Plagioclase feldspar, variety labradorite.

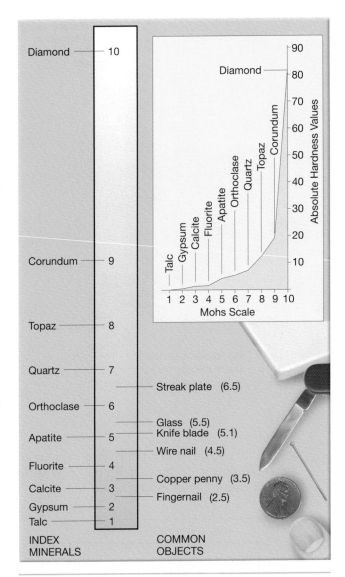

Figure 1.14 Mohs scale of hardness with the hardness of some common objects.

Table 1.2 **Hardness guide**

HARDNESS	DESCRIPTION
Less than 2.5	A mineral that can be scratched by your fingernail (hardness = 2.5).
2.5 to 5.5	A mineral that cannot be scratched by your fingernail (hardness = 2.5), and cannot scratch glass (hardness = 5.5).
Greater than 5.5	A mineral that scratches glass (hardness = 5.5).

Color

Color, although an obvious feature of minerals, may also be misleading. For example, slight impurities in a mineral may result in one sample of the mineral having one color while a different sample of the same mineral may have an entirely different color. *Thus, color is one of the least reliable physical properties.*

5. Observe the minerals in Figure 1.8. Both of the minerals shown are varieties of the same mineral, quartz. What is the reason for the variety of colors that quartz exhibits?

   _____

6. Examine the mineral specimens supplied by your instructor, and see if any appear to be the same mineral but with variable colors.

Streak

The **streak** of a mineral is the color of the fine powder of a mineral obtained by rubbing a corner across a piece of unglazed porcelain—called a *streak plate.* Whereas the color of a mineral may vary from sample to sample, its streak usually does not and is therefore the more reliable property (see Figure 1.7). In many cases, the color of a mineral's streak may not be the same as the color of the mineral. [*Note:* Minerals that have about the same hardness as, or are harder than, a streak plate (about 7 on Mohs scale of hardness), may not powder or produce a streak.]

7. Select three of the mineral specimens provided by your instructor. Do they exhibit a streak? If so, is the streak the same color as the mineral specimen?

   List your observations for each specimen in the following space.

	COLOR OF SPECIMEN	STREAK
Specimen 1:	_____	_____
Specimen 2:	_____	_____
Specimen 3:	_____	_____

Mohs scale in Figure 1.14, a hardness value can be assigned to a mineral. For example, a mineral that has a hardness greater than 5.5 will scratch glass. Table 1.2 can serve as a guide for determining the hardness of a mineral.

3. Test the hardness of several of the mineral specimens provided by your instructor by rubbing any two together to determine which are hard (the minerals that do the scratching) and which are soft (the minerals that are scratched). Doing this will give you an indication of what is meant by the term "relative hardness" of minerals.

4. Use the hardness guide in Table 1.2 to find an example of a mineral supplied by your instructor that falls in each of the three categories.

Figure 1.15 Atomic arrangement, crystal form, and cleavage of the mineral halite. Halite (NaCl) contains sodium and chlorine atoms arranged in a one-to-one ratio forming a cube. The internal, orderly arrangement of atoms produces the external cubic crystal form of the mineral. Planes of weak bonding between atoms in the internal crystalline structure are responsible for halite's cubic cleavage.

Crystal Form

Crystal form is the external appearance or shape of a mineral that results from the internal, orderly arrangement of atoms (Figure 1.15). Most inorganic substances consist of crystals. The flat external surfaces on a crystal are called *crystal faces*. A mineral that forms without space restrictions will exhibit well-formed crystal faces. However, most of the time, minerals must compete for space, and the result is a dense intergrown mass in which crystals do not exhibit their crystal form, especially to the unaided eye.

8. At the discretion of your instructor, you may be asked to grow crystals by evaporating prepared concentrated solutions. Following the specific directions of your instructor, and after you have completed your experiment(s), write a brief paragraph summarizing your observations.

_____

_____

_____

One of the most useful instruments for measuring the angle between crystal faces on large crystals is the **contact goniometer** (Figure 1.16).

9. The mineral shown in Figure 1.1 has a well-developed crystal form with six faces that intersect at

about 120° and come to a point. Two varieties of the same mineral are shown in Figure 1.8. Why do those in Figure 1.8 not exhibit crystal form?

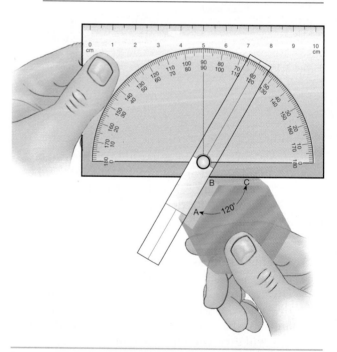

Figure 1.16 Contact goniometer. To use the instrument, hold the straight edge of the protractor in contact with one crystal face and the edge of the celluloid strip in contact with the other face. The angle is read where the fine line on the celluloid strip overlaps the degrees on the protractor. For example, the angle between the adjacent crystal faces on the mineral illustrated (angle ABC) is 120°.

10. Select one of the photographed minerals, other than Figure 1.1, that exhibits its crystal form and describe its shape.

   Figure _____ : _____

11. Observe the various size crystals of the mineral quartz on display in the lab. Use the contact goniometer to measure the angle between similar, adjacent crystal faces on several crystals. Then write a statement relating the angle between adjacent crystal faces to the size of the crystal.

   _____

   _____

Cleavage and Fracture

Cleavage is the tendency of some minerals to break along regular planes of weak bonding between atoms in the internal crystalline structure (see Figure 1.15). When broken, minerals that exhibit cleavage produce smooth, flat surfaces, called **cleavage planes**.

Cleavage is described by (1) noting the number of **directions of cleavage**, which is the number of different sets of planes that form the surfaces of a mineral crystal when it cleaves, and (2) the angle(s) at which the directions of cleavage meet (see Figure 1.15). Each cleavage plane of a mineral crystal that has a different orientation is counted as a different direction of cleavage. When two or more cleavage planes are parallel or line up with each other, they are counted only once, as one direction of cleavage. Minerals may have one, two, three, four, or more directions of cleavage (Figure 1.17).

Number of Cleavage Directions	Shape	Sketch	Directions of Cleavage	Sample
1	Flat sheets			 Muscovite
2 at 90°	Elongated form with rectangle cross section (prism)	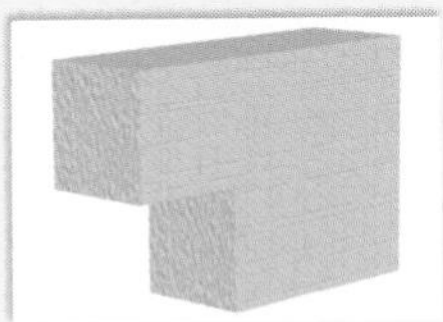	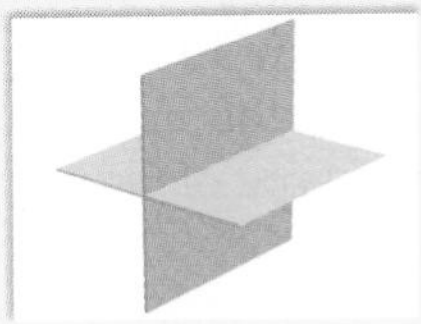	 Feldspar
2 not at 90°	Elongated form with parallelogram cross section (prism)	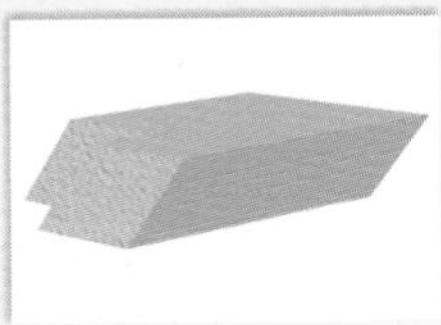		 Hornblende
3 at 90°	Cube	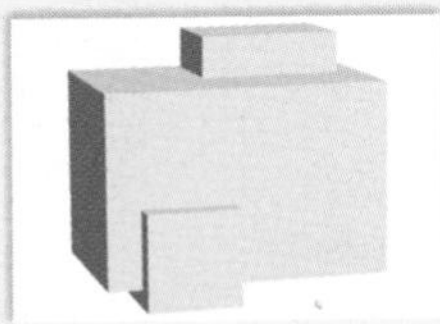	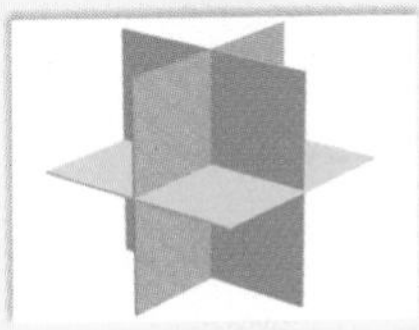	 Halite
3 not at 90°	Rhombohedron	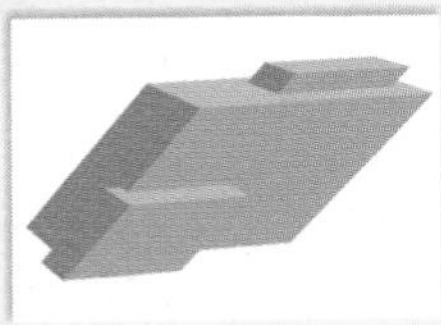		 Calcite
4	Octahedron	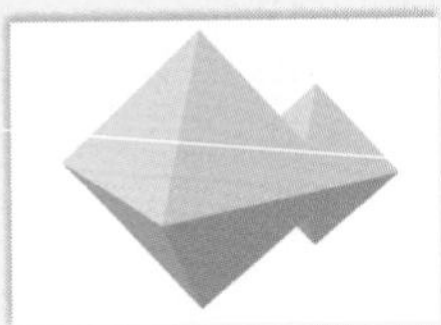	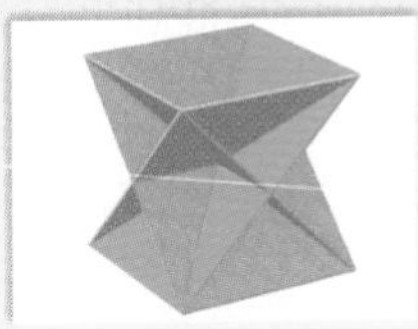	 Fluorite

Figure 1.17 Common cleavage directions of minerals.

Observe the minerals shown in Figures 1.5 and 1.15. These minerals have broken with regularity and exhibit cleavage. The specimens shown are in the form of a cube. Although there are six planes of cleavage surrounding each specimen, each exhibits only three directions of cleavage: top and bottom form one parallel set of planes—hence the first direction of cleavage; the two sides are a second parallel set—a second direction of cleavage; and the front and back form the third direction of cleavage. The cleavage of both minerals is described as three directions of cleavage that intersect at 90° (also commonly called *cubic cleavage*) (see Figure 1.17).

Cleavage and crystal form are *not* the same. Some mineral crystals cleave while others do not. Cleavage is determined by the bonds that hold atoms together, while crystal form results from the internal, orderly arrangement of the atoms. The best way to determine whether or not a mineral cleaves is to break it and carefully examine the results.

Minerals that do not exhibit cleavage when broken are said to **fracture** (see Figure 1.4). Fracturing can be irregular, splintery, or conchoidal (smooth curved surfaces resembling broken glass). Some minerals may cleave in one or two directions and also exhibit fracturing (see Figure 1.12).

12. The minerals shown in Figure 1.9 have one direction of cleavage. Describe the appearance of a mineral that exhibits this type of cleavage.

_____

13. Observe the photograph of calcite, the mineral on the right in Figure 1.2. Several smooth, flat planes result when the mineral is broken.

    a. How many planes of cleavage are present on the specimen?

    _____ planes of cleavage

    b. How many directions of cleavage are present on the specimen?

    _____ directions of cleavage

    c. The cleavage directions meet at (90° angles, angles other than 90°). Circle your answer.

14. Select one mineral specimen supplied by your instructor that exhibits cleavage. Describe its cleavage by completing the following statement.

    _____ directions of cleavage at _____ degrees

Specific Gravity

Specific gravity is a number that represents the ratio of the weight of a mineral to the weight of an equal volume of water. For example, the mineral quartz, Figures 1.1 and 1.8, has a specific gravity of 2.65; this means it weighs 2.65 times more than an equal volume of water. The mineral galena, Figure 1.5, with a specific gravity of 7.4, feels heavy when held in your hand. With a little practice, you can estimate the specific gravity of a mineral by hefting it in your hand. The average specific gravity of minerals is about 2.7, but some metallic minerals have a specific gravity two or three times greater than the average. (*Note:* Exercise 22, "The Metric System, Measurements, and Scientific Inquiry," contains a simple experiment for estimating the specific gravity of a solid.)

15. Find a mineral specimen supplied by your instructor that exhibits a high specific gravity by giving each mineral a heft in your hand.

Other Properties of Minerals

Luster, the ability to transmit light, hardness, color, streak, crystal form, cleavage or fracture, and specific gravity are the most basic and common physical properties used to identify minerals. However, other special properties can also be used to identify certain minerals. These other properties include:

Magnetism Magnetism is characteristic of minerals, such as magnetite, that have a high iron content and are attracted by a magnet. A variety of magnetite called *lodestone* is itself actually magnetic and will pick up paper clips (Figure 1.18).

Figure 1.18 Magnetite, variety lodestone, has polarity like a magnet and will attract iron objects.

Specimen Number	Luster	Hardness	Color	Streak	Fracture or Cleavage (number of directions and angle of intersection)	Other Properties	Name	Economic Use or Rock-forming

Figure 1.19 Mineral identification chart.

Specimen Number	Luster	Hardness	Color	Streak	Cleavage Fracture or (number of directions and angle of intersection)	Other Properties	Name	Economic Use or Rock-forming

Figure 1.19 Mineral identification chart (*continued*)

Taste   The mineral halite (Figure 1.2, center) has a "salty" taste.

> *CAUTION:* Do not taste any minerals or other materials without knowing it is *absolutely* safe to do so.

Feel   The mineral talc often feels "soapy," while the mineral graphite (Figure 1.4) has a "greasy" feel.

Striations   Striations are closely spaced, fine lines on the crystal faces of some minerals. They resemble the surface of a phonograph record but are straight. Certain plagioclase feldspar minerals often exhibit striations on one cleavage surface (see Figure 1.13).

Tenacity   Tenacity is the manner in which a substance resists breaking. Terms like *flexible* (a thin piece of plastic) and *brittle* (glass) are used to describe this property.

Reaction to Dilute Hydrochloric Acid   A very small drop of dilute hydrochloric acid, when placed on a freshly exposed surface of some minerals, will cause them to "fizz" (effervesce) as the gas carbon dioxide is released. The test is often used to identify a group of minerals called the *carbonate minerals.* The mineral calcite, Figure 1.2 (right), (chemical name: calcium carbonate) is the most common carbonate mineral and is frequently found in rocks.

> *CAUTION:* Hydrochloric acid can discolor, decompose, and disintegrate mineral and rock samples. Use the acid only after you have received specific instructions on its use from your instructor. Never taste minerals that have had acid placed on them.

16. Following the directions given by your instructor, examine the mineral specimens to determine if any exhibit one or more of the special properties listed above.

Identification of Minerals

Having investigated the physical properties of minerals, you are now prepared to proceed with the identification of the minerals supplied by your instructor.

To identify a mineral, you must first determine, using available tools, as many of its physical properties as you can. Next, knowing the properties of the mineral, you proceed to a mineral identification key, which often functions like an outline, to narrow down the choices and arrive at a specific name. *As you complete the exercise, remember that the goal is to learn the procedure for identifying minerals through observation and not simply to put a name on them.*

Arrange your mineral specimens by placing them on a numbered sheet of paper. Locate the mineral identification chart, Figure 1.19, and write the numbers of your mineral specimens, in order, under the column labeled "Specimen Number."

Using a Mineral Identification Key

Figure 1.20 is a mineral identification key that uses the property of luster as the primary division of minerals into two groups, those with metallic lusters and those with nonmetallic lusters. Color (either dark- or light-colored) is used as a secondary division for the nonmetallic minerals. Examine the mineral identification key closely to see how it is arranged.

17. Use the mineral identification key (Figure 1.20). What would be the name of the mineral with these properties: nonmetallic luster, light colored, softer than a fingernail, produces small, thin plates or sheets when scratched by a fingernail, white color, and a "soapy" feel?

    Mineral name:

    (Check your answer with your instructor before proceeding.)

18. Complete the mineral identification chart, Figure 1.19, by listing the properties of each of the mineral specimens supplied by your instructor. Use the mineral identification key, Figure 1.20, to determine the name of each of the minerals.

19. Use Table 1.1, *Mineral Uses,* to determine which of the minerals you identify have an economic use.

    List their use in the column to the right of their name on the mineral identification chart, Figure 1.19.

20. The mineral photographs, Figure 1.2 (right, calcite) and Figures 1.8 through 1.13, show some of the most common rock-forming minerals. If your mineral specimens include examples of these minerals, indicate that they are rock-forming in the column to the right of their name on your mineral identification chart.

Minerals on the Internet

Apply the concepts from this exercise to an investigation of the mineral resources in your home state by completing the corresponding online activity on the *Applications & Investigations in Earth Science* website at http://prenhall.com/earthsciencelab

METALLIC MINERALS

Hardness	Streak	Other Diagnostic Properties	Name (Chemical Composition)
Harder than glass	Black	Black; magnetic; hardness = 6; specific gravity = 5.2; often granular	Magnetite (Fe_3O_4)
	Greenish-black	Brass yellow; hardness = 6; specific gravity = 5.2; generally an aggregate of cubic crystals	Pyrite (FeS_2)-fool's gold
	Red-brown	Gray or reddish brown; hardness = 5–6; specific gravity = 5; platy appearance	Hematite (Fe_2O_3)
Softer than glass but harder than a finger nail	Greenish-black	Golden yellow; hardness = 4; specific gravity = 4.2; massive	Chalcopyrite ($CuFeS_2$)
	Gray-black	Silvery gray; hardness = 2.5; specific gravity = 7.6 (very heavy); good cubic cleavage	Galena (PbS)
	Yellow-brown	Yellow brown to dark brown; hardness variable (1–6); specific gravity = 3.5–4; often found in rounded masses; earthy appearance	Limonite ($Fe_2O_3 \cdot H_2O$)
	Gray-black	Black to bronze; tarnishes to purples and greens; hardness = 3; specific gravity = 5; massive	Bornite (Cu_5FeS_4)
Softer than your fingernail	Dark gray	Silvery gray; hardness = 1 (very soft); specific gravity = 2.2; massive to platy; writes on paper (pencil lead); feels greasy	Graphite (C)

NONMETALLIC MINERALS

	Hardness	Cleavage	Other Diagnostic Properties	Name (Chemical Composition)
Dark colored	Harder than glass	Cleavage Present	Greenish black to black; hardness = 5–6; specific gravity = 3.4; fair cleavage, two directions at nearly 90 degrees	Augite (Ca, Mg, Fe, Al silicate)
			Black to greenish black; hardness = 5–6; specific gravity = 3.2; fair cleavage, two directions at nearly 60 degrees and 120 degrees	Hornblende (Ca, Na, Mg, Fe, OH, Al silicate)
			Red to reddish brown; hardness = 6.5–7.5; conchoidal fracture; glassy luster	Garnet (Fe, Mg, Ca, Al silicate)
		Cleavage not prominent	Gray to brown; hardness = 9; specific gravity = 4; hexagonal crystals common	Corundum (Al_2O_3)
			Dark brown to black; hardness = 7; conchoidal fracture; glassy luster	Smoky quartz (SiO_2)
			Olive green; hardness = 6.5–7; small glassy grains	Olivine (Mg, Fe)$_2$SiO$_4$

Figure 1.20   Mineral identification key.

NONMETALLIC MINERALS

Hardness	Cleavage	Other Diagnostic Properties	Name (Chemical Composition)
Dark colored (continued) — Softer than glass but harder than a fingernail	Cleavage present	Yellow brown to black; hardness = 4; good cleavage in six directions, light yellow streak that has the smell of sulfur	Sphalerite (ZnS)
		Dark brown to black; hardness = 2.5–3, excellent cleavage in one direction; elastic in thin sheets; black mica	Biotite mica (K, Mg, Fe, OH, Al silicate)
	Cleavage absent	Generally tarnished to brown or green; hardness = 2.5; specific gravity = 9; massive	Native copper (Cu)
Softer than your fingernail	Cleavage not prominent	Reddish brown; hardness = 1–5; specific gravity = 4–5; red streak; earthy appearance	Hematite (Fe_2O_3)
		Yellow brown; hardness = 1–3; specific gravity = 3.5; earthy appearance; powders easily	Limonite ($Fe_2O_3 \cdot H_2O$)
Light Colored — Harder than glass	Cleavage present	Pink or white to gray; hardness = 6; specific gravity = 2.6; two directions of cleavage at nearly right angles	Potassium feldspar ($KAlSi_3O_8$) (pink)
			Plagioclase feldspar ($NaAlSi_3O_8$ to $CaAl_2Si_2O_8$) (white to gray)
	Cleavage absent	Any color; hardness = 7; specific gravity = 2.65; conchoidal fracture; glassy appearance; varieties: milky (white), rose (pink), smoky (gray), amethyst (violet)	Quartz (SiO_2)
Softer than glass but harder than a finger nail	Cleavage present	White, yellowish to colorless; hardness = 3; three directions of cleavage at 75 degrees (rhombohedral); effervesces in HCl; often transparent	Calcite ($CaCO_3$)
		White to colorless; hardness = 2.5; three directions of cleavage at 90 degrees (cubic); salty taste	Halite (NaCl)
		Yellow, purple, green, colorless; hardness = 4; white streak; translucent to transparent; four directions of cleavage	Fluorite (CaF_2)
Softer than your fingernail	Cleavage present	Colorless; hardness = 2–2.5; transparent and elastic in thin sheets; excellent cleavage in one direction; light mica	Muscovite mica (K, OH, Al silicate)
		White to transparent, hardness = 2; when in sheets; is flexible but not elastic; varieties: selenite (transparent, three directions of cleavage); satin spar (fibrous, silky luster); alabaster (aggregate of small crystals)	Gypsum ($CaSO_4 \cdot 2H_2O$)
	Cleavage not prominent	White, pink, green; hardness = 1–2; forms in thin plates; soapy feel; pearly luster	Talc (Mg silicate)
		Yellow; hardness = 1–2.5	Sulfur (S)
		White; hardness = 2; smooth feel; earthy odor; when moistened, has typical clay texture	Kaolinite (Hydrous Al silicate)
		Pale to dark reddish brown; hardness = 1–3; dull luster; earthy; often contains spheroidal-shaped particles; not a true mineral	Bauxite (Hydrous Al oxide)

Figure 1.20 Mineral identification key (*continued*)

Notes and calculations.

The Study of Minerals

Date Due: _____

Name: _____

Date: _____

Class: _____

After you have finished Exercise 1, complete the following questions. You may have to refer to the exercise for assistance or to locate specific answers. Be prepared to submit this summary/report to your instructor at the designated time.

1. Describe the procedure for identifying a mineral and arriving at its name.

_____

_____

_____

2. Name the physical property of a mineral that is described by each of the following statements.

PHYSICAL PROPERTY

Breaks along smooth planes: _____

Scratches glass: _____

Shines like a metal: _____

A red-colored powder on unglazed porcelain:

_____

3. Describe the shape of a mineral that has three directions of cleavage that intersect at 90°.

_____

4. Name two minerals you identified that have good cleavage. Describe the cleavage of each mineral.

MINERAL   **CLEAVAGE**

_____ : _____

_____ : _____

5. Select five minerals you identified, and list their names and physical properties.

_____ : _____

_____

_____

_____ : _____

_____

_____

_____ : _____

_____

_____

_____ : _____

_____

_____

_____ : _____

_____

_____

6. Name one mineral that you identified that has an economic use.

MINERAL   **MINED FOR**

_____ : _____

17

7. List the name and hardness of two minerals you identified.

   MINERAL   **HARDNESS**

   _____ : _____

   _____ : _____

8. How many directions of cleavage do the feldspar minerals—potassium feldspar and plagioclase feldspar—have?

   _____ directions of cleavage

9. What was your conclusion concerning the angles between similar crystal faces on different size crystals of the same mineral?

   _____

10. List the name(s) of the minerals you identified that had a special property such as magnetism or feel. Write the special property that you observed next to the name of the mineral.

    MINERAL   **SPECIAL PROPERTY**

    _____ : _____

    _____ : _____

    _____ : _____

11. What physical property most distinguishes biotite mica from muscovite mica?

    _____

12. Selecting from the minerals illustrated in Figures 1.1 through 1.13, list, by name, one mineral that exhibits each of the following:

    one direction of cleavage: _____

    striations: _____

    multiple colors: _____

    cubic cleavage: _____

    nonmetallic, vitreous luster: _____

    fracture: _____

    metallic luster: _____

Common Rocks

To an Earth scientist, rocks represent much more than usable substances. They are the materials of the Earth; understanding their origin and how they change allows us to begin to understand Earth and its processes. It is often said that "the history of Earth is written in the rocks"—we just have to be smart enough to read the "words."

In this exercise, you will investigate some of the common rocks that are found on and near Earth's surface. The criteria used to classify a rock as being of either igneous, sedimentary, or metamorphic origin are examined, as well as the procedure for identifying rocks within each of these three families.

Objectives

After you have completed this exercise, you should be able to:

1. Examine a rock and determine if it is an igneous, sedimentary, or metamorphic rock.
2. List and define the terms used to describe the textures of igneous, sedimentary, and metamorphic rocks.
3. Name the dominant mineral(s) found in the most common igneous, sedimentary, and metamorphic rocks.
4. Use a classification key to identify a rock.
5. Recognize and name some of the common rocks by sight.

Materials

metric ruler   hand lens

Materials Supplied by Your Instructor

igneous rocks   dilute hydrochloric acid
sedimentary rocks   streak plate

metamorphic rocks   glass plate
hand lens or binocular   copper penny
  microscope

Terms

rock	weathering	composition
rock cycle	sediment	detrital material
igneous rock	lithification	chemical material
magma	metamorphic	foliation
sedimentary	rock	texture
rock		

Introduction

Most **rocks** are aggregates (mixtures) of minerals. However, there are some rocks that are composed essentially of one mineral found in large impure quantities. The rock limestone, consisting almost entirely of the mineral calcite, is a good example.

Rocks are classified into three types, based on the processes that formed them. One of the most useful devices for understanding rock types and the geologic processes that transform one rock type into another is the **rock cycle**. The cycle, shown in Figure 2.1, illustrates the various Earth materials and uses arrows to indicate chemical and physical processes. As you examine the rock cycle and read the following definitions, notice the references to the origin of each rock type.

The three types of rock are igneous, sedimentary, and metamorphic.

Igneous Igneous rocks (Figures 2.2–2.9) are the solidified products of once molten material called **magma**. The distinguishing feature of most igneous rocks is an interlocking arrangement of mineral crystals that forms as the molten material cools and crystals grow. *Intrusive* igneous rocks form below the surface of Earth, while those that form at the surface from lava are termed *extrusive*.

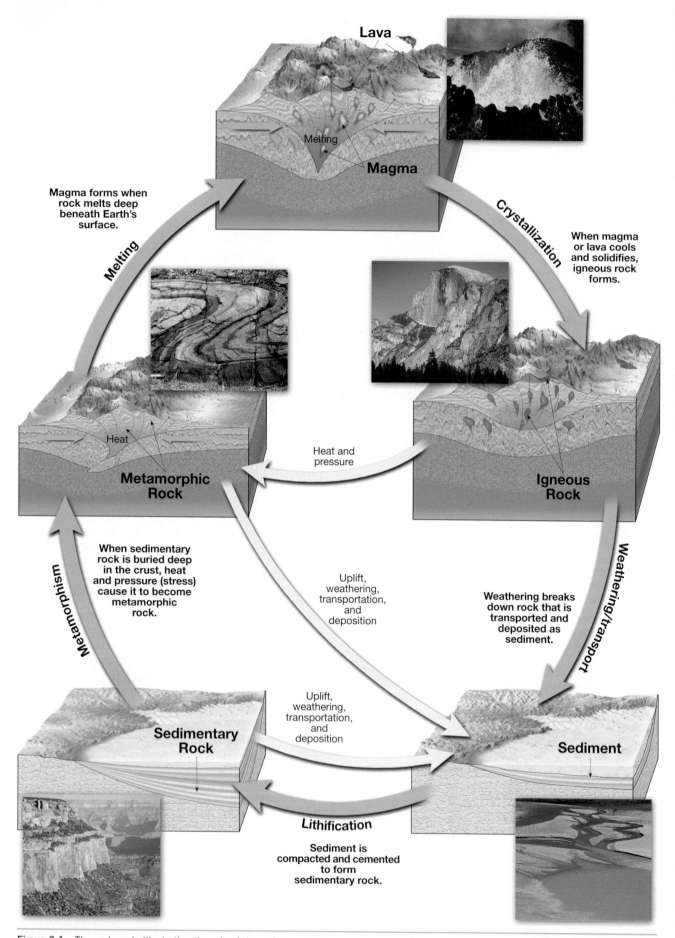

Figure 2.1 The rock cycle illustrating the role of the various geologic processes that act to transform one rock type into another.

Sedimentary These rocks (Figures 2.10–2.17) form at or near Earth's surface from the accumulated products of **weathering**, called **sediment**. These products may be solid particles or material that was formerly dissolved and then precipitated by either inorganic or organic processes. The process of **lithification** transforms the sediment into hard rock. Since sedimentary rocks form at, or very near. Earth's surface, they often contain organic matter, or fossils, or both. The layering (or bedding) that develops as sediment is sorted by, and settled out from, a transporting material (usually water or air) helps make sedimentary rocks recognizable.

Metamorphic These rocks (Figures 2.18–2.25) form below Earth's surface where high temperatures, pressures, and/or chemical fluids change preexisting rocks without melting them.

Minerals are identified by using their physical and chemical properties. However, rock types and the names of individual rocks are determined by describing their *textures* and *compositions*. The key to success in rock identification lies in learning to accurately determine and describe these properties.

Texture refers to the shape, arrangement, and size of mineral grains in a rock. The shape and arrangement of mineral grains help determine the type (igneous, sedimentary, or metamorphic) of rock. Mineral grain size is often used to separate rocks within a particular type. Each rock type uses different terms to describe its textures.

Composition refers to the minerals that are found in a rock. Often the larger mineral grains can be identified by sight or by using their physical properties. In some cases, small mineral grains may require the use of a hand lens or microscope. Occasionally, very small grains cannot be identified with the normal magnification of a microscope. Practice and increased familiarity with the minerals will make this assessment easier.

Comparing Igneous, Sedimentary, and Metamorphic Rocks

One of the first steps in the identification of rocks is to determine the rock type. Each of the three rock types has a somewhat unique appearance that helps to distinguish one type from the other.

Examine the specimens of the three rock types supplied by your instructor, as well as the photographs of the rocks in Figures 2.2–2.25. Then answer the following questions.

1. Which two of the three rock types appear to be made primarily of intergrown crystals?

   _____ rocks and _____ rocks

2. Which one of the two rocks types you listed in question 1 has the mineral crystals aligned or arranged so that they are oriented in the same direction in a linear, linelike manner?

   _____

3. Which one of the two rock types you listed in question 1 has the mineral crystals in most of the rocks arranged in a dense interlocking mass with no alignment?

   _____

4. Of the three rock types, (igneous, sedimentary, metamorphic) rocks often contain haphazardly arranged pieces or fragments, rather than crystals. Circle your answer.

Igneous Rock Identification

Igneous rocks form from the cooling and crystallization of magma. The interlocking network of mineral crystals that develop as the molten material cools gives most igneous rocks their distinctive crystalline appearance.

Textures of Igneous Rocks

The rate of cooling of the magma determines the size of the interlocking crystals found in igneous rocks. The slower the cooling rate, the larger the mineral crystals. The five principal textures of igneous rocks are:

Coarse Grained (or *phaneritic*) The majority of mineral crystals are of a uniform size and large enough to be identifiable without a microscope. This texture occurs when magma cools slowly inside Earth.

Fine Grained (or *aphanitic*) Very small crystals, which are generally not identifiable without strong magnification, develop when molten material cools quickly on, or very near, the surface of Earth.

Porphyritic Two very contrasting sizes of crystals are caused by magma having two different rates of cooling. The larger crystals are termed *phenocrysts;* and the smaller, surrounding crystals are termed *groundmass* (or *matrix*).

Glassy No mineral crystals develop because of very rapid cooling. This lack of crystals causes the rock to have a glassy appearance. In some cases, rapidly escaping gases may produce a frothy appearance similar to spun glass.

Fragmental The rock contains broken, angular fragments of rocky materials produced during an explosive volcanic eruption.

Examine the igneous rock photographs in Figures 2.2–2.9. Then answer the following questions.

5. The igneous rock illustrated in Figure 2.2 is made of large mineral crystals that are all about the same size. The rock formed from magma that cooled (slowly, rapidly) (inside, on the surface of) Earth. Circle your answers.

Igneous Rocks

Figure 2.2 Granite, a common coarse-grained, intrusive igneous rock.

Figure 2.6 Basalt, a fine-grained igneous rock.

Figure 2.3 Rhyolite, a fine-grained, extrusive rock.

Figure 2.7 Gabbro, a coarse-grained, intrusive igneous rock.

Figure 2.4 Diorite, a coarse-grained igneous rock.

Figure 2.8 Obsidian, an igneous rock with a glassy texture.

Figure 2.5 Andesite porphry, an igneous rock with a porphyritic texture.

Figure 2.9 Pumice, a glassy rock containing numerous tiny voids.

Sedimentary Rocks

Figure 2.10 Conglomerate, a detrital sedimentary rock.

Figure 2.11 Sandstone, a common detrital sedimentary rock.

Figure 2.12 Shale, a detrital sedimentary rock composed of very fine grains.

Figure 2.13 Breccia, a detrital sedimentary rock containing large, angular fragments.

Figure 2.14 Fossiliferous limestone, a biochemical sedimentary rock.

Figure 2.15 Coquina, a biochemical limestone consisting of visible shells and shell fragments, loosely cemented.

Figure 2.16 Rock salt, a chemical sedimentary rock formed as water evaporates.

Figure 2.17 Bituminous coal, a sedimentary rock composed of altered plant remains.

Metamorphic Rocks

Figure 2.18 Slate, a fine-grained, foliated metamorphic rock.

Figure 2.19 Phyllite, a foliated metamorphic rock with barely visible grains.

Figure 2.20 Schist, a foliated metamorphic rock with visible grains (variety: garnet-mica schist).

Figure 2.21 Gneiss, a foliated-banded metamorphic rock that often forms during intensive metamorphism.

Figure 2.22 Schist, variety mica schist.

Figure 2.23 Marble, a nonfoliated metamorphic rock that forms from the metamorphism of the sedimentary rock limestone.

Figure 2.24 Quartzite, a nonfoliated metamorphic rock composed of fused quartz grains.

Figure 2.25 Anthracite coal, often called hard coal, forms from the metamorphism of bituminous coal.

6. The rock shown in Figure 2.6 is made of mineral crystals that are all small and not identifiable without a microscope. The rock formed from magma that cooled (slowly, rapidly) (inside, on/near the surface of) Earth. Circle your answers.

7. The igneous rock in Figure 2.5 has a porphyritic texture. The large crystals are called _____, and the surrounding, smaller crystals are called _____.

8. The rocks in Figures 2.2 and 2.3 have nearly the same mineral composition. What fact about the mineral crystals in the rocks makes their appearances so different? What caused this difference?

_____

_____

Select a coarse-grained rock from the igneous rock specimens supplied by your instructor and examine the mineral crystals closely using a hand lens or microscope.

9. Sketch a diagram showing the arrangement of the mineral crystals in the igneous rock specimen you examined in the space provided below. Indicate the scale of your sketch by writing the appropriate length within the ( ) provided on the bar scale.

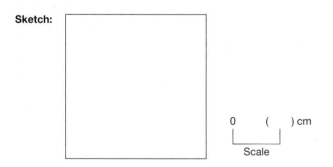

Sketch:

0   (   ) cm

Scale

Composition of Igneous Rocks

The specific mineral composition of an igneous rock is ultimately determined by the chemical composition of the magma from which it crystallized. However, the minerals found in igneous rocks can be arranged into four groups. Each group can be identified by observing the proportion of dark-colored minerals compared to light-colored minerals. The four groups are

> **Felsic** (or *granitic*)—composed mainly of the light-colored minerals quartz and potassium feldspars. Dark-colored minerals account for less than 15% of the minerals in rocks found in this group.

> **Intermediate** (or *andesitic*)—a mixture of both light-colored and dark-colored minerals. Dark minerals comprise about 15% to 45% of these rocks.

> **Mafic** (or *basaltic*)—dark-colored minerals such as pyroxene and olivine account for over 45% of the composition of these rocks.

> **Ultramafic**—composed almost entirely of the dark-colored minerals pyroxene and olivine, these rocks are rarely observed on Earth's surface. However, the ultramafic rock peridotite is believed to be a major constituent of Earth's upper mantle.

10. Estimate the percentage of dark minerals contained in the igneous rock in Figure 2.4. (You may find the color index at the top of Figure 2.26, *Igneous & Rock Identification Key*, helpful.) The rock's color is (light, medium, dark, very dark). Circle your answer.

11. The rocks shown in Figures 2.3 and 2.6 have the same texture. What fact about the mineral crystals makes their appearances so different?

_____

Using an Igneous Rock Identification Key

The name of an igneous rock can be found by first determining its texture and color (an indication of mineral composition), identifying visible mineral grains, and then using an igneous rock identification key such as the one shown in Figure 2.26 to determine the name.

For example, the igneous rock shown in Figure 2.2 has a coarse-grained texture and is light-colored (quartz and potassium feldspar dominant). Intersecting the light-colored column with the coarse-grained row on the igneous rock identification key, Figure 2.26, determines that the name of the rock is "granite."

12. Place each of the igneous rocks supplied by your instructor on a numbered piece of paper. Then complete the igneous rock identification chart, Figure 2.27, for each rock. Use the igneous rock identification key, Figure 2.26, to determine each specimen's name.

Sedimentary Rock Identification

Sedimentary rocks, Figures 2.10–2.17, form from the accumulated products of weathering called *sediment*. Sedimentary rocks can be made of either, or a combination of, detrital or chemical material.

> **Detrital material** consists of mineral grains or rock fragments derived from the process of mechanical weathering that are transported and deposited as solid particles (sediment). Rocks formed in this manner are called *detrital sedimentary rocks*. The mineral pieces that make

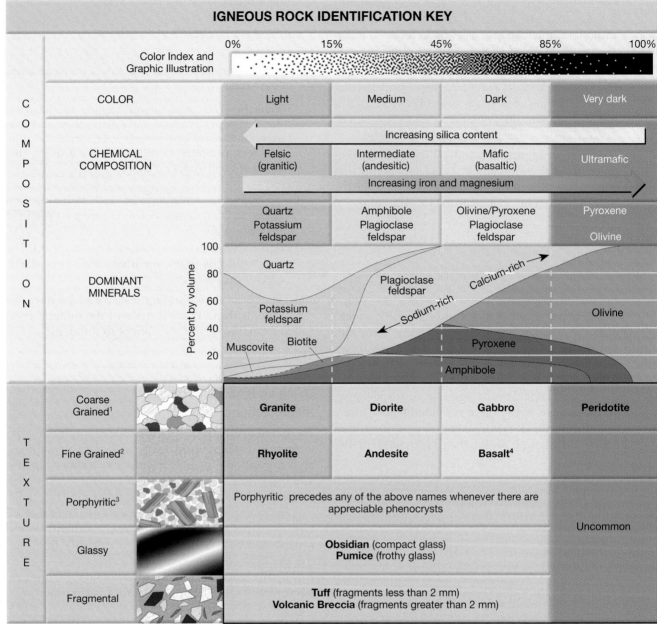

IGNEOUS ROCK IDENTIFICATION KEY

[1] Also called *phaneritic.* Crystals generally 1-10 mm (1 cm). The term *pegmatite* is added to the rock name when crystals are greater than 1 cm; e.g. *granite-pegmatite.*

[2] Also called *aphanitic.* Crystals generally less than 1 mm.

[3] For example, a granite with phenocrysts is called *porphyritic granite.*

[4] Basalt with a cinder-like appearance that develops from gas bubbles trapped in cooling lava (a texture referred to as *vesicular*) is called *scoria.*

Figure 2.26 Igneous rock identification key. Color, with associated mineral composition, is shown along the top axis. Each rock in a column has the color and composition indicated at the top of the column. Texture is shown along the left side of the key. Each rock in a row has the texture indicated for that row. To determine the name of a rock, intersect the appropriate column (color & mineral composition) with the appropriate row (texture) and read the name at the place of intersection.

up a detrital sedimentary rock are called *grains* (or *fragments* if they are pieces of rock). The identification of a detrital sedimentary rock is determined primarily by the size of the grains or fragments. Mineral composition of the rock is a secondary concern.

Chemical material was previously dissolved in water and later precipitated by either inorganic or organic processes. Rocks formed in this manner are called *chemical sedimentary rocks.* If the material is the result of the life processes of water-dwelling organisms—for example, the formation of a shell—it is said to be of biochemical origin. Mineral composition is the primary consideration in the identification of chemical sedimentary rocks.

Sedimentary rocks come in many varieties that have formed in many different ways. For the purpose of examination, this investigation divides the sedimentary rocks into the two groups, *detrital* and *chemical*, based upon the type of material found in the rock.

Specimen Number	Texture	Color (light-intermediate-dark)	Dominant Minerals	Rock Name

Figure 2.27   Igneous rock identification chart.

Examining Sedimentary Rocks

Examine the sedimentary rock specimens supplied by your instructor. Separate those that are made of pieces or fragments of mineral, rock material, or both. They are the detrital sedimentary rocks. Do *not* include any rocks that have abundant shells or shell fragments. You may find the photographs of the detrital sedimentary rocks in Figures 2.10–2.13 helpful. The remaining sedimentary rocks, those with shells or shell fragments and those that consist of crystals, are the chemical rocks.

Pick up each detrital rock specimen and rub your finger over it to feel the size of the grains or fragments.

13. How many of your detrital specimens feel rough like sand? How many feel smooth like mud or clay?

_____ specimens feel rough and _____ feel smooth.

Use a hand lens or microscope to examine the grains or fragments of several coarse detrital rock specimens. Notice that they are not crystals.

14. Sketch the magnified pieces and surrounding material, called *cement* (or *matrix*), of a coarse detrital rock in the space provided on the following page. Indicate the scale of your sketch by writing

the appropriate length within the ( ) provided on the bar scale.

Sketch:

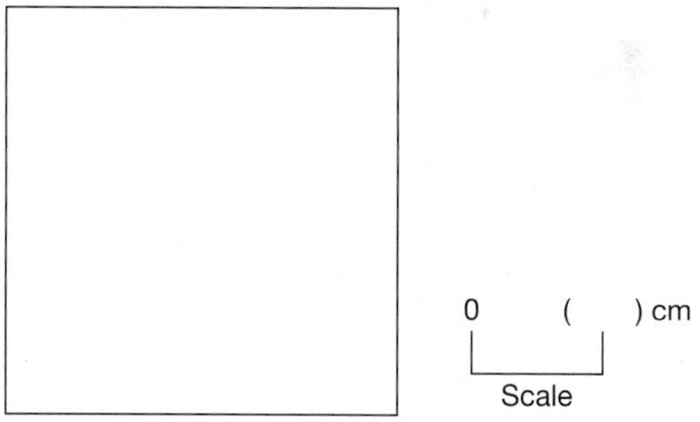

0   (   ) cm

Scale

a. Observe the material surrounding the grains or fragments in the rock specimen closely with a hand lens or microscope. The material is (course, fine). Circle your answer.

b. Write a brief description of the detrital rock specimen you have examined.

_____

_____

Two of the minerals that often comprise the grains of detrital sedimentary rocks are quartz, a hard (hardness = 7) mineral with a glassy luster, and clay, a soft, fine mineral that consists of microscopic platy particles. The difference in appearance and hardness of quartz and clay is helpful in distinguishing them.

15. How many of your detrital specimens are made of quartz, and how many appear to be made of clay?

    _____ specimens have quartz grains and

    _____ have clay grains.

As a result of their method of formation, many chemical sedimentary rocks are fine-to-coarse crystalline, while others consist of shells or shell fragments.

16. How many of your chemical sedimentary rocks are crystalline, and how many contain abundant shells or shell fragments?

    _____ specimens are crystalline and

    _____ contain shells or shell fragments.

Limestones, Figures 2.14 and 2.15, are the most abundant chemical sedimentary rocks. They have several origins and many different varieties; however, one thing that all limestones have in common is that they are made of the calcium carbonate mineral called *calcite*. Calcite can precipitate directly from the sea to form limestone or can be used by marine organisms to make shells. After the organisms die, the shells become sediment and eventually the sedimentary rock limestone.

Calcite is a mineral that reacts with dilute hydrochloric acid and effervesces (fizzes) as carbon dioxide gas is released. Most limestones react readily when a small drop of acid is placed on them, thus providing a good test for identifying the rock. Many limestones also contain fragments of seashells, which also aid in their identification.

17. Follow the directions of your instructor to test the specified sedimentary rock(s) with the dilute hydrochloric acid provided and observe the results. (*Note:* Several detrital sedimentary rocks have calcite surrounding their grains or fragments (calcite cement) that will effervesce with acid and give a *false* test for limestone. Observe the acid reaction closely.)

Using a Sedimentary Rock Identification Key

The sedimentary rock identification key in Figure 2.28 divides the sedimentary rocks into detrital and chemical types. Notice that the primary subdivisions for the detrital rocks are based upon grain size, whereas composition is used to subdivide the chemical rocks.

Detrital Sedimentary Rocks

Texture (particle size)		Sediment Name	Rock Name
Coarse (over 2 mm)		Gravel (rounded particles)	Conglomerate
		Gravel (angular particles)	Breccia
Medium (1/16 to 2 mm)		Sand (if abundant feldspar is present the rock is called **Arkose**)	Sandstone
Fine (1/16 to 1/256 mm)		Mud	Siltstone
Very fine (less than 1/256 mm)		Mud	Shale

Chemical Sedimentary Rocks

Composition	Texture	Rock Name		
Calcite, $CaCO_3$ (effervesces in HCl)	Fine to coarse crystalline	Crystalline Limestone		
		Travertine		
	Visible shells and shell fragments loosely cemented	Coquina	Biochemical	Limestone
	Various size shells and shell fragments cemented with calcite cement	Fossiliferous Limestone		
	Microscopic shells and clay	Chalk		
Quartz, SiO_2	Very fine crystalline	**Chert** (light colored) **Flint** (dark colored)		
Gypsum $CaSO_4 \bullet 2H_2O$	Fine to coarse crystalline	Rock Gypsum		
Halite, NaCl	Fine to coarse crystalline	Rock Salt		
Altered plant fragments	Fine-grained organic matter	Bituminous Coal		

Figure 2.28 Sedimentary rock identification key. Sedimentary rocks are divided into two groups, detrital and chemical, depending upon the type of material that composes them. Detrital rocks are further subdivided by the size of their grains, while the subdivision of the chemical rocks is determined by composition.

Specimen Number	Detrital or Chemical	Texture (grain size)	Sediment Name or Composition	Rock Name

Figure 2.29 Sedimentary rock identification chart.

18. Place each of the sedimentary rocks supplied by your instructor on a numbered piece of paper. Then complete the sedimentary rock identification chart, Figure 2.29, for each rock. Use the sedimentary rock identification key, Figure 2.28, to determine each specimen's name.

Sedimentary Rocks and Environments

Sedimentary rocks are extremely important in the study of Earth's history. Particle size and the materials from which they are made often suggest something about the place, or environment, in which the rock formed. The fossils that often are found in a sedimentary rock also provide information about the rock's history.

Reexamine the sedimentary rocks and think of them as representing a "place" on Earth where the sediment was deposited.

19. Figure 2.11 is the rock sandstone that formed from sand. Where on Earth do you find sand, the primary material of sandstone, being deposited today?

_____

_____

Figure 2.30 shows a few generalized environments (places) where sediment accumulates. Often, an environment is characterized by the type of sediment and life forms associated with it.

20. Use Figure 2.30 to name the environment(s) where, in the past, the sediment for the following sedimentary rocks may have been deposited.

	ORIGINAL SEDIMENT	ENVIRONMENT(S)
Sandstone:	(sand)	_____
Shale:	(mud)	_____
Limestone:	(coral, shells)	_____

Metamorphic Rock Identification

Metamorphic rocks were previously igneous, sedimentary, or other metamorphic rocks that were changed by any combination of heat, pressure, and chemical fluids during the process of **metamorphism**. They are most often located beneath sedimentary rocks on the continents and in the cores of mountains.

During metamorphism new minerals may form, and/or existing minerals can grow larger as metamorphism becomes more intense. Frequently, mineral crystals that are elongated (like hornblende) or have a sheet structure (like the micas—biotite and muscovite) become oriented perpendicular to compressional forces. The resulting parallel, linear alignment of mineral crystals perpendicular to compressional forces (differential stress) is called **foliation** (Figure 2.31). Foliation is unique to many metamorphic rocks and gives them a layered or banded appearance.

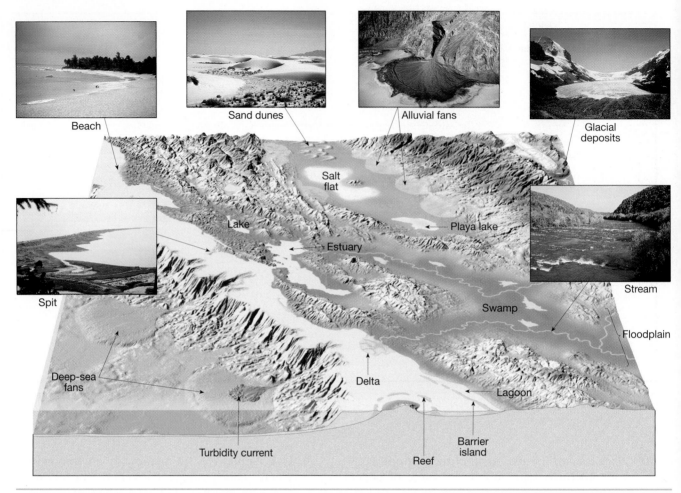

Beach

Sand dunes

Alluvial fans

Glacial deposits

Salt flat

Lake

Estuary

Playa lake

Stream

Spit

Swamp

Floodplain

Deep-sea fans

Delta

Lagoon

Turbidity current

Reef

Barrier island

Figure 2.30 Generalized illustration of sedimentary environments. Although many environments exist on both the land and in the sea, only some of the most important are represented in this idealized diagram. (Photos by E. J. Tarbuck, except alluvial fan, by Marli Miller)

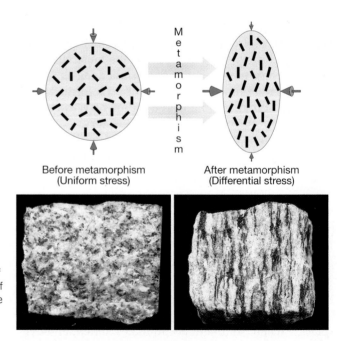

Before metamorphism (Uniform stress)

Metamorphism

After metamorphism (Differential stress)

Figure 2.31 Under directed pressure, planar minerals, such as the micas, become reoriented or recrystallized so that their surfaces are aligned at right angles to the stress. The resulting planar orientation of mineral grains is called **foliation** and gives the rock a foliated texture. If the coarse-grained igneous rock (granite) on the left underwent intense metamorphism, it could end up closely resembling the metamorphic rock on the right (gneiss). (Photos by E. J. Tarbuck)

Metamorphic rocks are divided into two groups based on texture—foliated and nonfoliated. These textural divisions provide the basis for the identification of metamorphic rocks.

Foliated Metamorphic Rocks

The mineral crystals in foliated metamorphic rocks are either elongated or have a sheet structure and are arranged in a parallel or "layered" manner. *During metamorphism, increased heat and pressure can cause the mineral crystals to become larger and the foliation more obvious.* (Figure 2.32) The metamorphic rocks in Figures 2.18–2.22 exhibit foliated textures.

21. From the rocks illustrated in Figures 2.18 and 2.20, the (slate, schist) resulted from more intensive heat and pressure. Circle your answer.

22. From the metamorphic rocks in Figures 2.19 and 2.21, the (phyllite, gneiss) shows the minerals separated into light and dark bands. Circle your answer. (The foliated-banded texture of the rock that you have selected often results

from the most intensive heat and pressure during metamorphism.)

Select several of the foliated metamorphic rock specimens supplied by your instructor that have large crystals and examine them with a hand lens or microscope.

23. Sketch the appearance of the magnified crystals of one foliated metamorphic rock in the space provided below. Indicate the scale of your sketch by writing the appropriate length within the ( ) provided on the bar scale.

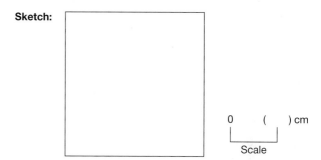

Sketch:

0   ( )cm

Scale

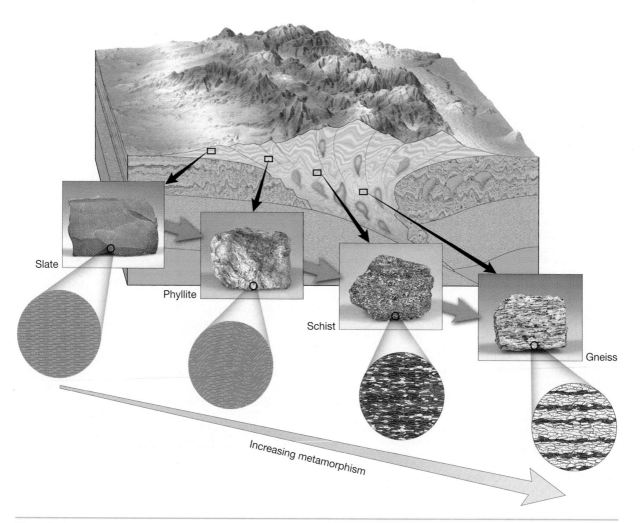

Slate

Phyllite

Schist

Gneiss

Increasing metamorphism

Figure 2.32 Idealized illustration showing the effect of increasing metamorphism in foliated metamorphic rocks. (Photos by E. J. Tarbuck)

Nonfoliated Metamorphic Rocks

Nonfoliated metamorphic rocks are most often identified by determining their mineral composition. The minerals that comprise them, most often calcite or quartz, are neither elongated nor sheet structured and therefore cannot be as easily aligned. Hence, no foliation develops during metamorphism.

24. Examine the nonfoliated metamorphic rocks supplied by your instructor to determine if any are composed of calcite or quartz. Hardness and the reaction to dilute hydrochloric acid often provide a clue.

CAUTION: Follow the directions of your instructor when using acid to test for calcite.

Using a Metamorphic Rock Identification Key

A metamorphic rock identification key is presented in Figure 2.33. To use the key, first determine a rock's texture, foliated or nonfoliated, and then proceed to further subdivisions to arrive at a name. The names of the medium or coarse foliated rocks are often modified with the mineral composition placed in front of the name, e.g., "mica schist."

25. Place each of the metamorphic rocks supplied by your instructor on a numbered piece of paper. Then complete the metamorphic rock identification chart, Figure 2.34, for each rock. Use the metamorphic rock identification key, Figure 2.33 to determine each specimen's name.

Rocks on the Internet

Associated with igneous rock, the most abundant rock on Earth, are often geologic hazards related to volcanic activity. Investigate this potentially destructive geologic process by completing the corresponding online activity on the *Applications & Investigations in Earth Science* website at http://prenhall.com/earthsciencelab

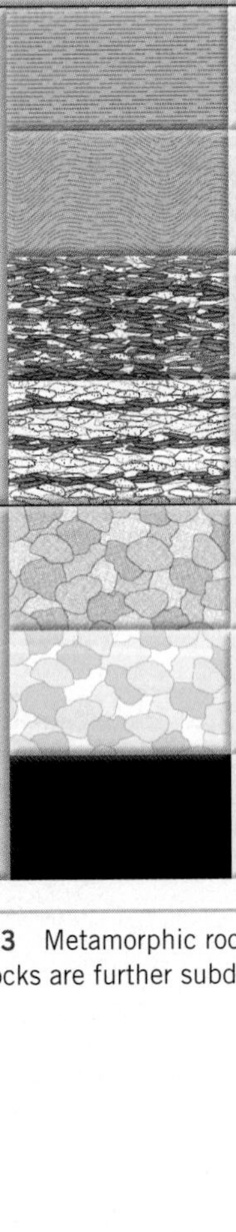

Texture	Grain Size	Rock Name		Comments	Parent Rock
Foliated	Very fine	Slate	Increasing Metamorphism	Excellent rock cleavage, smooth dull surfaces	Shale, mudstone, or siltstone
	Fine	Phyllite		Breaks along wavey surfaces, glossy sheen	Slate
	Medium to Coarse	Schist		Micas dominate, scaly foliation	Phyllite
	Medium to Coarse	Gneiss		Compositional banding due to segregation of minerals	Schist, granite, or volcanic rocks
Nonfoliated	Medium to coarse	Marble		Interlocking calcite or dolomite grains	Limestone, dolostone
	Medium to coarse	Quartzite		Fused quartz grains, massive, very hard	Quartz sandstone
	Fine	Anthracite		Shiny black organic rock that may exhibit conchoidal fracture	Bituminous coal

Figure 2.33 Metamorphic rock identification key. Metamorphic rocks are divided into the two textual groups, foliated and nonfoliated. Foliated rocks are further subdivided based upon the size of the mineral grains.

Specimen Number	Foliated or Nonfoliated	Grain Size	Composition (if identifiable)	Rock Name

Figure 2.34   Metamorphic rock identification chart.

Notes and calculations.

Common Rocks

Date Due: _____

Name: _____

Date: _____

Class: _____

After you have finished Exercise 2, complete the following questions. You may have to refer to the exercise for assistance or to locate specific answers. Be prepared to submit this summary/report to your instructor at the designated time.

1. Write a brief definition of each of the three rock types.

   Igneous rocks: _____

   _____

   _____

   Sedimentary rocks: _____

   _____

   _____

   Metamorphic rocks: _____

   _____

   _____

2. What unique factor about the arrangement of mineral crystals occurs in many metamorphic rocks?

   _____

3. Describe the procedure you would follow to determine the name of a specific igneous rock.

   _____

   _____

   _____

4. Describe the basic difference between detrital and chemical sedimentary rocks.

   _____

   _____

5. List the *texture* and mineral *composition* of each of the following rocks.

	TEXTURE	MINERAL COMPOSITION
Granite:	_____	_____
Marble:	_____	_____
Sandstone:	_____	_____

6. What are two possible environments for the origin of the sedimentary rock sandstone?

   _____

   _____

   _____

7. Of the three rock types, which one is most likely to contain fossils? Explain the reason for your choice.

   _____

   _____

   _____

8. What factor determines the size of the crystals in igneous rocks?

   _____

9. What is a good chemical test to determine the primary mineral in limestone?

   _____

10. What factor(s) determine(s) the size of crystals in metamorphic rocks?

    _____

    _____

11. If the sedimentary rock limestone is subjected to metamorphism, what metamorphic rock will likely form?

   _____

12. With reference to the rock cycle, describe the processes and changes that an igneous rock will undergo as it is changed first to a sedimentary rock, which then becomes a metamorphic rock.

   _____

   _____

   _____

13. Select two igneous, two sedimentary, and two metamorphic rocks that you identified, and write a brief description of each.

   Rock type: _____

   Rock name: _____

   Description: _____

   _____

   _____

   Rock type: _____

   Rock name: _____

   Description: _____

   _____

   _____

   Rock type: _____

   Rock name: _____

   Description: _____

   _____

Rock type: _____

Rock name: _____

Description: _____

_____

_____

Rock type: _____

Rock name: _____

Description: _____

_____

_____

Rock type: _____

Rock name: _____

Description: _____

_____

_____

14. Referring to Figure 2.35, list each rock's name and write a brief description of each.

   A: _____

   _____

   _____

   _____

   B: _____

   _____

   _____

   _____

   C: _____

   _____

   _____

   _____

A.   B.   C.

Figure 2.35 Three rock specimens for use with question 14.

Introduction to Aerial Photographs and Topographic Maps

Aerial photographs, satellite images, and topographic maps are important research tools that provide insight into the various processes that shape the surface of the land. Each is an indispensable method for reducing vast amounts of data to a scale that can be easily managed. The ability of an Earth scientist to effectively interpret and use these tools is essential to identifying and understanding Earth features.

Objectives

After you have completed this exercise, you should be able to:

1. Use a stereoscope to view a stereogram, a pair of aerial photographs.

2. Explain what a topographic map is and how it can be used to study landforms.

3. Use map scales to determine distances.

4. Determine the latitude and longitude of a place from a topographic map.

5. Use the Public Land Survey system to locate features.

6. Explain how contour lines are drawn and be able to use contours to determine elevation, relief, and slope of the land.

7. Construct a simple contour map.

8. Construct a topographic profile.

Materials

ruler   hand lens

Materials Supplied by Your Instructor

stereoscope   string
topographic map
United States and
    world wall maps

Terms

topographic map	Public Land	contour line
stereoscope	Survey	contour
stereogram	base line	interval
datum	principal	index contour
quadrangle	meridian	bench mark
magnetic	township	slope
declination	range	relief
map scale	congressional	topographic
fractional scale	township	profile
graphic scale	section	

Aerial Photographs

Aerial photographs are useful for geological, environmental, agricultural, and related studies. Photographs of the same feature, when taken sequentially and overlapped, can be viewed in three dimensions through a viewer called a **stereoscope**.

To view a stereoscopic aerial photograph, called a **stereogram**, the stereoscope is placed directly over the line separating any two photos of the same feature (Figure 3.1). As you look through a stereoscope, it may have to be moved around slightly until the image appears in three dimensions. The observed heights will be vertically exaggerated, and the difference in heights you see through the stereoscope will not be the same as the actual difference in heights on the land.

To provide some practice viewing stereograms, obtain a stereoscope from your instructor, unfold it, and center it over the line that separates the two aerial photographs of the volcanic cone in Figure 3.2. As you view the photographs, adjust the stereoscope until the cone appears in three dimensions. You may have to be patient until your eyes focus.

Use the stereogram in Figure 3.2 to answer questions 1–5.

1. Identify and label the crater at the summit of the volcano.

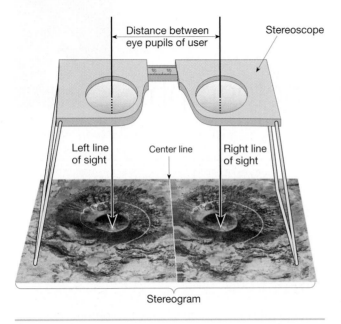

Figure 3.1 Aligning a stereoscope to view a stereogram.

2. Outline and label the lava flow, located at the intersection of the two coordinates, 1.5 and A.8, at the base of the volcano.

3. What is the white, curved feature that extends from the base of the cone to its summit?

_____

4. Mark the highest point on the volcano with an "X."

5. Assume the summit of the volcanic cone is 1500 feet above the surrounding land. While viewing the cone through the stereoscope, draw lines around the volcano at approximately 400-foot intervals above the local surface.

In addition to aerial photographs, beginning in the early 1970s, the United States began launching several satellites that systematically collect images of Earth's surface using a variety of remote sensing techniques. The ability of the images to be computer manipulated

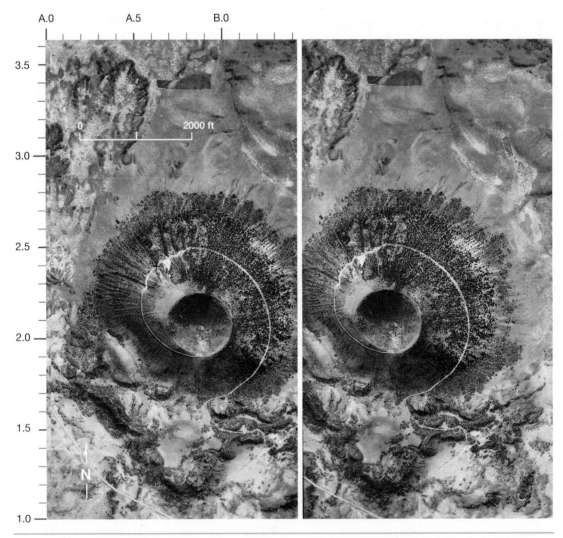

Figure 3.2 Stereogram of Mt. Capulin, a volcanic cinder cone located in northeastern New Mexico. The stereogram is composed from two overlapping aerial photographs taken from an altitude of approximately 16,000 feet. To view the three-dimensional image of the volcano, center a stereoscope over the line that separates the two photographs. Then, while looking through the stereoscope, adjust the stereoscope until the image appears in three dimensions. (Courtesy of U.S. Geological Survey)

Figure 3.3 Satellite image of a portion of the delta of the Mississippi River in May 2001. The image covers an area of 54 × 57 kilometers. For the past 600 years or so, the main flow of the river has been along its present course, extending southeast from New Orleans. During that span, the delta advanced into the Gulf of Mexico at a rate of about 10 kilometers (6 miles) per century. For a more detailed investigation of satellite images see the Internet activities at the URL listed at the end of this exercise. (Photo courtesy of NASA)

and enhanced has made a tremendous amount of new data available to Earth scientists (Figure 3.3). (Further investigations of satellite imagery can be found at the URL indicated at the end of this exercise.)

Topographic Maps

One type of map, the **topographic map**, is most useful when investigating the many kinds of landforms that exist on Earth's surface.

Topography means "the shape of the land." Each topographic map shows, to scale, the width, length, and variable height of the land above a **datum** or reference plane—generally average sea level. The maps, which are also referred to as **quadrangles**, are two-dimensional representations of the three-dimensional surface of Earth. Their primary value to the Earth scientist is for determining locations, landform types, elevations, and other physical data.

Topographic maps have been produced by the United States Geological Survey (USGS) since the late 1800s. Today, a vast area of the United States has been accurately portrayed on these commercially available maps.

To facilitate their use, topographic maps follow a similar format. In addition to standard colors and symbols, each contains information about where the

area mapped is located, the date when the mapping was done or revised, scale, north arrow, and the names of adjoining quadrangle maps.

To help understand the basics of topographic maps, obtain a copy of a topographic map from your instructor and examine it. You will use this map to answer specific groups of questions that follow. PLEASE DO NOT WRITE OR MARK ON THE MAP.

General Map Information

Every topographic map contains useful information printed in its margin. Locate and record the following information for your map.

Each topographic map is assigned a name for reference. The name of a topographic map is located in the upper-right corner of the map.

6. What is the name of your map?

    Map name: _____

Notice the small reference map and compass arrow in the lower margin of the map.

7. In what part of the state (north, southwest, etc.) is the area covered by your map located?

    _____

The names of adjoining maps are given along the four margins and four corners of the map.

8. What is the name of the map that adjoins the northeast corner of your map?

    Adjoining map: _____

Information about when the area was surveyed and the map published is provided in the margin of the map.

9. When was the area surveyed? When was the map published? If the map has been revised, when was the revision completed?

    Surveyed: _____ Published: _____ Revised: _____

Since the geographic North Pole and North Magnetic Pole of Earth do not coincide, the north arrow on a topographic map often shows the difference between true north (TN) and magnetic north (MN), the direction a compass would point, for the area represented. This difference in degrees is called the **magnetic declination**.

10. What is the magnetic declination of the area shown on your map?

     Magnetic declination: _____

Map Colors and Symbols

Each symbol and color used on a U.S. Geological Survey topographic map has a meaning. Refer to the inside cover of this manual and carefully examine the standard U.S. Geological Survey topographic map symbols.

Using the standard map symbols as a guide, locate examples of various types of roads, buildings, and streams on the topographic map supplied by your instructor.

11. In general, what color(s) are used for the following types of features?

Highways and roads: _____

Buildings: _____

Urban areas: _____

Wooded areas: _____

Water features: _____

Map Scale

Many people have built or seen scale model airplanes or cars that are miniature representations of the actual objects. Maps are similar in that they are "scale models" of Earth's surface. Each map will have a **map scale** that expresses the relation between distance on the map to the true distance on Earth's surface. Different map scales depict an area on Earth with more or less detail. On a topographic map, scale is usually indicated in the lower margin and is expressed in two ways.

Fractional scale (e.g. 1/24,000 or 1:24,000) means that a distance of 1 unit on the map represents a distance of 24,000 of the *same* units on the surface of Earth. For example, one inch on the map equals 24,000 inches on Earth, or one centimeter on the map equals 24,000 centimeters on Earth. Maps with small fractional scales (fractions with large numbers in the denominator, e.g., 1/250,000) cover large areas. Those with large fractional scales (fractions with small numbers in the denominator, e.g., 1/1,000) cover small areas. The United States Geological Survey publishes maps at various scales to meet both the need for broad coverage and detail (see Figures 3.4 and 3.5).

Graphic, or **bar, scale** is a bar that is divided into segments that show the relation between distance on the map to actual distance on Earth (Figure 3.6). Scales showing miles, feet, and kilometers are generally included. The left side of the bar is often divided into fractions to allow for more accurate measurement of distance. The graphic scale is more useful than the fractional scale for measuring distances between points. Graphic scales can be used to make your own "map ruler" for measuring distances on the map using a piece of paper or string.

12. Examine your topographic map as well as the large wall maps in the laboratory and write out the fractional scale for each in the following space. Then answer questions 12a and 12b.

Topographic map: _____ : _____

Wall map of the United States (if available):

_____ : _____

World map (if available):

_____ : _____

a. Which of the three maps has the smallest scale (largest denominator in the fractional scale)?

_____

b. Which of the three maps covers more square miles?

_____

Figure 3.4 Standard U.S. Geological Survey topographic map scales, sizes, and coverage.

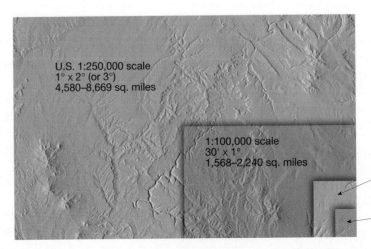

U.S. 1:250,000 scale
1° x 2° (or 3°)
4,580–8,669 sq. miles

1:100,000 scale
30' x 1°
1,568–2,240 sq. miles

1:62,500 scale
15' series (15' x 15')
197–282 sq. miles

1:24,000 scale
7.5' series (7.5' x 7.5')
49–70 sq. miles

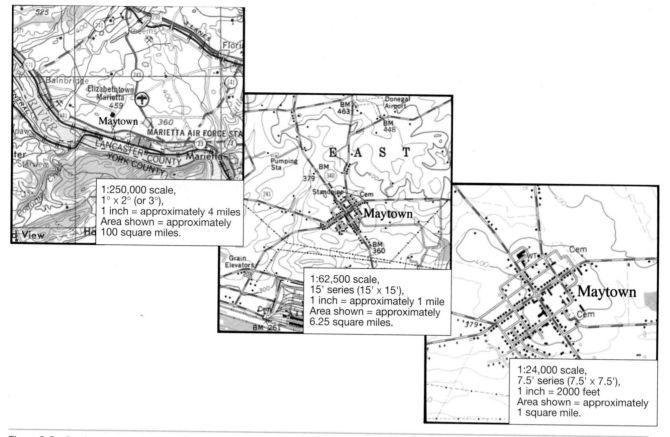

Figure 3.5 Portions of three topographic maps of the same area showing the effect that different map scales have on the detail illustrated.

13. Depending upon the map scale, one inch on a topographic map represents various distances on Earth. Convert the following scales.

SCALE	1 INCH ON THE MAP REPRESENTS
1 : 24,000	_____ feet on Earth
1 : 63,360	_____ mile(s) on Earth
1 : 250,000	_____ miles on Earth

14. Use the graphic scale provided on your topographic map to construct a "map ruler" in miles, and measure the following distances that are represented on the map.

Width of the map along the south edge = _____ miles

Length of the map along the east edge = _____ miles

15. How many square miles are represented on your topographic map? (HINT: The area of a rectangle is calculated using the formula, area = width × length)

Map area equals _____ square miles

Location

One of the most useful functions of a topographic map is determining the precise location of a feature on Earth's surface. Two frequently used methods for designating location are 1) latitude and longitude to determine the location of a point and 2) the **Public Land Survey** system (PLS) to define an area. Because topographic maps

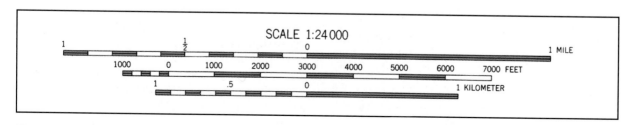

Figure 3.6 Typical graphic scale.

are very accurate, both methods of location can be used to provide information helpful to engineers, surveyors, realtors, and others. A third method, the Universal Transverse Mercator (UTM) grid, is investigated at the website located at the URL listed at the end of this exercise.

Latitude and Longitude. Topographic maps are bounded by parallels of latitude on the north and south, and meridians of longitude on the east and west. The latitudes and longitudes covered by the quadrangles are printed at the four corners of the map in degrees (°), minutes ('), and seconds (") and are indicated at intervals along the margins. Maps that cover 15 minutes of latitude and 15 minutes of longitude are called *15-minute series topographic maps*, and although no longer produced by the USGS are still available. A $7\frac{1}{2}$-minute series topographic map covers $7\frac{1}{2}$-minutes of latitude and $7\frac{1}{2}$-minutes of longitude (see Figure 3.4). (*Note*: There are 60 minutes of arc in one degree and 60 seconds of arc in one minute of arc. Therefore, $\frac{1}{2}$-minute is the same as 30 seconds.) A more complete examination of latitude and longitude can be found in Exercise 21 "Location and Distance on Earth."

Use the topographic map supplied by your instructor to answer questions 16–22. PLEASE DO NOT WRITE OR MARK ON THE MAPS.

16. What are the latitudes of the southern edge and northern edge of the map to the nearest $\frac{1}{2}$ minutes of latitude?

    Latitude of southern edge: _____

    Latitude of northern edge: _____

17. How many total minutes of latitude does the map cover?

    _____ minutes of latitude

18. What are the longitudes of the eastern edge and western edge of the map to the nearest $\frac{1}{2}$ minutes of longitude?

    Longitude of eastern edge: _____

    Longitude of western edge: _____

19. How many total minutes of longitude does the map cover?

    _____ minutes of longitude

20. The map is a _____ -minute series topographic map because it covers _____ minutes of latitude and _____ minutes of longitude.

21. The total minutes of latitude and total minutes of longitude covered by the map are equal. Why is the appearance of the map rectangular rather than square?

    _____

    _____

22. Your instructor will supply you with the names of two features (school, church, etc.) located on the map. Write the name of each feature, as well as its latitude and longitude to the nearest minute, in the following spaces.

    Feature name: _____

    Latitude: _____ Longitude: _____

    Feature name: _____

    Latitude: _____ Longitude: _____

Public Land Survey. The Public Land Survey (PLS) provides a precise method for identifying the location of land in most states west of the Appalachian Mountains by establishing a grid system that systematically subdivides the land area (Figure 3.7). The Public Land Survey begins at an initial point (generally there are one or more initial points for each state that utilizes the PLS). An east-west line, called a **base line**, and a north-south line, called a **principal meridian**, extend through the initial point and provide the basis of the grid (see Figures 3.7 and 3.8).

Horizontal lines at six-mile intervals that parallel the base line establish east-west tracts, called **townships**. Each township is numbered north and south from the base line. The first horizontal six-mile-wide tract north of the base line is designated Township One North (T1N), the second T2N, etc. Vertical lines at six-mile intervals that parallel the principal meridian define north-south tracts, called **ranges**. Each range is numbered east and west of the principal meridian. The first vertical six-mile wide tract west of the principal meridian is designated Range One West (R1W), the second R2W, etc. *On a topographic map, the townships and ranges covered by the map are printed in red along the margins.*

The intersection of a township and a range defines a six-mile-by-six-mile rectangle, called a **congressional township**, which may or may not coincide with a civil township. Each congressional township is identified by referring to its township and range numbers. For example, in Figure 3.7A, the shaded congressional township would be identified as T1N, R4W.

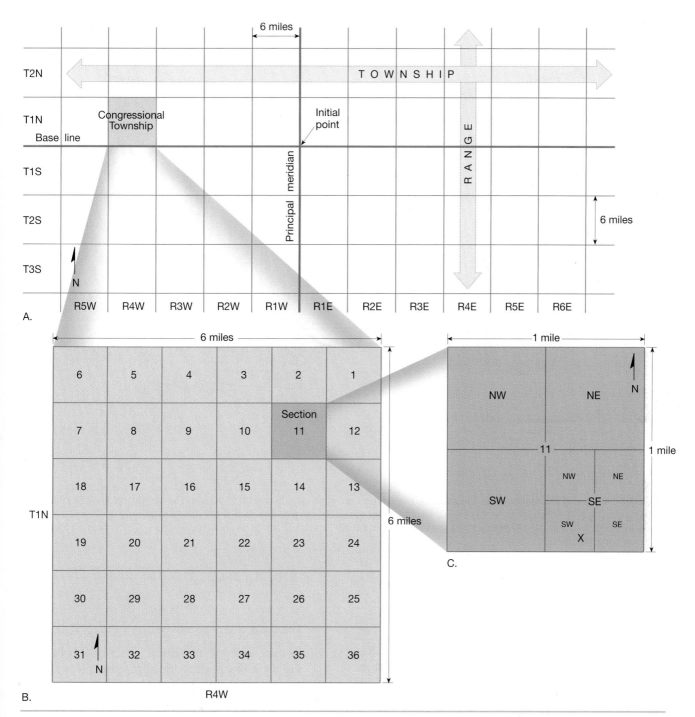

Figure 3.7 The Public Land Survey system (PLS).

Each congressional township is divided into 36 one-mile-square parcels of land, called **sections**, with each section containing 640 acres. Sections are numbered beginning with number one in the northeast corner of the congressional township and ending with number 36 in the southeast corner (Figure 3.7B). The shaded section of land in Figure 3.7B would be designated as Section 11, T1N, R4W. *On a topographic map, the sections are outlined and their numbers are printed in red.*

For more detailed descriptions, sections may be subdivided into halves, quarters, or quarters of a quarter (Figure 3.7C). Each of these subdivisions are identified by their compass position. For example, the forty acre area designated with the letter X in Figure 3.7C would be described as the SW$\frac{1}{4}$ (southwest $\frac{1}{4}$), of the SE$\frac{1}{4}$ (southeast $\frac{1}{4}$) of Section 11. Hence, the complete locational description of the area marked with the letter X would be SW$\frac{1}{4}$, SE$\frac{1}{4}$, Sec. 11, T1N,

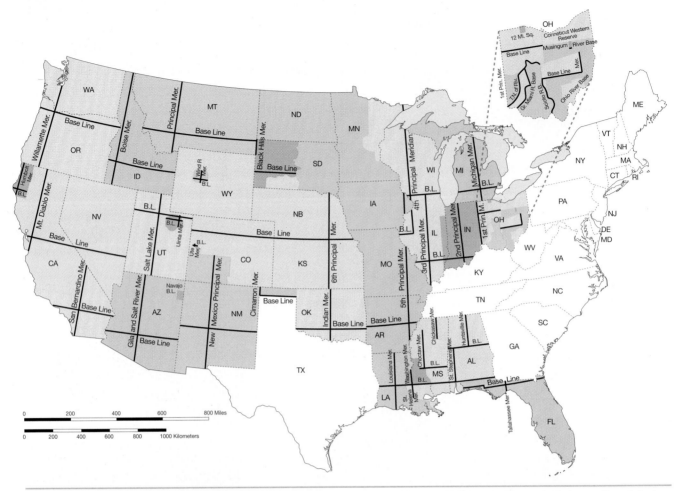

Figure 3.8 Regions in the conterminous United States that utilize the Public Land Survey (PLS) system. The unshaded areas (the original thirteen states and five others, plus Texas) were, for the most, settled prior to the establishment of the PLS in the late 1700s and utilize a variety of less structured systems (often referred to as "metes and bounds" descriptions) for designating land. (U.S. Department of Interior, Bureau of Land Management)

R4W. *By convention, in the description the smallest subdivision is given first and the township number precedes the range number.*

Figure 3.9 illustrates a hypothetical area that has been surveyed using the Public Land Survey system. Figure 3.9A is a township and range diagram, 3.9B represents a congressional township within the township and range system, and Figure 3.9C is a section of the congressional township. Use Figures 3.9A–3.9C to complete questions 23–25.

23. Use the PLS system to label the townships, ranges, and sections in Figure 3.9A and 3.9B with their proper designation.

24. Follow each of the letters. A through D, through Figure 3.9C–3.9A and write the PLS location description of each in the following spaces. As an example, letter A has already been done.

A: <u>NW</u> $\frac{1}{4}$, <u>SW</u> $\frac{1}{4}$, Sec. <u>8</u>, T <u>3N</u>, R <u>4W</u>

B: ____ $\frac{1}{4}$, ____ $\frac{1}{4}$, Sec. ____, T____, R____

C: ____ $\frac{1}{4}$, ____ $\frac{1}{4}$, Sec. ____, T ____, R ____

D: ____ $\frac{1}{4}$, ____ $\frac{1}{4}$, Sec. ____, T ____, R ____

25. Locate each of the areas described below on Figure 3.9 by placing the appropriate letter in the proper places in Figure 3.9A–3.9C.

E: SW$\frac{1}{4}$, SW$\frac{1}{4}$, Sec. 5, T5N, R3E

F: SE$\frac{1}{4}$, NE$\frac{1}{4}$, Sec. 34, T4S, R7W

Use the topographic map supplied by your instructor to answer questions 26–29. PLEASE DO NOT WRITE OR MARK ON THE MAPS.

26. List the townships and ranges represented on the map.

Townships: _____

Ranges: _____

27. To reach the principal meridian that was used to survey the land represented on the map, people

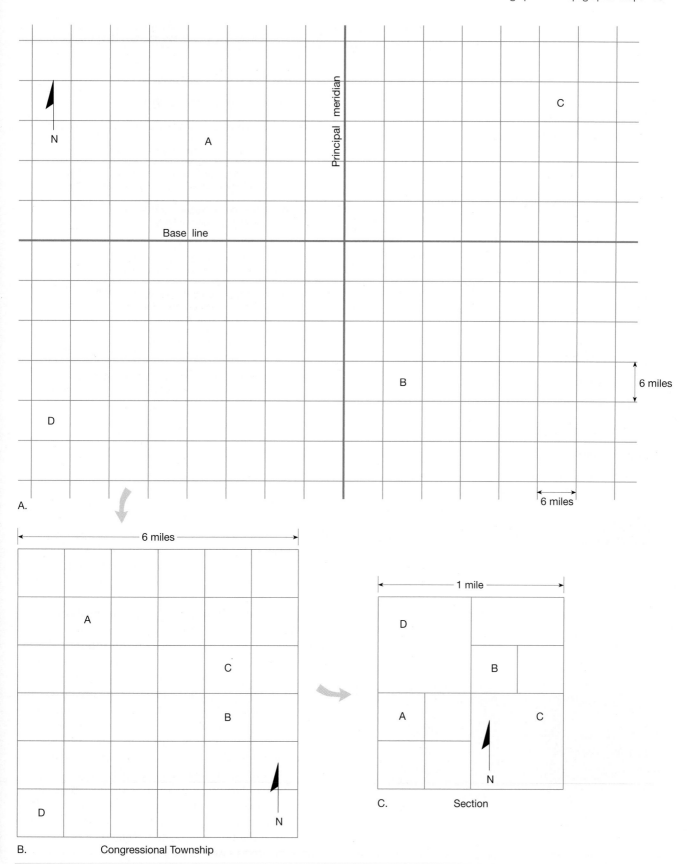

Figure 3.9 Hypothetical Public Land Survey system showing the locations of various points.

living within the area would have to travel (east-ward, westward). To reach the base line, they would travel to the (north, south). Circle your answers.

28. What is the section, township, and range at each of the following locations on the map?

    Exact center of the map:

    Sec. _____ , T _____ , R _____

    Extreme NE corner of the map:

    Sec. _____ , T _____ , R _____

29. Your instructor will supply you with the names of three features (school, church, etc.) located on the map. Using the PLS system, write each of the feature's complete location to the nearest $\frac{1}{4}$ of a $\frac{1}{4}$ section in the following spaces.

    Feature name: _____

    Location: _____

    Feature name: _____

    Location: _____

    Feature name: _____

    Location: _____

Contour Lines

Depicting the height or elevation of the land, thereby showing the shape of landforms, is what makes a topographic map unique. A **contour line** is a line drawn on a topographic map that connects all points that have equal elevations above or below a datum or reference plane on Earth's surface (Figure 3.10). The reference

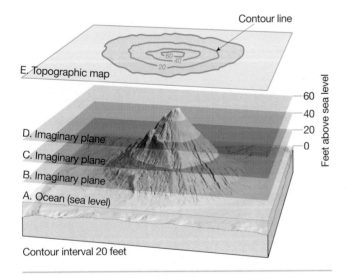

Figure 3.10 Schematic illustration showing how contour lines are determined when topographic maps are constructed. **A** is the ocean surface. **B** is an imaginary plane 20 feet above the ocean that intersects the land. **C** is an imaginary plane 40 feet above the ocean that intersects the land. **D** is an imaginary plane 60 feet above the ocean that intersects the land. **E** is the topographic map that results when the contour lines that mark where the imaginary planes intersect the surface are drawn on a map.

GENERAL RULES FOR CONTOUR LINES

1. A contour line connects points of equal elevation.

2. A contour line never branches or splits.

3. Steep slopes are shown by closely spaced contours.

4. Contour lines never cross, except to show an overhanging cliff. (To show an overhanging cliff, the hidden contours are dashed. Contour lines can also merge to form a single line along a vertical cliff.)

5. Hills are represented by a concentric series of closed contour lines.

6. A concentric series of closed contours with hachure marks on the downhill side represents a closed depression.

7. When contour lines cross streams or dry stream channels, they form a V that points upstr eam.

8. Contour lines that occur on opposite sides of a valley always occur in pairs.

9. Topographic maps published by the U.S. Geological Survey are contoured in feet or meters referenced to sea level.

Figure 3.11 Some general rules for contour lines.

plane from which elevations are measured for most topographic maps is mean (average) sea level. The datum for a map is usually indicated in the lower, central margin of the map with the phrase, "Datum is ..."

Contour lines must conform to certain guidelines. Figure 3.11 presents some of the general rules that apply to contour lines.

The **contour interval** (CI) is the vertical difference in elevation between adjacent contour lines. All contour lines are multiples of the contour interval. For example, for a contour interval of 20 feet, the lines may read 420', 440', 460', etc. Most maps use the smallest contour interval possible to provide the greatest detail for the surface that is being mapped. The contour interval of a topographic map is usually indicated in the lower central margin of the map with the phrase, "Contour interval ..." *The contour interval should always be known before using a topographic map.*

To help determine the elevation of the contour lines, on most topographic maps every fifth contour line, called an **index contour**, is printed as a bold line and the elevation of the line is indicated. Reference points of elevation, called **bench marks** (BM), are also often present on the map and can be used to establish elevations.

Contour lines that are close together indicate a steep **slope** (vertical change in elevation per horizontal distance, usually expressed in feet/mile or meters/kilometer), while widely spaced lines show a gradual slope. Consequently, the "shading" that results from closely spaced contour lines allows for the recognition of such features as hills, valleys, ridges, etc.

Relief is defined as the difference in elevation between two points on a map. *Total relief* is the difference between the highest and lowest points on a map. *Local relief* refers to the difference in elevation between two specified points, for example, a hill and nearby valley.

Examining Contour Lines

Figure 3.12 shows a contour map of volcanic cones, along with a stereoscopic contour map of the same

features. The large volcano illustrated is very similar to the one you observed in the stereogram, Figure 3.2.

30. Examine the stereoscopic contour map in Figure 3.12 by centering your stereoscope over the center line and observing the three-dimensional image. Before continuing the exercise, compare the stereoscopic image closely with the contour map. Examine the features noted in the caption on both maps, paying particular attention to the use of *hachure marks* on contours to show a depression.

A. Contour Map

Scale 1:17000
Contour interval 20 feet

Figure 3.12 Contour map and stereoscopic contour map of the same volcanic cones. The two maps show a large, steep-sided cinder cone (**A**) with a well-developed crater (**C**) at its summit. Ridges (**E**) and depressions (**F**) are evident on the lava flows (**D**), while water erosion has carved gullies (**B**) on the sides of the cones. (From Horace MacMahan, Jr., *Stereogram Book of Contours*, p. 18. Copyright (c) 1972, Hubbard Scientific Company. Reprinted by permission of American Educational Products—Hubbard Scientific)

Center line

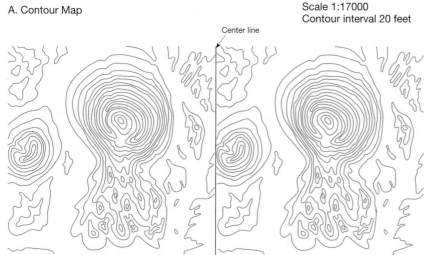

B. Stereoscopic Contour Map

Figure 3.13 shows both a perspective view and contour map of a hypothetical area situated along an ocean coast. The elevations in feet above mean sea level of several contour lines and points are identified on the map for reference. Use Figure 3.13 to answer questions 31–37.

31. What is the contour interval that has been used on the map?

    Contour interval: _____ feet

32. Indicate the two areas on the map that have the steepest slopes by writing the word "steep" on the map. What characteristic of the contour lines shows that the slopes are steep?

    _____

33. Notice what happens to the contour lines as they cross a stream. The "peak" formed by a contour line as it crosses a stream points (upstream, downstream). Circle your answer.

34. What are the elevations of the points designated with the following letters?

    Point A: _____ feet

    Point B: _____ feet

    Point C: _____ feet

35. The approximate elevation of the church is (12, 22, 32) feet. Circle your answer.

36. What is the total relief shown on the map?

    Highest elevation (_____ ft) − lowest elevattion (_____ ft) = total relief (_____ ft)

A. Perspective aerial view

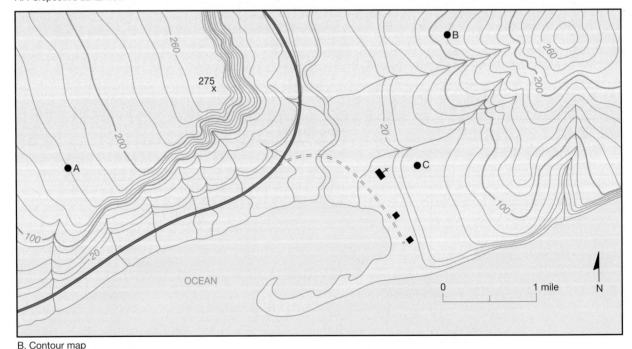

B. Contour map

Figure 3.13 Perspective (A) and map (B) view of a hypothetical coastal area. All elevations are in feet above mean sea level. (After U.S. Geological Survey)

37. What is the slope of the mountain located on the east side of the diagram from its summit, directly south to the ocean?

Slope = _____ feet/mile

Use the topographic map supplied by your instructor to answer questions 38–43. PLEASE DO NOT WRITE OR MARK ON THE MAPS.

38. What is the datum that has been used for determining the elevations on the map?

Datum: _____

39. What is the contour interval of the map?

Contour interval: _____ feet

40. What are the lowest and highest elevations found on the map?

Lowest elevation: _____ feet

Highest elevation: _____ feet

41. What is the elevation of the exact center of the map?

Elevation: _____ feet

42. Your instructor will supply you with the names of three features (school, church, etc.) located on the map. Write the elevation of each feature in the following spaces.

Feature name: _____

Elevation: _____

Feature name: _____

Elevation: _____

Feature name: _____

Elevation: _____

43. After examining the contour lines, etc., write a brief description of the slope of the land represented by the map.

_____

_____

_____

Constructing a Contour Map

Originally, contour maps were constructed by first surveying an area and establishing the elevations of several points in the field. The surveyor then sketched contour lines on the map by estimating their location between the points of known elevation. Today, topographic maps are made from stereoscopic aerial photographs that are computer processed to determine elevations and contours.

44. To help understand the process of drawing a contour map, using a pencil, complete the contour map shown in Figure 3.14. The points illustrated are of known elevation. The 100-foot contour line has been drawn to provide a reference. Using a 20-foot contour interval, draw a contour line for each 20-foot change in elevation below and above 100 feet—for example, 80 feet, 60 feet, 120 feet, etc. You will have to estimate the elevations between the points. Label each of the lines with the proper elevation.

45. In Figure 3.14, in general, the land slopes toward the (north, south). Circle your answer.

46. After examining the contour lines and elevations in Figure 3.14, show the directions that the streams are flowing by drawing arrows on the map.

47. What is the average slope of the stream on the west side of the map you drew in Figure 3.14?

Slope = _____ feet/mile

Drawing a Topographic Profile

Topographic maps, like most other maps, depict Earth's surface viewed from above. Often a topographic profile or "side-view" will provide a more useful representation of the elevations and slopes of an area. To change an overhead, or map, view into a profile, follow the steps illustrated in Figure 3.15.

48. Use the horizontal line and vertical scale in Figure 3.16 to construct a west-east profile along the profile line indicated on the contour map you have drawn in Figure 3.14. Follow the guidelines for preparing a topographic profile in Figure 3.15.

Air Photos, Satellite Images, and Maps on the Internet

Apply the concepts from this exercise in an examination of the aerial/satellite photographs and topographic maps that are available for your area by completing the corresponding online activity on the *Applications & Investigations in Earth Science* website at http://prenhall.com/earthsciencelab

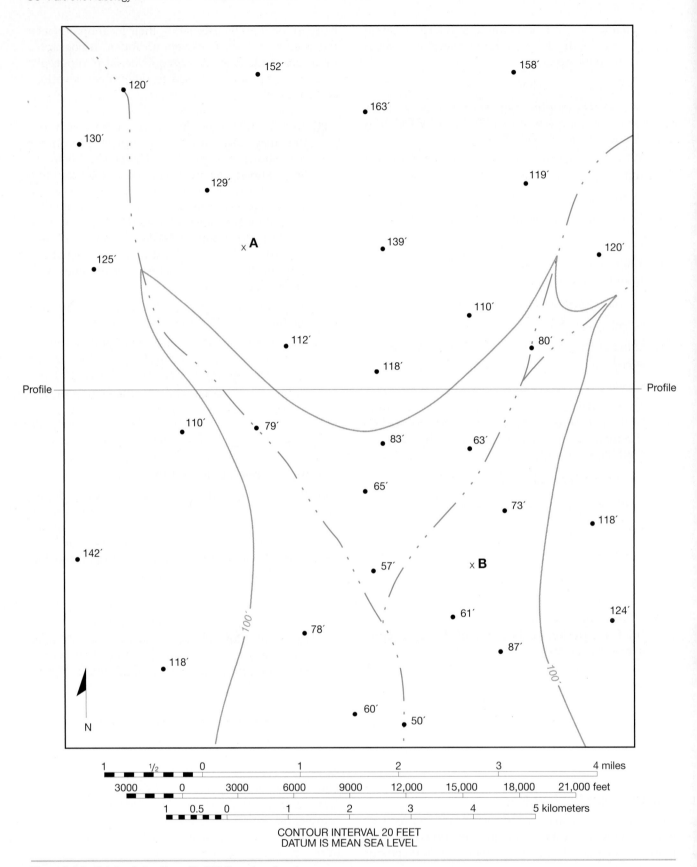

Figure 3.14 Points of elevation with 100-ft contour line drawn.

Step 1

Figure 3.15 Construction of a topograph profile. **Step 1**. On the topographic map, draw a line along which the profile is to be constructed. Label the line A–A′.

As shown in **Step 2**, lay a piece of paper along line A–A′. Mark each place where a contour line intersects the edge of the paper and note the elevation of the contour line by each mark.

In **Step 3**, on a separate piece of paper, draw a horizontal line slightly longer than your profile line, A–A′. Select a vertical scale for your profile that begins slightly below the lowest elevation along the profile and extends slightly beyond the highest elevation. Mark this scale off on either side of the horizontal line. Lay the marked paper edge (from Step 2) along the horizontal line. Wherever you have marked a contour line on the edge of the paper, place a dot directly above the mark at an elevation on the vertical scale equal to that of the contour line. Connect the dots on the profile with a smooth line to see the finished product. (Note: Since you have more or less arbitrarily selected the vertical scale for the profile, *the finished profile may be somewhat vertically exaggerated and not the same as you would see it from the ground.*)

Step 2

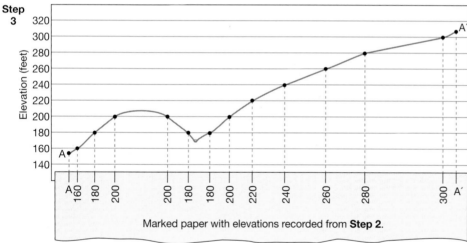

Step 3

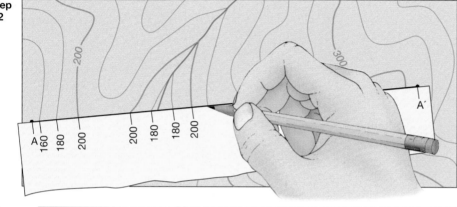

Marked paper with elevations recorded from **Step 2**.

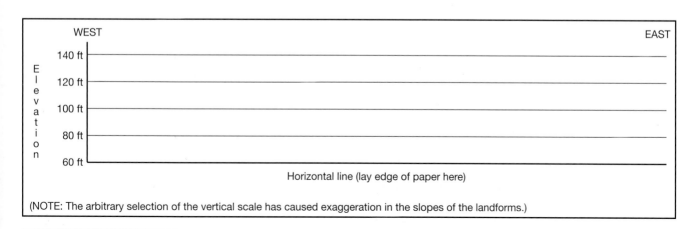

(NOTE: The arbitrary selection of the vertical scale has caused exaggeration in the slopes of the landforms.)

Figure 3.16 West–east topographic profile along the profile line on Figure 3.14.

Notes and calculations.

Introduction to Aerial Photographs and Topographic Maps

Date Due: _____

Name: _____

Date: _____

Class: _____

After you have finished Exercise 3, complete the following questions. You may have to refer to the exercise for assistance or to locate specific answers. Be prepared to submit this summary/report to your instructor at the designated time.

1. Use Figure 3.17, a portion of the Ontario, California, topographic map, to answer questions 1a.–1i.

a. One inch on the map is approximately _____ mile(s).

b. The fractional scale of the map is (1:12,000; 1:24,000; 1:62,500; 1:250,000). Circle your answer.

c. What is the direction and shortest distance a hiker would have to travel to reach the "Big

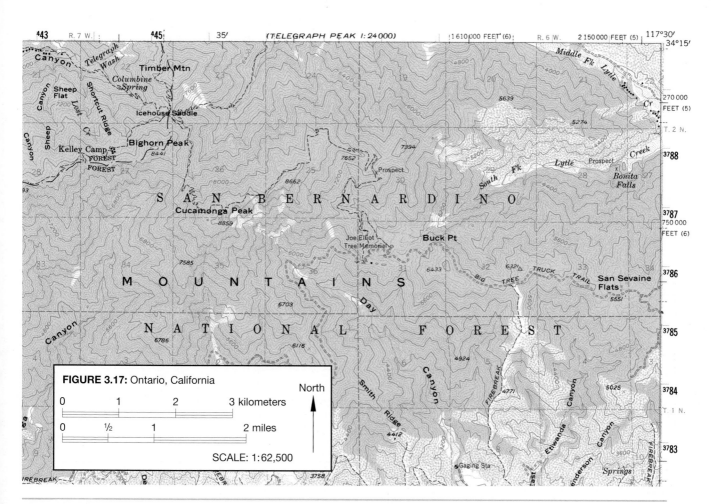

FIGURE 3.17: Ontario, California

0   1   2   3 kilometers

0   ½   1   2 miles

North

SCALE: 1:62,500

Figure 3.17   Portion of the Ontario, CA, topographic map to be used with question 1.

Tree Truck Trail" from Kelley Camp?

_____

_____

d. What is the approximate elevation of Kelley Camp?

_____

e. What is the contour interval of the map?

_____

f. Toward what direction does the stream in Day Canyon flow? How did you arrive at your answer?

_____

_____

g. Is the slope of the stream in Day Canyon steeper near Cucamonga Peak or the gaging station? How did you arrive at your answer?

_____

_____

h. Portions of which townships and ranges are covered by the map?

_____

i. Give the complete PLS location of Kelley Camp to the nearest $\frac{1}{4}$ of a $\frac{1}{4}$ section.

_____

2. What is the latitude and longitude to the nearest minute of the *exact center* of the topographic map supplied by your instructor?

Latitude: _____ Longitude: _____

3. The topographic map supplied by your instructor is a _____ -minute series topographic map, which means that it:

_____

4. What are the numbers of the townships and ranges covered by the topographic map supplied by your instructor?

Townships: _____

Ranges: _____

5. Use the Public Land Survey system to give the name and location of a feature your instructor requested that you locate on your topographic map in question 29.

Name of feature: _____

Location:

_____ $\frac{1}{4}$, _____ $\frac{1}{4}$, Sec. _____, T _____, R _____

6. What was your calculated slope for the mountain in question 37?

Slope: _____ feet/mile

7. What was the elevation of a feature your instructor requested that you determine on your topographic map in question 42?

Feature name: _____

Elevation: _____

8. In Figure 3.18, sketch a copy of the west-east topographic profile you constructed in question 48. Label the appropriate elevations on the vertical axis of your sketch.

WEST   EAST

Figure 3.18   West–east topographic profile along the profile line on Figure 3.14.

Shaping Earth's Surface
Running Water and Groundwater

The study of the processes that modify Earth's surface is of major significance to the Earth scientist. By understanding those processes and the features they produce, scientists gain insights into the geologic history of an area and make predictions concerning its future development.

Some of the agents that are responsible for modifying the surface of Earth are running water (Figure 4.1), groundwater, glacial ice, wind, and volcanic activity. Each produces a unique landscape with characteristic features that can be recognized on topographic maps. Exercises 4 and 5 examine several of these agents, the variety of landforms associated with them, and some of the consequences of human interaction with these natural systems.

Objectives

After you have completed this exercise, you should be able to:

1. Sketch, label, and discuss the complete hydrologic cycle.
2. Explain the relation between infiltration and runoff that occurs during a rainfall.
3. Discuss the effect that urbanization has on the runoff and infiltration of an area.
4. Identify on a topographic map the following features that are associated with rivers and valleys: rapids, meanders, floodplain, oxbow lake, and backswamps.
5. Explain the occurrence, fluctuation, use, and misuse of groundwater supplies.
6. Identify on a topographic map the following features associated with karst landscapes: sinkholes, disappearing streams, and solution valleys.

Materials

calculator   hand lens
ruler

Materials Supplied by Your Instructor

graduated measuring   coarse sand, fine sand,
  cylinder (100 ml)   soil

Figure 4.1 Mountain stream. (Photo by E.J. Tarbuck)

small funnel   stereoscope
cotton   string
beaker (100 ml)

Terms

hydrologic cycle	permeability	aquifer
infiltration	hydrograph	karst topography
groundwater	discharge	disappearing
runoff	base level	stream
erosion	meander belt	solution valley
evaporation	zone of saturation	sinkhole
transpiration	water table	cave
porosity	zone of aeration	cavern

Introduction

On Earth, the water is constantly being exchanged between the surface and atmosphere. The **hydrologic cycle**, illustrated in Figure 4.2, describes this continuous movement of water from the oceans to the atmosphere, from the atmosphere to the land, and from the land back to the sea.

A portion of the precipitation that falls on land will soak into the ground via **infiltration** and become **groundwater**. If the rate of rainfall is greater than the surface's ability to absorb it, the additional water flows over the surface and becomes **runoff**. Runoff initially flows in broad sheets; however, it soon becomes confined and is channeled to form streams and rivers. **Erosion** by both groundwater and runoff wears down the land and modifies the shape of Earth's surface. Eventually runoff and groundwater from the continents return to the sea or the atmosphere, continuing the endless cycle.

Examining the Hydrologic Cycle

Figure 4.2 illustrates Earth's water balance, a quantitative view of the hydrologic cycle. Although the figure correctly implies a uniform exchange of water between Earth's atmosphere and surface on a worldwide basis, factors such as climate, soil type, vegetation, and urbanization often produce local variations.

Use Figure 4.2 as a reference to answer questions 1–6.

1. On a worldwide basis, more water is evaporated into the atmosphere from the (oceans, land). Circle your answer.

2. Approximately what percent of the total water evaporated into the atmosphere comes from the oceans?

$$\text{Percent from oceans} = \frac{\text{ocean evaporation}}{\text{total evaporation}} \times 100$$

= _____ %

Notice in the figure that more water evaporates from the oceans than is returned directly to them by precipitation.

3. Since sea level is not dropping, what are the other sources of water for the oceans in addition to precipitation?

_____

Over most of Earth, the quantity of precipitation that falls on the land must eventually be accounted for by the sum total of **evaporation**, **transpiration** (the release of water vapor by vegetation), **runoff**, and **infiltration**.

4. Define each of the following four variables.

Evaporation: _____

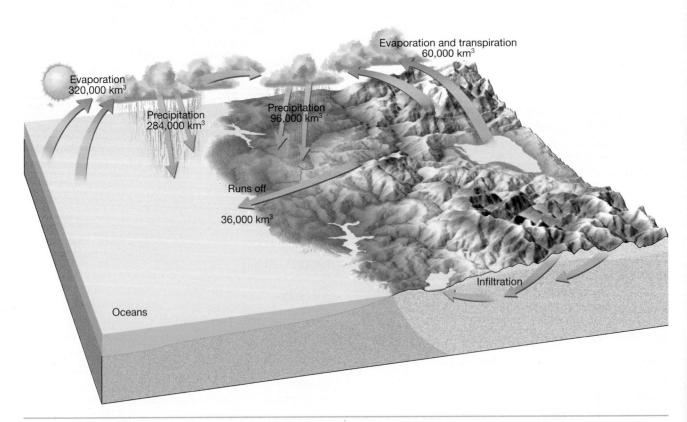

Figure 4.2 Earth's water balance, a quantitative view of the hydrologic cycle.

Transpiration: _____

Runoff: _____

Infiltration: _____

5. On a worldwide basis, about (35, 55, 75) percent of the precipitation that falls on the land becomes runoff. Circle your answer.

6. At high elevations or high latitudes, some of the water that falls on the land does not immediately soak in, run off, evaporate, or transpire. Where is this water being temporarily stored?

_____

_____

Infiltration and Runoff

During a rainfall most of the water that reaches the land surface will infiltrate or run off. The balance between infiltration and runoff is influenced by factors such as the **porosity** and **permeability** of the surface material, slope of the land, intensity of the rainfall, and type and amount of vegetation. After infiltration saturates the land and the ground contains all the water it can hold, runoff will begin to occur on the surface.

7. Describe the difference between the terms *porosity* and *permeability*. Is it possible for a substance to have a high porosity and a low permeability? Why?

_____

_____

_____

_____

Permeability Experiment

To gain a better understanding of how the permeability of various Earth materials affects the flow of groundwater, examine the equipment setup in Figure 4.3 and conduct the following experiment by completing each of the indicated steps.

Step 1. Obtain the following equipment and materials from your instructor:
   graduated measuring cylinder
   beaker
   small funnel
   piece of cotton
   samples of coarse sand, fine sand, and
      soil (enough of each to fill the funnel
      approximately two-thirds full)

Step 2. Place a small wad of cotton in the neck of the funnel.

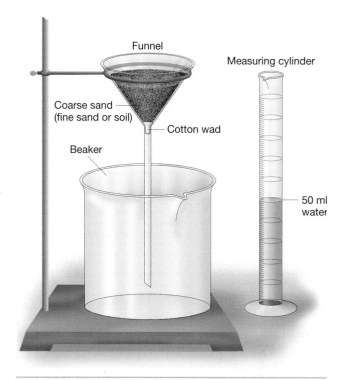

Figure 4.3 Equipment setup for permeability experiment.

Step 3. Fill the funnel above the cotton about two-thirds full with coarse sand.

Step 4. With the bottom of the funnel placed in the beaker, measure the length of time that it takes for 50 ml of water to drain through the funnel filled with coarse sand. Record the time in the data table, Table 4.1.

Step 5. Using the measuring cylinder, measure the amount (in milliliters) of water that has drained into the beaker and record the measurement in the data table.

Step 6. Empty and clean the measuring cylinder, funnel, and beaker.

Step 7. Repeat the experiment two additional times, using fine sand and then soil. Record the results of each experiment at the appropriate place in the data table, Table 4.1. (*Note:* In each case, fill the funnel with the material to the same level that was used for the coarse sand and use the same size wad of cotton.)

Step 8. Clean the glassware and return it to your instructor, along with any unused sand and soil.

Table 4.1 **Data table for permeability experiment**

	Length of time to drain 50 ml of water through funnel	Milliliters of water drained into beaker
Coarse sand	seconds	ml
Fine sand	seconds	ml
Soil	seconds	ml

8. Questions 8a–8c refer to the permeability experiment.

   a. Of the three materials you tested, the (coarse sand, fine sand, soil) has the greatest permeability. Circle your answer.

   b. Suggest a reason why different amounts of water were recovered in the beaker for each material that was tested.

   _____

   c. Write a brief statement summarizing the results of your permeability experiment.

   _____

   _____

9. What will be the effect of each of the following conditions on the relation between infiltration and runoff?

Highly permeable surface material: _____

_____

Steep slope: _____

_____

Gentle rainfall: _____

_____

Dense ground vegetation: _____

_____

10. What will be the relation between infiltration and runoff in a region with a moderate slope that has a permeable surface material covered with sparse vegetation?

_____

_____

Infiltration and Runoff in Urban Areas

In urban areas much of the land surface has been covered with buildings, concrete, and asphalt. The consequence of covering large areas with impervious materials is to alter the relation between runoff and infiltration of the region.

   Figure 4.4 shows two hypothetical **hydrographs** (plots of stream flow, or runoff, over time) for an area before and after urbanization. The amount of precipitation the area receives is the same after urbanization as before. Runoff is evaluated by measuring the stream **discharge**, which is the volume of water flowing past a given point per unit of time, usually measured in cubic feet per second. Use Figure 4.4 to answer questions 11–14.

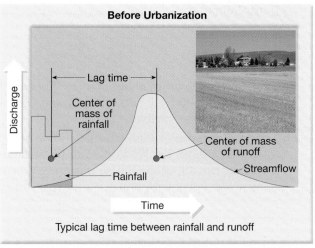

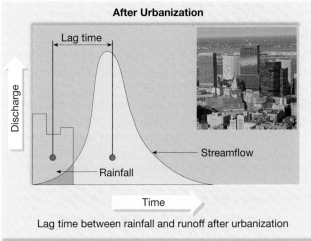

Figure 4.4 The effect of urbanization on stream flow before urbanization (top) and after urbanization (bottom). (After L. B. Leopold, U.S. Geological Survey Circular 559, 1968)

11. As illustrated in Figure 4.4, urbanization (increases, decreases) the peak, or maximum, stream flow. Circle your answer.

12. What is the effect that urbanization has on the lag time between the time of the rainfall and the time of peak stream discharge?

_____

13. Total runoff occurs over a (longer, shorter) period of time in an area that has been urbanized. Circle your answer.

14. Based on what you have learned from the hydrographs, explain why urban areas often experience flash-flooding during intense rainfalls.

_____

_____

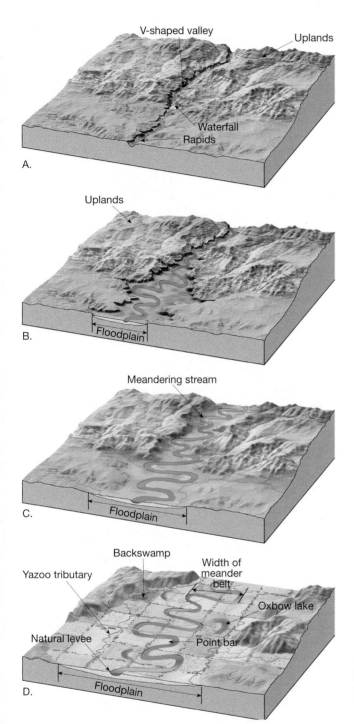

Figure 4.5 Common features of valleys. **A.** Near the headwaters. **B.** and **C.** In the middle. **D.** At the mouth. (After Ward's Natural Science Establishment, Inc., Rochester, New York)

Running Water

Of all the agents that shape Earth's surface, running water is the most important. Rivers and streams are responsible for producing a vast array of erosional and depositional landforms in both humid and arid regions. As illustrated in Figure 4.5, many of these features are associated with the *headwaters* of a river, while others typically are found near the *mouth*.

An important factor that governs the flow of a river is its **base level**. Base level is the lowest point to which a river or stream may erode. The ultimate base level is sea level. However, lakes, resistant rocks, and main rivers often act as temporary, or local, base levels that control the erosional and depositional activities of a river for a period of time.

Often the *head,* or source area, of a river is well above base level. At the headwaters, rivers typically have steep slopes and downcutting prevails. As the river deepens its valley it may encounter rocks that are resistant to erosion and form *rapids* and *waterfalls.* In arid areas rivers often erode narrow valleys with nearly vertical walls. In humid regions the effect of mass wasting and slope erosion caused by heavy rainfall produce typical V-shaped valleys (Figure 4.5A).

In humid regions downstream from the headwaters, the gradient or slope of a river decreases while its discharge increases because of the additional water being added by tributaries. As the level of the channel begins to approach base level, the river's energy is directed from side to side and the channel begins to follow a sinuous path, or *meanders* (Figure 4.6). Lateral erosion by the meandering river widens the valley floor and a *floodplain* begins to form (Figures 4.5B and 4.5C).

Near the mouth of a river where the channel is nearly at base level, maximum discharge occurs and meandering often becomes very pronounced. Widespread lateral erosion by the meandering river produces a floodplain that is often several times wider than the river's *meander belt*. Features such as *oxbow lakes, natural levees, backswamps* or *marshes*, and *yazoo tributaries* commonly develop on broad floodplains (Figure 4.5D).

Figure 4.6 This high-attitude image shows incised meanders of the Delores River in western Colorado. (Courtesy of USDA–ASCS)

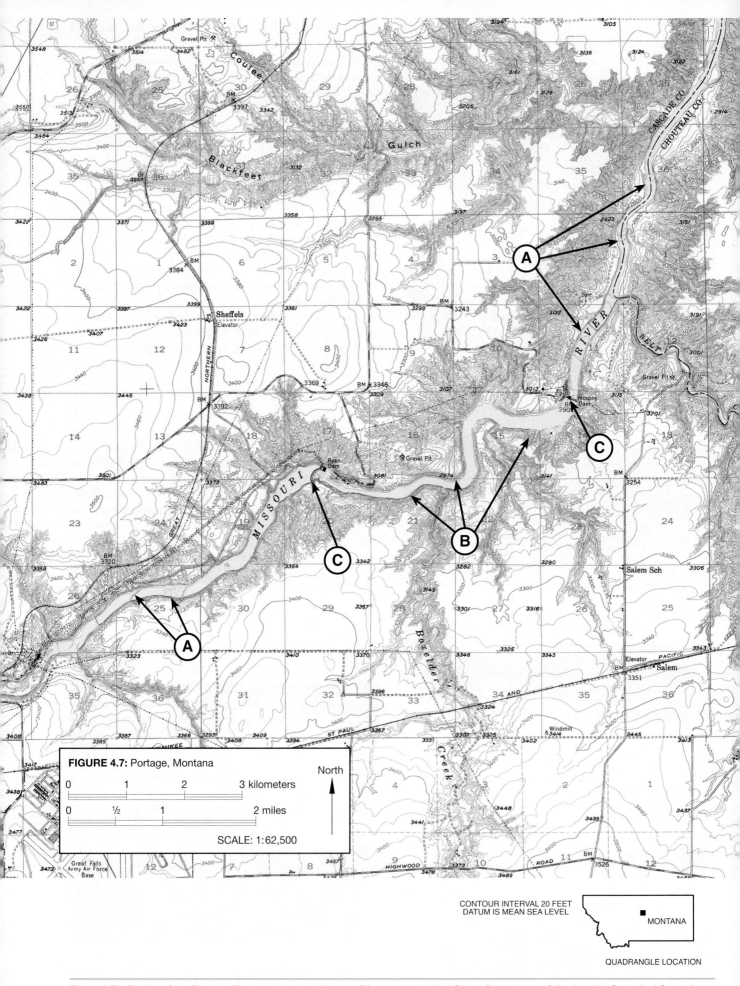

Figure 4.7 Portion of the Portage, Montana, topographic map. (Map source: United States Department of the Interior, Geological Survey)

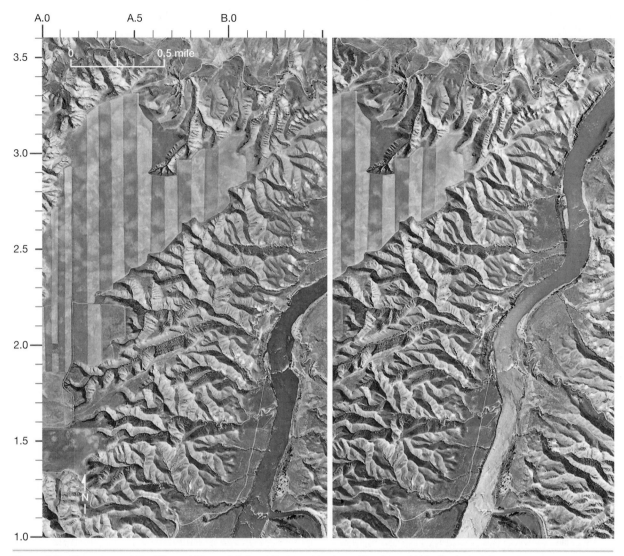

Figure 4.8 Stereogram of the Missouri River, vicinity of Portage, Montana. (Courtesy of U.S. Geological Survey)

Questions 15 to 25 refer to the Portage, Montana, topographic map, Figure 4.7, and stereogram of a portion of the same region, Figure 4.8. On the map, notice the rapids indicated by A and the steep-sided valley walls of the Missouri River indicated by B.

15. Compare the aerial photograph to the map. Then, on the topographic map, outline the area that is shown in the photo.

16. Use the map to determine the approximate total *relief* (vertical distance between the lowest and highest points of the area represented).

    Highest elevation (_____ ft) − lowest elevation (_____ ft) = total relief (_____ ft).

17. On Figure 4.9, draw a north-south topographic profile through the center of the map along a line from north of Blackfeet Gulch to south of the Missouri River. Indicate the appropriate elevations on the vertical axis of the profile. Label

Blackfeet Gulch and the Missouri River on the profile. (*Note:* Exercise 3 contains a detailed explanation for constructing topographic profiles.)

18. Label the upland areas between stream valleys on the topographic profile in Figure 4.9 with the word "upland."

19. The upland areas are (broad and flat, narrow ridges). Circle your answer.

20. Approximately what percentage of the area shown on the map is stream valley and what percentage upland?

    Stream valley: _____ %

    Upland: _____ %

21. Approximately how deep would the Missouri River have to erode to reach ultimate base level?

    _____ ft

NORTH   SOUTH

Figure 4.9 North-south topographic profile through the center of the Portage, Montana, map.

22. It appears that the Missouri River and its tributaries are, for the most part, actively (eroding, depositing) in the area. Circle your answer.

23. With increasing time, as the tributaries erode and lengthen their courses near the headwaters, what will happen to the upland areas?

_____

_____

Notice the dams located along the Missouri River at C.

24. What effect have the dams had on the width of the river, upriver from their locations?

_____

_____

25. Assuming that climate, base level, and other factors remain unchanged, how might the area look millions of years from now?

_____

_____

Questions 26 to 31 refer to the Angelica, New York, topographic map, Figure 4.10.

26. What is the approximate total relief shown on the map?

_____ feet of total relief

27. Draw an arrow on the map indicating the direction that the main river, the Genesee, is flowing. (*Hint*: Use the elevations of the contour lines on the floodplain to determine your answer.)

28. What is the approximate *gradient* (the slope of a river; generally measured in feet per mile) of the Genesee River?

Average gradient = _____ ft/mile

29. The Genesee River (follows a straight course, meanders from valley wall to valley wall). Circle your answer.

30. Most of the areas separating the valleys on the Angelica map are (very broad and flat, relatively narrow ridges). Circle your answer.

31. Assume that erosion continues in the region without interruption. How might the appearance of the area change over a span of millions of years?

_____

_____

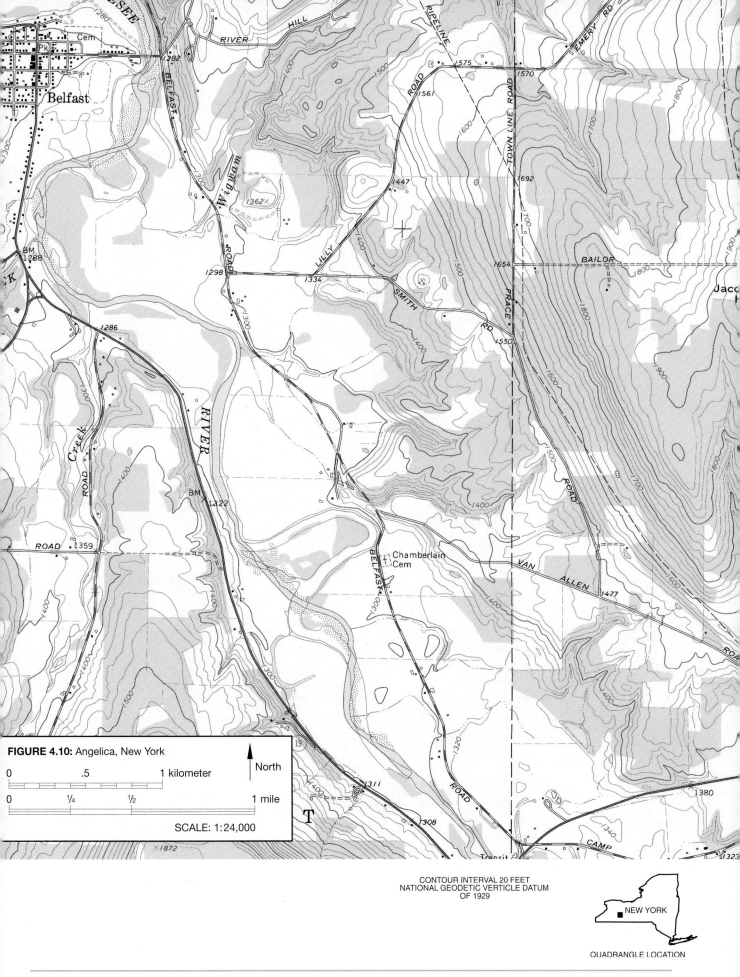

FIGURE 4.10: Angelica, New York

0   .5   1 kilometer

0   1/4   1/2   1 mile

North

SCALE: 1:24,000

CONTOUR INTERVAL 20 FEET
NATIONAL GEODETIC VERTICLE DATUM
OF 1929

NEW YORK

QUADRANGLE LOCATION

Figure 4.10   Portion of the Angelica, New York, topographic map. (Map source: United States Department of the Interior, Geological Survey)

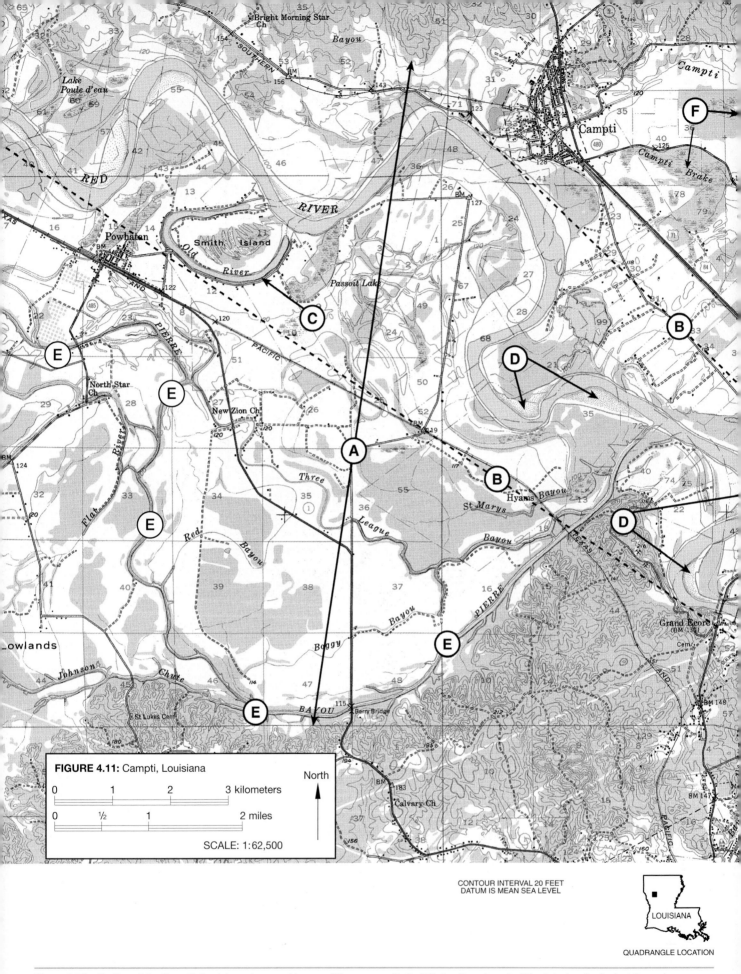

Figure 4.11 Portion of the Campti, Louisiana, topographic map. (Map source: United States Department of the Interior, Geological Survey)

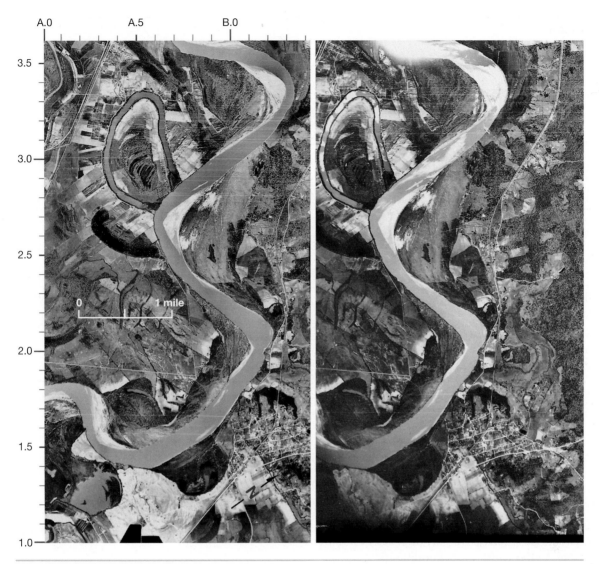

Figure 4.12   Stereogram of the Campti, Louisiana, area. (Courtesy of U.S. Geological Survey)

Questions 32–39 refer to the Campti, Louisiana, topographic map, Figure 4.11, and stereogram of the same area, Figure 4.12. On the map, A indicates the width of the floodplain of the Red River and the dashed lines, B, mark the two sides of the meander belt of the river.

32. Approximately what percentage of the map area is floodplain?

Floodplain = _____ % of the map area

33. In Figure 4.13, draw a north-south topographic profile along a line from the south edge of the City of Campti to south of Bayou Pierre. Indicate the appropriate elevations on the vertical axis of the profile. Label the floodplain area and Bayou Pierre on the sketch.

34. Approximately how many feet is the floodplain above ultimate base level?

_____ feet above ultimate base level

35. Using Figure 4.5 as a reference, identify the type of feature found at each of the following lettered positions on the map. Also, write a brief statement describing how each feature forms.

Letter C (in particular, *Old River*): _____

_____

Letter D: _____

_____

Letter E: _____

_____

Letter F: _____

_____

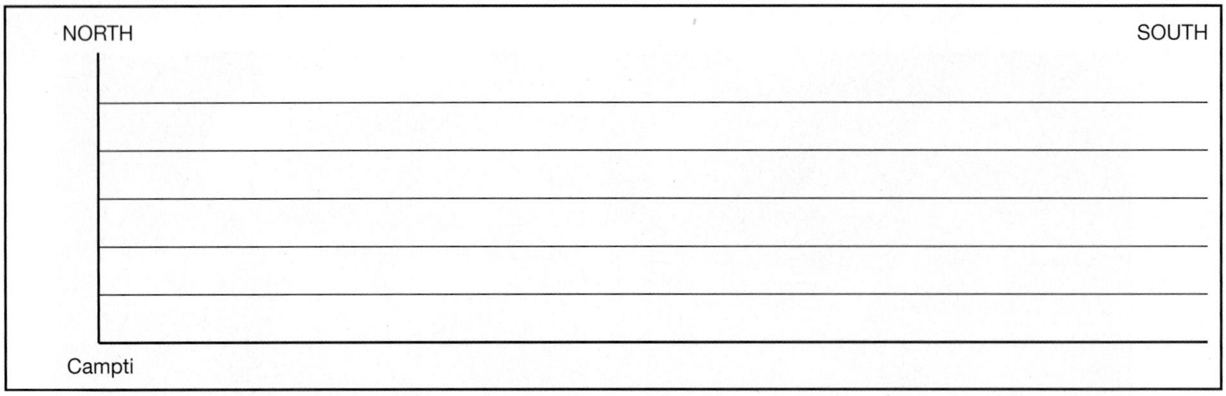

NORTH   SOUTH

Campti

Figure 4.13   North-south topographic profile of the Campti map.

36. Identify and label examples of a point bar, cut-bank, and an oxbow lake on the stereogram.

37. Write a statement that compares the width of the meander belt of the Red River to the width of its floodplain.

_____

_____

38. (Downcutting, Lateral erosion) is the dominant activity of the Red River. Circle your answer.

39. Assuming that erosion by the Red River continues without interruption, what will eventually happen to the width of its floodplain?

_____

_____

_____

Answer questions 40–42 by comparing the Portage, Angelica, and Campti topographic maps.

40. On which of the three maps is the gradient of the main river the steepest?

_____

41. Which of the three areas has the greatest total relief (vertical distance between the lowest and highest elevations)?

_____

42. Choosing from the three topographic maps, write the name of the map that is best described by each of the following statements.

Primarily floodplain: _____

_____

River valleys separated by broad, relatively flat upland areas: _____

_____

Most of the area consists of steep slopes:

_____

Greatest number of streams and tributaries:

_____

Poorly drained lowland area with marshes and swamps: _____

_____

Active downcutting by rivers and streams:

_____

Surface nearest to base level: _____

_____

Groundwater

As a resource, groundwater supplies much of our water needs for consumption, irrigation, and industry. On the other hand, as a hazard, groundwater can damage building foundations and aid the movement of materials during landslides and mudflows. In many areas, overuse and contamination of this valuable resource threaten the supply. One of the most serious problems faced by many localities is land subsidence caused by groundwater withdrawal.

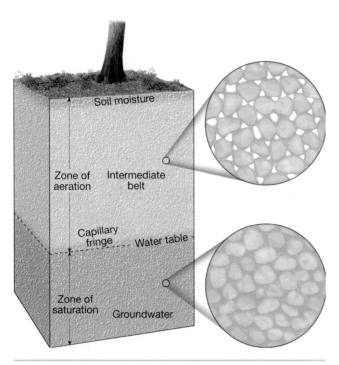

Figure 4.14 Idealized distribution of groundwater.

Water Beneath the Surface

Groundwater is water that has soaked into Earth's surface and occupies all the pore spaces in the soil and bedrock in a zone called the **zone of saturation**. The upper surface of this saturated zone is called the **water table**. Above the water table in the **zone of aeration**, the pore spaces of the materials are unsaturated and mainly filled with air. (Figure 4.14)

Figure 4.15 illustrates a profile through the subsurface of a hypothetical area. Use Figures 4.14 and 4.15 to answer questions 43–50.

43. Label the zone of saturation, zone of aeration, and water table on Figure 4.15.

44. Describe the shape of the water table in relation to the shape of the land surface.

_____

_____

45. What is the relation of the surface of the water in the stream to the water table?

_____

_____

46. What is the lowered surface in the water table around the well called? What has caused the lowering of the surface of the water table around the well? What will make it larger or smaller?

_____

_____

_____

47. At point *A* on Figure 4.15, sketch a small, impermeable pocket of clay that intersects the valley wall.

48. Describe what will happen to water that infiltrates to the depth of the clay pocket at point *A*.

_____

_____

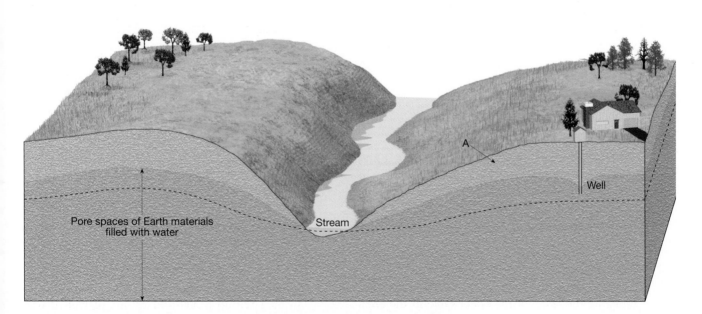

Figure 4.15 Earth's subsurface showing saturated and unsaturated materials.

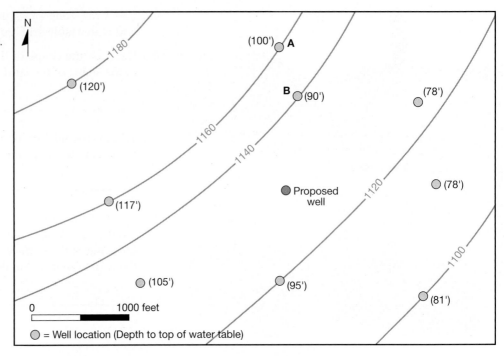

Figure 4.16 Hypothetical topographic map showing the location of several water walls.

The dashed line in Figure 4.15 represents the level of the water table during the dry season when infiltration is no longer replenishing the groundwater.

49. What is the consequence of the lower elevation of the water table during the dry season on the operation of the well? How might the problem have been avoided?

50. What are two main sources of pollutants that can contaminate groundwater supplies?

Groundwater Movement

Figure 4.16 is a hypothetical topographic map showing the location of several water wells. The numbers in parentheses indicate the depth of the water table below the surface in each well.

Questions 51–53 refer to Figure 4.16.

51. Begin by calculating the elevation of the water table at each indicated well location. Then, using a colored pencil, draw smooth 10-foot contours that illustrate the slope of the water table in the area. Using a different colored pencil, draw arrow(s) on

the map that indicate the direction of the slope of the water table.

a. What is the average amount of slope of the water table in the area? Toward which direction does the water table slope?

b. Referring to the site of the proposed water well, at approximately what depth below the surface should the well drill the water table?

52. Assume that a dye was put into well A at 1 PM on May 10, 1990 and detected in well B at 8 AM on October 1, 1998. What was the velocity of the groundwater movement between the two wells in centimeters per day?

53. Use a different colored pencil to draw dashed 10-ft contour lines on the map that illustrate the configuration of the water table after well B was pumped for a sufficient period of time to lower the water table 22 feet at its location. Assume that an area within a 500-foot radius of well B was affected by the pumping.

The Problem of Ground Subsidence

As the demand for freshwater increases, surface subsidence caused by the withdrawal of groundwater from **aquifers** presents a serious problem for many areas. Several major urban areas such as Las Vegas, Houston-Galveston, Mexico City, and the Central Valley of California are experiencing subsidence caused by overpumping wells (Figure 4.17). In Mexico City alone, compaction of the subsurface material resulting from the reduction of fluid pressure as the water table is lowered has caused as much as seven meters of subsidence. Fortunately, in many areas an increased reliance on surface water and replenishing the groundwater supply has slowed the trend.

A classic example of land subsidence caused from groundwater withdrawal is in the Santa Clara Valley, which borders the southern part of San Francisco Bay in California. The graph presented in Figure 4.18 illustrates the relation between ground subsidence in the valley and the level of water in a well in the same area.

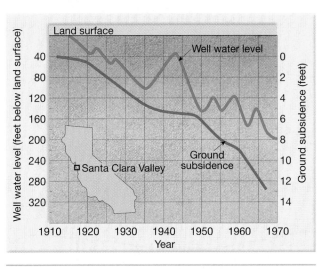

Figure 4.18 Ground subsidence and water level in a well in the Santa Clara Valley, California. (Data courtesy of U.S. Geological Survey)

Figure 4.17 The marks on this utility pole indicate the level of the surrounding land in preceding years. Between 1925 and 1975 this part of the San Joaquin Valley, CA subsided almost 9 meters because of the withdrawal of groundwater and the resulting compaction of sediments. (Photo courtesy of U.S. Geological Survey)

Questions 54–58 refer to Figure 4.18.

54. What is the general relation between the ground subsidence and level of water in the well illustrated on the graph?

_____

_____

55. What was the total ground subsidence and total drop in the level of water in the well during the period shown on the graph?

    Total ground subsidence = _____ ft

    Total drop in well level = _____ ft

56. During the period shown on the graph, on an average, about (1 ft, 5 ft, 10 ft) of land subsidence occurred with each 20-ft decrease in the level of water in the well. Circle your answer.

57. The ground subsidence that took place during the twenty years before 1950 was (less, greater) than the subsidence that took place between 1950 and 1970. Circle your answer.

58. Notice that minimal subsidence took place between 1935 and 1950. After referring to the well water level during the same period of time, suggest a possible reason for the reduced rate of subsidence between 1935 and 1950.

_____

_____

Figure 4.20 Portion of the Mammoth Cave, Kentucky, topographic map. (Map source: United States Department of the Interior, Geological Survey)

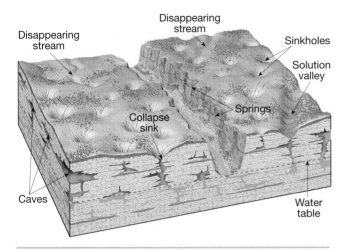

Figure 4.19 Generalized features of an advanced stage of karst topography.

Examining a Karst Landscape

Landscapes that are dominated by features that form from groundwater dissolving the underlying rock are said to exhibit **karst topography** (Figure 4.19). On the surface, karst topography is characterized by irregular terrain, springs, **disappearing streams, solution valleys**, and depressions called **sinkholes** (Figure 4.19). Beneath the surface, dissolution of soluble rock may result in **caves** and **caverns**.

One of the classic karst regions in the United States is the Mammoth Cave, Kentucky, area. Locate and examine the Mammoth Cave, Kentucky, topographic map, Figure 4.20. An insoluble sandstone layer is the surface rock that forms the upland area in the northern quarter of the map. Underneath the sandstone layer is a soluble limestone. Erosion has removed all the sandstone in the southern three fourths of the map and exposed the limestone. On the limestone surface, numerous sinkholes, indicated by closed contour lines with hachures, are present, as well as several disappearing streams (letter A).

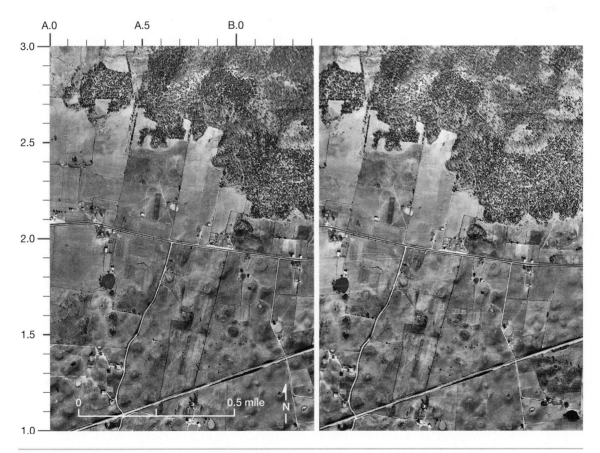

Figure 4.21 Stereogram of the Mammoth Cave, Kentucky, area. (Courtesy of the U.S. Geological Survey)

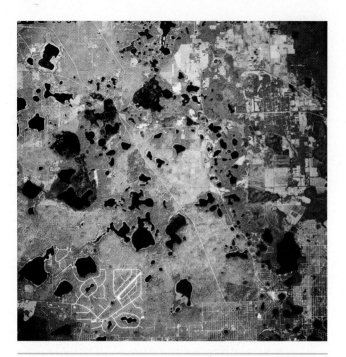

Figure 4.22 This high-altitude infrared image shows an area of karst topography in central Florida. The numerous lakes occupy sinkholes. (Courtesy of USDA–ASCS)

Use the Mammoth Cave topographic map, Figure 4.20, the stereogram of the same area, Figure 4.21, and Figure 4.22, to answer questions 59–63.

59. On the topographic map, outline the area that is shown on the stereogram.

60. What does the absence of water in the majority of sinkholes indicate about the depth of the water table in the area?

_____

61. Examine both the stereogram and map. Then describe the difference in appearance between the northern quarter and southern three-fourths of the mapped area.

_____

_____

_____

_____

62. Describe what is happening to Gardner Creek in the area indicated with the letter B on the map.

_____

63. List two ways that sinkholes commonly form.

a. _____

b. _____

Running Water and Groundwater on the Internet

Apply the concepts from this exercise to investigate the hydrology of a river and the groundwater resources in your home state by completing the corresponding online activity on the *Applications & Investigations in Earth Science* website at http://prenhall.com/earthsciencelab

SUMMARY/REPORT PAGE

EXERCISE 4

Shaping Earth's Surface
Running Water and Groundwater

Date Due: ______________________

Name: ______________________

Date: ______________________

Class: ______________________

After you have finished Exercise 4, complete the following questions. You may have to refer to the exercise for assistance or to locate specific answers. Be prepared to submit this summary/report to your instructor at the designated time.

1. Write a statement that describes the movement of water through the hydrologic cycle, citing several of the processes that are involved.

____

____

____

2. Assume you are assigned a project to determine the quantity of infiltration that takes place in an area. What are the variables you must measure or know before you can arrive at your answer?

____

____

3. Write a brief paragraph summarizing the results of your permeability experiment in question 8 of the exercise.

____

____

____

4. Describe the effects that urbanization has on the stream flow of a region.

____

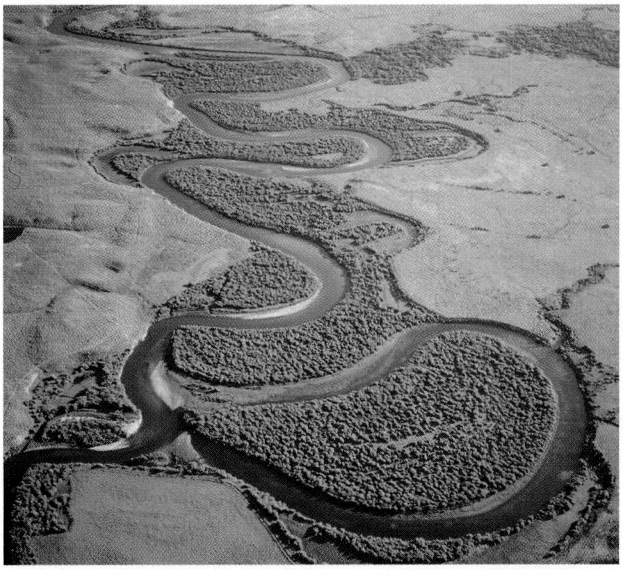

Figure 4.23 River and valley features. (Photo by Michael Collier)

5. On Figure 4.23, identify and label as many features of the river and valley as possible. Write a brief paragraph describing the area and its relation to base level.

____

____

____

____

____

____

6. Refer to the proportion of water that either infiltrates or runs off. Why does a soil-covered hillside with sparse vegetation often experience severe soil erosion? What are some soil conservation methods that could be used to reduce the erosion?

_____

_____

_____

7. Name and describe two features you would expect to find on the floodplain of a widely meandering river near its mouth.

Feature	Description
_____	_____
_____	_____

8. Assume you have decided to drill a water well. What are at least two factors concerning the water table and zone of saturation that should be considered prior to drilling?

_____

_____

9. What is the average slope of the water table illustrated on Figure 4.16?

_____

10. What was the velocity of the groundwater movement between wells A and B in Figure 4.16?

_____

11. How might a rapidly growing urban area that relies on groundwater as a freshwater source avoid the problem of land subsidence from groundwater withdrawal?

_____

_____

12. Name and describe two features you would expect to find in a region with karst topography.

Feature	Description
_____	_____
_____	_____

Shaping Earth's Surface
Arid and Glacial Landscapes

The previous exercise explored the hydrologic cycle and the role of running water and groundwater in shaping the landscape in humid regions. However, when taken together, the dry regions of the world and those areas whose surfaces have been modified by glacial ice also comprise a significant portion of Earth's surface (Figure 5.1). Since desert or near-desert conditions and glaciated regions prevail over a large area of Earth, an understanding of the landforms and processes that shaped these regions is essential to the Earth scientist.

Objectives

After you have completed this exercise, you should be able to:

1. Locate the desert and steppe regions of North America.
2. Describe the evolution of the landforms that exist in the mountainous desert areas of the Basin and Range region of the western United States.
3. Describe the different types of glacial deposits and the features they compose.
4. Identify and explain the formation of the features commonly found in areas where the landforms are the result of deposition by continental ice sheets.
5. Describe the evolution and appearance of a glaciated mountainous area.
6. Identify and explain the formation of the features caused by alpine glaciation.

Materials

calculator   ruler   hand lens

Materials Supplied by Your Instructor

stereoscope   string

Terms

desert	bajada	till
steppe	playa lake	stratified drift
flash flood	inselberg	moraine
Basin and Range	pediment	Pleistocene epoch
fault-block	alpine glacier	arête
mountains	continental ice	cirque
alluvial fan	sheet	horn
	drift	hanging valley

Figure 5.1 Glacial striations in bedrock, Yellowknife, Canada NWT. (Photo courtesy of Dr. Richard Waller, Keele University, UK)

Desert Landscapes

Arid **(desert)** and semiarid **(steppe)** climates cover about 30 percent of Earth's land area (Figure 5.2). At first glance, many desert landscapes with their angular hills and steep canyon walls may appear to have been shaped by processes other than those that are responsible for landforms in regions with an abundance of water. However, as striking as the contrasts may be, running water is still the dominant agent responsible for most of the erosional work in deserts. Wind erosion, although more significant in dry areas than elsewhere, is only of secondary importance.

The distinct effects that running water has on humid and dry areas are the result of the same processes operating under different climatic conditions. Precipitation in the dry climates is minimal, often sporadic, and frequently comes in the form of torrential downpours that last only a short time. Consequently, in desert areas **flash floods** occur, and few streams or rivers reach the sea because the water often evaporates and infiltrates into the ground.

Evolution of a Mountainous Desert Landscape

Mountainous desert landscapes have developed in response to a variety of geologic processes. A classic region for studying the effects of running water in dry areas is the western United States. Throughout much of this **Basin and Range** region, which includes southeastern California, Nevada, western Utah, southern Oregon, southern Arizona and New Mexico, the erosion of mountain ranges and subsequent deposition of sediment in adjoining basins have produced a landscape characterized by several unique landforms (Figure 5.3).

In a large area of the Basin and Range region of the western United States **fault-block mountains** have formed as large blocks of Earth's crust have been forced upward (Figure 5.3A). The infrequent and intermittent precipitation in this desert region typically results in streams that carry their eroded material from the mountains into interior basins. **Alluvial fans** and **bajadas** often form as streams deposit sediment on the less steep slopes at the base of the mountains (Figure 5.3B). On rare occasions when streams flow across the alluvial fans, a shallow **playa lake** may develop near the center of a basin.

Continuing erosion in the mountains and deposition in the basins may eventually fill the basin and only isolated peaks, called **inselbergs**, surrounded by gently sloping sediment, remain. As the front of the mountain is worn back by erosion, a broad, sloping bedrock surface called a **pediment**, covered by a thin layer of sediment, often forms at its base (Figure 5.3C). In the final stages, even the inselbergs will disappear, and all that remains is a nearly flat, sediment-covered surface underlain by the erosional remnants of mountains.

Use Figure 5.2 to answer questions 1 and 2.

1. Where are the desert and steppe regions of North America located?

   Desert: _____

   Steppe: _____

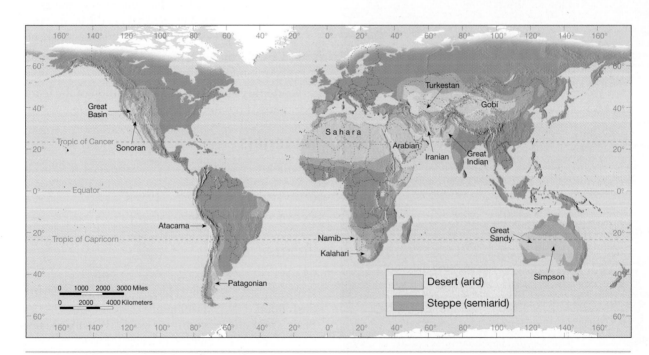

Figure 5.2 Global arid and semiarid climates cover about 30% of Earth's land surface. No other climate group covers so large an area. Low-latitude dry climates are the result of the global distribution of air pressure and winds. Middle-latitude deserts and steppes exist principally because they are sheltered in the deep interiors of large landmasses.

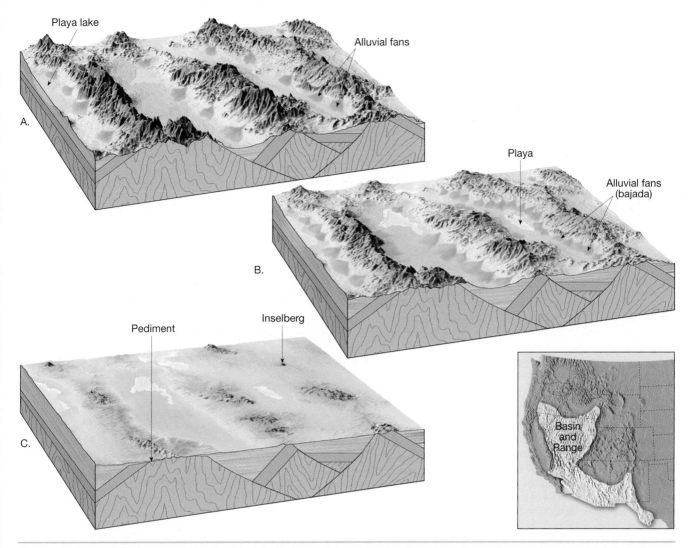

Figure 5.3 Stages of landscape evolution in a block-faulted, mountainous desert such as the Basin and Range region of the West. **A.** Early stage; **B.** Middle stage; **C.** Late stage.

2. Using an "X" to mark your selection(s), indicate which of the following statements are commonly held misconceptions concerning the world's dry lands.

_____ The world's dry lands are always hot.

_____ Desert landscapes are almost completely covered with sand dunes.

_____ The dry regions of the world encompass about 30 percent of Earth's land surface.

_____ Dry lands are practically all lifeless.

Figure 5.4 is a portion of the Antelope Peak, Arizona, topographic map that illustrates many of the features of the mountainous desert landscapes found in the western United States. Use the map and accompanying stereogram of the area (Figure 5.5) to answer questions 3–13. You may find the diagrams in Figure 5.3 helpful.

3. On the map, outline the area that is illustrated in the stereogram.

Use a stereoscope to examine the stereogram, Figure 5.5.

4. The vegetation in the area is (dense, sparse), and there are (few, many) dry stream courses. Circle your answers.

5. By examining the map, determine the total relief of the map area.

Total relief = _____ ft

6. (Continuously flowing, Intermittent) streams dominate the area shown on the map. Circle your answer.

7. On the map, of the two lines, A or B, (A, B) follows the steepest slope. Circle your answer.

8. By drawing arrows on the map, indicate the directions that intermittent streams will flow as they leave the mountains.

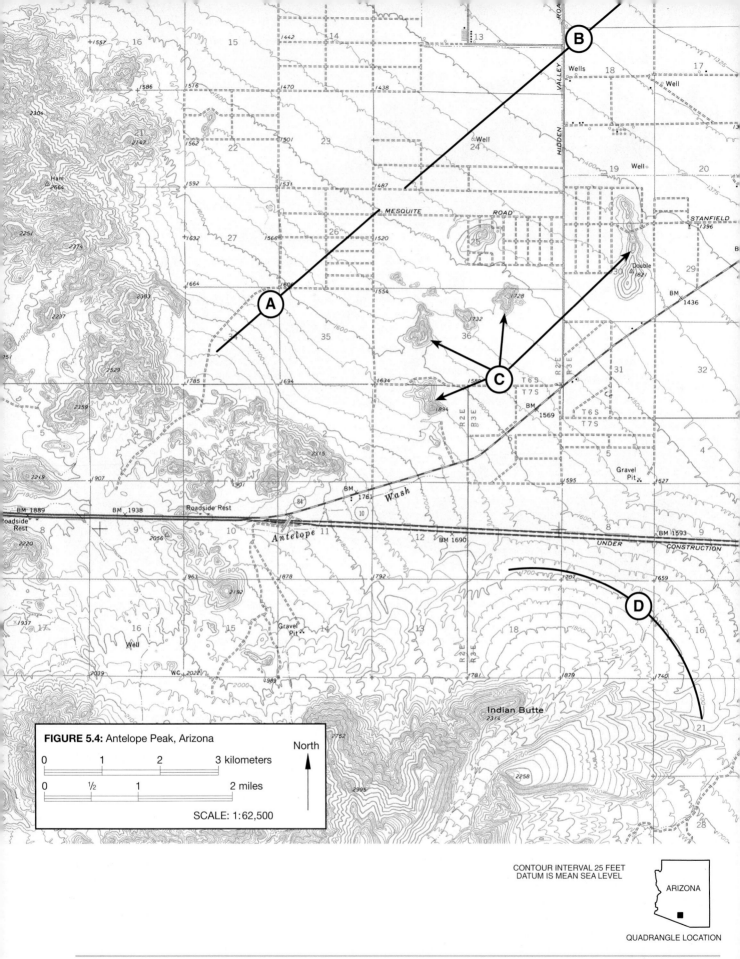

FIGURE 5.4: Antelope Peak, Arizona

North

```
0        1        2        3 kilometers
0      ½       1              2 miles
```

SCALE: 1:62,500

CONTOUR INTERVAL 25 FEET
DATUM IS MEAN SEA LEVEL

ARIZONA

QUADRANGLE LOCATION

Figure 5.4   Portion of the Antelope Peak, Arizona, topographic map. (Map source: United States Department of the Interior, Geological Survey)

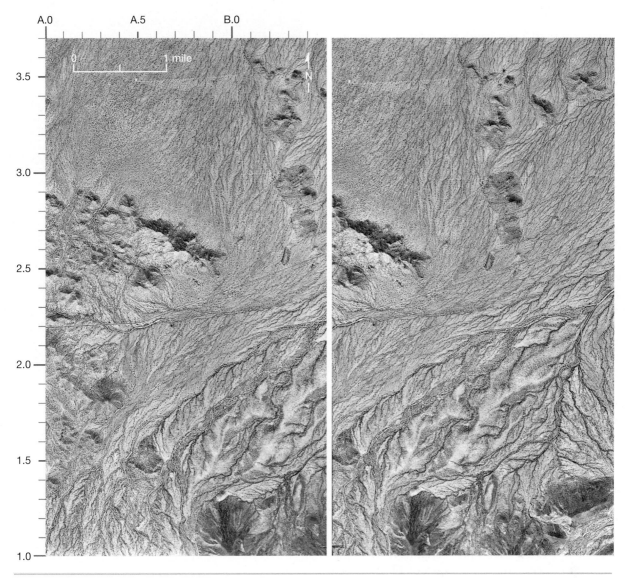

Figure 5.5 Stereogram of the Antelope Peak, Arizona, area. (Courtesy of the U.S. Geological Survey)

9. Where on the map is the most likely place that surface water may accumulate? Label the area "possible lake."

10. Identify the features indicated on the map at the following letters and briefly describe how each formed.

   Letter C: _____

   _____

   Letter D: _____

   _____

   The area at letter A on the map is a bedrock surface covered by a thin layer of sediment.

11. The feature labeled A is called a(n) _____.

12. Briefly describe how the Antelope Peak area may have looked millions of years ago.

   _____

   _____

   _____

13. Assume that erosion continues in the area without interruption. How might the area look millions of years from now?

   _____

   _____

   _____

Glacial Landscapes

Slightly more than 2 percent of the world's water is in the form of glacial ice that covers nearly 10 percent of Earth's land area. However, to an Earth scientist, glaciers represent more than storehouses of fresh water in the hydrologic cycle. Like the other agents that modify the surface, glaciers are dynamic forces capable of eroding, transporting, and depositing sediment.

Literally thousands of glaciers exist on Earth today. They occur in regions where, over long periods of time, the yearly snowfall has exceeded the quantity lost by melting or evaporation. **Alpine**, or **valley, glaciers** form from snow and ice at high altitudes. At high latitudes, enormous **continental ice sheets** cover much of Greenland and Antarctica.

Glacial erosion and deposition leave an unmistakable imprint on Earth's surface. In regions once covered by continental ice sheets, glacially scoured surfaces and subdued terrain dominated by glacial deposits are the rule (see Figure 5.1). By contrast, erosion by alpine glaciers in mountainous areas tends to accentuate the irregularity of the topography, often resulting in spectacular scenery characterized by sharp, angular features.

Glacial Deposits and Depositional Features

The general term **drift** applies to all sediments of glacial origin, no matter how, where, or in what form they were deposited. There are two types of glacial drift: (1) **till**, which is characteristically unsorted sediment deposited directly by the glacier (Figure 5.6), and (2) **stratified drift**, which is material that has been sorted and deposited by glacial meltwater.

The most widespread depositional features of glaciers are **moraines**, which are ridges of till that form along the edges of glaciers and layers of till that accumulate on the ground as the ice melts and recedes. There are several types of moraines, some common only to alpine glaciers, as well as other kinds of glacial depositional features.

Figure 5.7 illustrates a hypothetical retreating glacier. Use Figure 5.7 to answer questions 14–16.

14. Draw a large arrow on Figure 5.7 that indicates the direction of glacial ice movement in the area. Label the arrow "ice flow."

15. On Figure 5.7, label an example of a terminal moraine, recessional moraine, and ground moraine.

Features of Continental Ice Sheets

During a division of Earth's history called the **Pleistocene epoch**, continental ice sheets, as well as alpine glaciers, were considerably more extensive over Earth's surface than they are today. At one time, these thick sheets of ice covered all of Canada, portions of

Close up
of cobble

Figure 5.6 Glacial till is an unsorted mixture of many different sediment sizes. A close examination often reveals cobbles that have been scratched as they were dragged along by the glacier. (Photos by E. J. Tarbuck)

Alaska, and much of the northern United States as well as extensive areas of northern Europe and Asia (Figure 5.8). Today, the impact that these ice sheets had on the landscape is still very obvious.

Use Figure 5.8 as a reference to answer question 16.

16. By listing state abbreviations, indicate the geographic area of the continental United States that Pleistocene glaciers covered during their maximum extent.

_____

_____

While alpine glaciers change the shape of the land surface primarily by erosion, landforms produced by

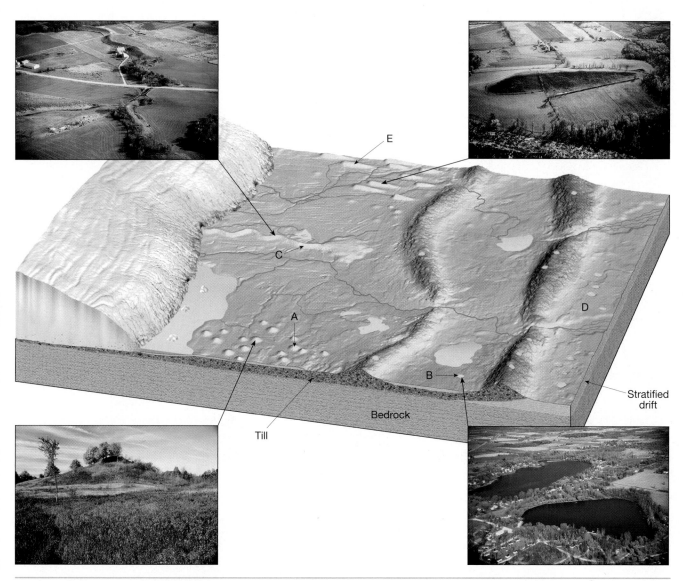

Figure 5.7 Characteristic depositional features of glaciers. (Drumlin photo courtesy of Ward's Natural Science Establishment; Kame, Esker, and Kettle photos by Richard P. Jacobs/JLM Visuals)

Figure 5.8 Maximum extent of glaciation in the Northern Hemisphere during the Pleistocene epoch.

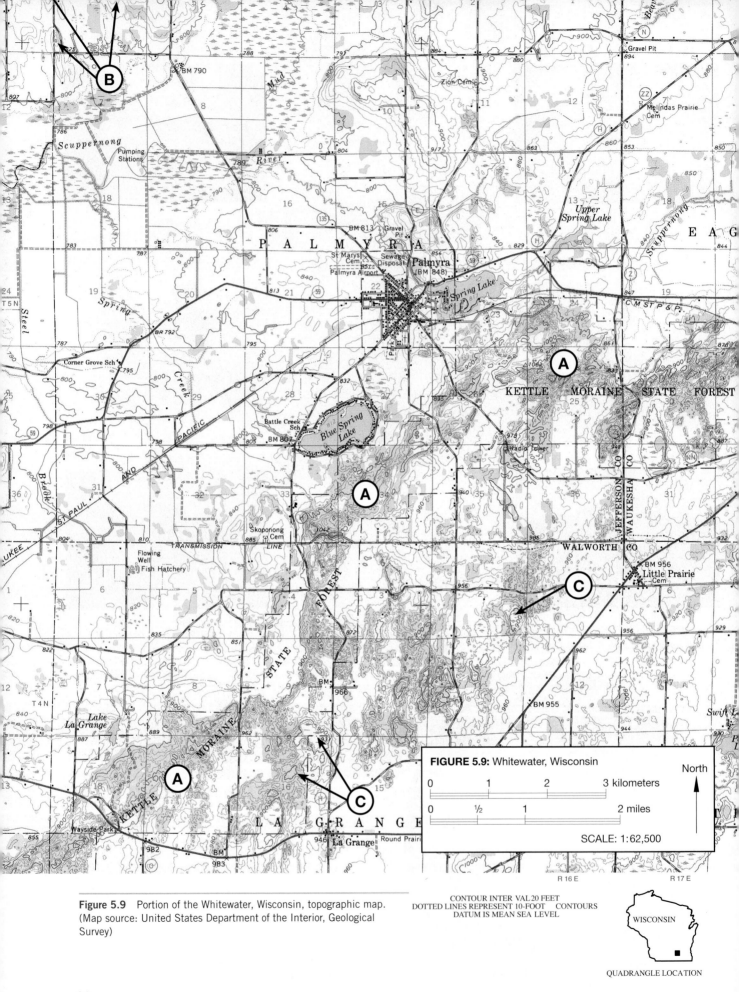

Figure 5.9 Portion of the Whitewater, Wisconsin, topographic map. (Map source: United States Department of the Interior, Geological Survey)

FIGURE 5.9: Whitewater, Wisconsin

North

0   1   2   3 kilometers

0   ½   1   2 miles

SCALE: 1:62,500

CONTOUR INTER VAL 20 FEET
DOTTED LINES REPRESENT 10-FOOT   CONTOURS
DATUM IS MEAN SEA LEVEL

WISCONSIN

QUADRANGLE LOCATION

continental ice sheets, especially those that covered portions of the United States during the Pleistocene epoch, are essentially depositional in origin. Some of the most extensive areas of glacial deposition occurred in the north-central United States. Here, glacial drift covers the surface and landscapes are dominated by moraines, outwash plains, kettles, and other depositional features.

17. Briefly describe each of the following glacial depositional features and select the letter on Figure 5.7 that indicates an example of each.

Drumlin: _____

_____ Letter: _____

Esker: _____

_____ Letter: _____

Kame: _____

_____ Letter: _____

Kettle: _____

_____ Letter: _____

Outwash plain: _____

_____ Letter: _____

Figure 5.9 is a portion of the Whitewater, Wisconsin, topographic map, which illustrates many of the depositional features that are typical of continental glaciation. Use the map and the accompanying stereogram of the area (Figure 5.10) to answer questions 18–30.

18. After examining the map and stereogram, draw a line on the map that outlines the area illustrated on the photograph.

19. The general topography of the land in the southeast quarter of the region is (higher, lower) in elevation and (more, less) irregular than the land in the northwest. Circle your answers.

20. What features on the map indicate that portions of the area are poorly drained? Where are these features located?

_____

_____

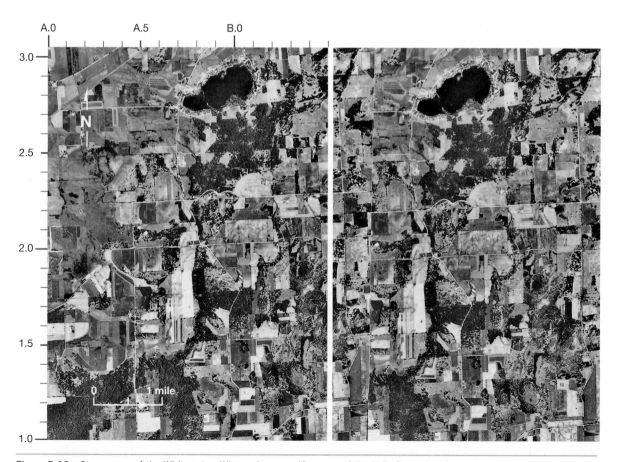

Figure 5.10 Stereogram of the Whitewater, Wisconsin, area. (Courtesy of the U.S. Geological Survey)

NORTHWEST   SOUTHEAST

Scuppernong River   Little Prairie

Figure 5.11   Northwest-southeast topographic profile of the Whitewater map.

Examine the elevations of the feature indicated with the letter As on the map that, in general, coincides with Kettle Moraine State Forest. Compare the elevations to those found to the northwest and southeast of the feature.

21. In Figure 5.11, sketch a northwest-southeast topographic profile along a line that extends from the Scuppernong River to near the city of Little Prairie. Indicate the appropriate elevations on the vertical axis of the profile.

22. The area that coincides with Kettle Moraine State Forest is (higher, lower) in elevation than the land to the northwest and southeast. Circle your answer.

23. The feature labeled with As on the map is a long ridge composed of till called a (kettle, moraine, drumlin). Circle your answer.

24. The streamlined, asymmetrical hills composed of till, labeled B, are what type of feature?

_____

25. Examine the shape of the features labeled B on the map and in Figure 5.7. How can the features be used to determine the direction of ice flow in a glaciated area?

_____

26. Use the features labeled B as a guide to draw an arrow on the map that indicates the direction of ice flow in the region.

27. Where on the map is the likely location of the outwash plain? Identify and label the area "outwash plain."

_____

28. Identify and label the ground moraine area on the map.

29. What term is applied to the numerous almost circular depressions designated with the letter Cs on the map?

_____

30. What is the probable origin of the material that is being mined in the gravel pits north and northeast of Palmyra?

_____

_____

_____

Features of Alpine or Valley Glaciation

As they flow, alpine glaciers often exaggerate the already irregular topography of a region by eroding the mountain slopes and deepening the valleys. Figure 5.12 illustrates the changes that a formerly unglaciated mountainous area (Figure 5.12A) experiences as the result of alpine glaciation. Many of the landforms produced by glacial erosion, such as **arêtes**, **cirques**, **horns** and **hanging valleys** (Figure 5.13) are identified in Figure 5.12C.

Questions 31–33 refer to Figure 5.12.

31. How has glaciation changed the shape and depth of the main valley?

_____

_____

Prior to glaciation, tributary streams were adjusted to the depth of the main valley.

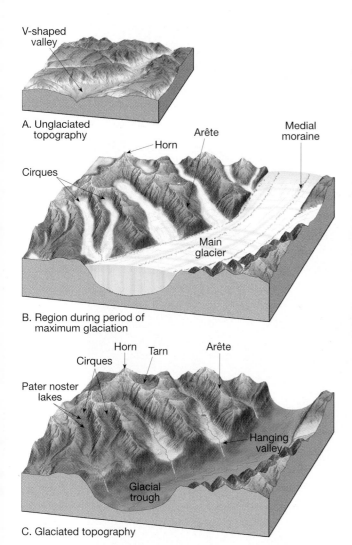

Figure 5.13 Bridalveil Falls in Yosemite National Park cascades from a hanging valley into the glacial trough below. (Photo by E. J. Tarbuck)

Figure 5.12 Landforms created by alpine glaciers. **A.** Landscape prior to glaciation; **B.** During glaciation; **C.** After glaciation.

Figure 5.14 is an oblique aerial photograph of an area experiencing alpine glacial erosion on Mont Blanc, France. Questions 34 and 35 refer to the figure.

32. What has been the consequence of glacial erosion on the gradients or slopes of tributary streams?

_____

_____

_____

33. Use your own words to describe how the appearance of the area has changed from what it was prior to glaciation.

_____

_____

_____

34. Draw arrows on the photograph that indicate the directions that the glaciers are flowing.

35. Give the name of the glacial feature described by each of the following statements as well as the letter on the photograph that labels an example of the feature. Use Figure 5.12C as a reference.

a. Sinuous, sharp-edged ridge:

Name: _____

Letter of Example: _____

b. Hollowed-out, bowl-shaped depression that is the glacier's source and the area of snow accumulation and ice formation:

Name: _____

Letter of Example: _____

c. Moraine formed along the side of a valley:

Name: _____

Letter of Example: _____

d. Moraine formed when two valley glaciers coalesce to form a single ice stream:

Name: _____

Letter of Example: _____

Figure 5.14 Oblique aerial photograph of alpine glaciers and glacial features, Mont Blanc, France. (Photo courtesy of U.S.G.S.)

Figure 5.15 is a portion of the Holy Cross, Colorado, topographic map, a mountainous area that underwent alpine glaciation in the past. Questions 36–41 refer to the map.

36. Following line A on the map, sketch a topographic profile of the valley of Lake Fork from Sugar Loaf Mtn. to Bear Lake on Figure 5.16. Indicate the appropriate elevations along the vertical axis of the profile.

37. Describe the shape of the profile of the valley of Lake Fork. The valley is called a glacial

_____.

38. Identify the type of glacial feature indicated on the map at each of the following letters. Use Figure 5.12C as a reference.

Letter B: _____

Letter C: _____

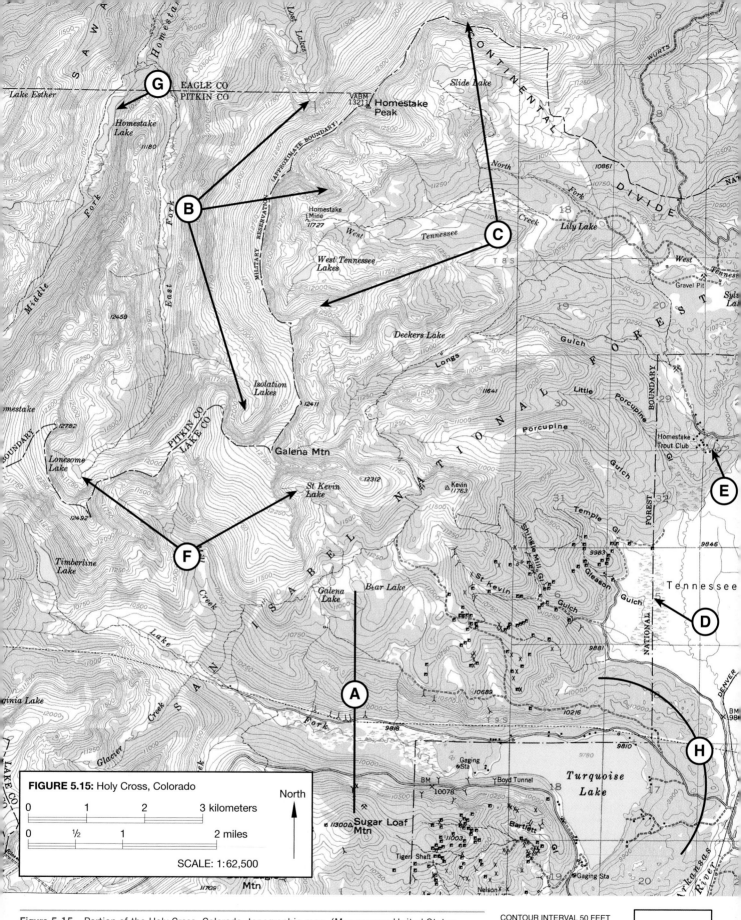

Figure 5.15 Portion of the Holy Cross, Colorado, topographic map. (Map source: United States Department of the Interior, Geological Survey)

FIGURE 5.15: Holy Cross, Colorado

North

0   1   2   3 kilometers

0   ½   1   2 miles

SCALE: 1:62,500

CONTOUR INTERVAL 50 FEET
DATUM IS MEAN SEA LEVEL

COLORADO

QUADRANGLE LOCATION

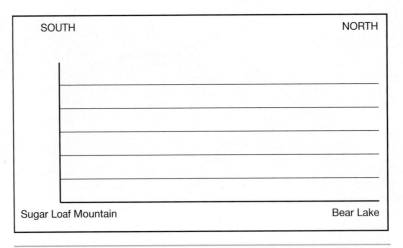

SOUTH   NORTH

Sugar Loaf Mountain   Bear Lake

Figure 5.16   Topographic profile of the valley of Lake Fork on the Holy Cross map.

39. Letter (D, E, F, G) on the map indicates a *tarn*(s), a lake that forms in a cirque. Circle your answer.

    The feature marked H on the map is composed of glacial till.

40. What type of glacial feature is designated H? How did it form?

    _____

    _____

41. What is the reason for the formation of Turquoise Lake?

    _____

    _____

Desert and Glaciers on the Internet

Investigate both desert and glacial regions by completing the corresponding online activity on the *Applications & Investigations in Earth Science* website at http://prenhall.com/earthsciencelab

Shaping Earth's Surface
Arid and Glacial Landscapes

Date Due: _____

Name: _____

Date: _____

Class: _____

After you have finished Exercise 5, complete the following questions. You may have to refer to the exercise for assistance or to locate specific answers. Be prepared to submit this summary/report to your instructor at the designated time.

1. What area of the United States is characterized by fault-block mountains with interior drainage into adjoining basins?

_____

2. Describe the sequence of geologic events that have produced the landforms in the Antelope Peak area of Arizona.

_____

_____

3. What type of feature is located at each of the following letters on the Antelope Peak, Arizona, topographic map, Figure 5.4?

Letter C: _____

Letter D: _____

4. Toward what direction does the pediment slope on the Antelope Peak, Arizona, topographic map?

_____

5. What is the reason for so many dry stream channels in the Antelope Peak, Arizona, area?

_____

6. If you were working in the field, explain how you might determine whether a glacial feature is a recessional moraine or an esker.

_____

_____

7. In the following space, sketch a map-view (the area viewed from above) of the Whitewater, Wisconsin, topographic map, Figure 5.9. Show and label the outwash plain, end moraine, area containing drumlins, and the area containing kettles and kettle lakes.

N

8. Assume you are hiking in the mountains. You suspect that the area was glaciated in the past. Describe some of the features you would look for to confirm your suspicion.

_____

_____

9. What was your conclusion as to the reason for the formation of Turquoise Lake on the Holy Cross, Colorado, topographic map, Figure 5.15?

_____

10. On Figure 5.17, identify and label the alpine glacial features.

Figure 5.17 Athabaska Glacier in Canada's Jasper National Park. (Photo by David Barnes/The Stock Market)

Determining Geologic Ages

The recognition of the vastness of geologic time and the ability to establish the sequence of geologic events that have occurred at various places at different times are among the great intellectual achievements of science. To accomplish the task of deciphering Earth's history, geologists have formulated several laws, principles, and doctrines that can be used to place geologic events in their proper sequence (Figure 6.1). Also, using the principles that govern the radioactive decay of certain elements, scientists are now able to determine the age of many Earth materials with reasonable accuracy. In this exercise you will investigate some of the techniques and procedures used by Earth scientists in their search to interpret the geologic history of Earth.

Objectives

After you have completed this exercise, you should be able to:

1. List and explain each of the laws, principles, and doctrines that are used to determine the relative ages of geologic events.

2. Determine the sequence of geologic events that have occurred in an area by applying the techniques and procedures for relative dating.

3. Explain the methods of fossilization and how fossils are used to define the ages of rocks and correlate rock units.

4. Explain how the radioactive decay of certain elements can be used to determine the age of Earth materials.

5. Apply the techniques of radiometric dating to determine the numerical age of a rock.

6. Describe the geologic time scale and list in proper order some of the major events that have taken place on Earth since its formation.

Figure 6.1 In any sequence of underformed sedimentary rocks, the oldest rock is always at the bottom and the youngest is at the top. (Photo by E. J. Tarbuck)

Materials

ruler   calculator

Materials Supplied by Your Instructor

fossils and fossil questions   meterstick or metric
  (optional)   tape measure
5-meter length of adding-
  machine paper

Terms

relative dating	unconformity	radiometric
uniformitarianism	cross-cutting	date
original	fossil	half-life
horizontality	fossil succession	eon
superposition		era
inclusion		

Relative Dating

Relative dating, the placing of geologic events in their proper sequence or order, does not tell how long ago something occurred, only that it preceded one event and followed another. Several logical doctrines, laws, and principles govern the techniques used to establish the relative age of an object or event.

Doctrine of Uniformitarianism

First proposed by James Hutton in the late 1700s, this doctrine states that the physical, chemical, and biological laws that operate today have operated throughout Earth's history. Although geologic processes such as erosion, deposition, and volcanism are governed by these unchanging laws, their rates and intensities may vary. The doctrine is often summarized in the statement, "The present is the key to the past."

Principle of Original Horizontality

Sediment, when deposited, forms nearly horizontal layers. Therefore, if we observe beds of sedimentary rocks that are folded or inclined at a steep angle, the implication is that some deforming force took place after the sediment was deposited (Figure 6.2).

Law of Superposition

In any sequence of undeformed sedimentary rocks (or surface deposited igneous rocks such as lava flows and layers of volcanic ash), the oldest rock is always at the bottom and the youngest is at the top. Therefore, each layer of rock represents an interval of time that is more recent than that of the underlying rocks (see Figure 6.1).

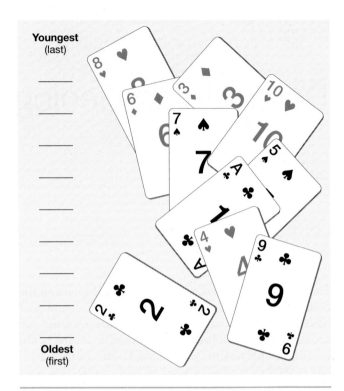

Figure 6.3 Sequence of playing cards illustrating the law of superposition.

Assume the playing cards shown in Figure 6.3 are layers of sedimentary rocks viewed from above.

1. In the space provided in Figure 6.3, list the order, first (oldest) to last (youngest), in which the cards were laid down.

2. Were you able to place all of the cards in sequence? If not, which one(s) could not be "relative" dated and why?

_____

_____

Figure 6.4 illustrates a geologic cross section, a side view, of the rocks beneath the surface of a hypothetical region. Use Figure 6.4 to answer questions 3 and 4.

3. Of the two sequences of rocks, A–D and E–G, (A–D, E–G) was disturbed by crustal movements after its deposition. Circle your answer. What law or principle did you apply to arrive at your answer?

_____

4. Apply the law of superposition to determine the relative ages of the *undisturbed* sequence of sedimentary rocks. List the letter of the oldest rock layer first.

Oldest _____ Youngest

Figure 6.2 Uplifted and tilted sedimentary strata in the Canadian Rockies. (Photo by E. J. Tarbuck)

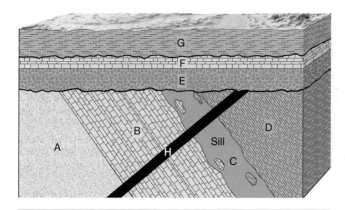

Figure 6.4 Geologic block diagram of a hypothetical region showing igneous intrusive features (C and H) and sedimentary rocks.

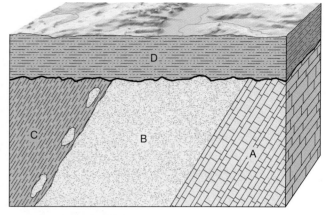

Figure 6.6 Geologic block diagram showing sedimentary rocks.

Inclusions

Inclusions are pieces of one rock unit that are contained within another unit (Figure 6.5). The rock mass adjacent to the one containing the inclusions must have been there first in order to provide the rock fragments. Therefore, the rock containing the inclusions is the younger of the two.

Refer to Figure 6.6 to answer questions 5 and 6. The sedimentary layer B is a sandstone. Letter C is the sedimentary rock, shale.

5. Identify and label the inclusions in Figure 6.6.
6. Of the two rocks B and C, rock (B, C) is older. Circle your answer.

Unconformities

As long as continuous sedimentation occurs at a particular place, there will be an uninterrupted record of the material and life forms. However, if the sedimentation process is suspended by an emergence of the area from below sea level, then no sediment will be deposited

Figure 6.5 Inclusions are fragments of one rock enclosed within another. (Photo by E. J. Tarbuck)

and an erosion surface will develop. The result is that no rock record will exist for a part of geologic time. Such a gap in the rock record is termed an **unconformity**. An unconformity is typically shown on a cross-sectional (side view) diagram by a wavy line ($\sim\sim\sim$). Several types of unconformities are illustrated in Figure 6.7.

7. Identify and label an example of an angular unconformity and a disconformity in Figure 6.4.

Principle of Cross-Cutting Relationships

Whenever a fault or intrusive igneous rock cuts through an existing feature, it is younger than the structure it cuts. For example, if a basalt dike cuts through a sandstone layer, the sandstone had to be there first and, therefore, is older than the dike (Figure 6.8).

Figure 6.9 is a geologic cross section showing sedimentary rocks (A, B, D, E, F, and G), an igneous intrusive feature called a *dike* (C), and a *fault* (H). Use Figure 6.9 to answer questions 8–11.

8. The igneous intrusion C is (older, younger) than the sedimentary rocks B and D. Circle your answer.
9. Fault H is (older, younger) than the sedimentary beds A–E.
10. The relative age of fault H is (older, younger) than the sedimentary layer F.
11. Did the fault occur before or after the igneous intrusion? Explain how you arrived at your answer.

_____

_____

12. Refer to Figure 6.4. The igneous intrusion H is (older, younger) than rock layer E and (older, younger) than layer D. Circle your answers.
13. Refer to Figure 6.4. What evidence supports the conclusion that the igneous intrusive feature

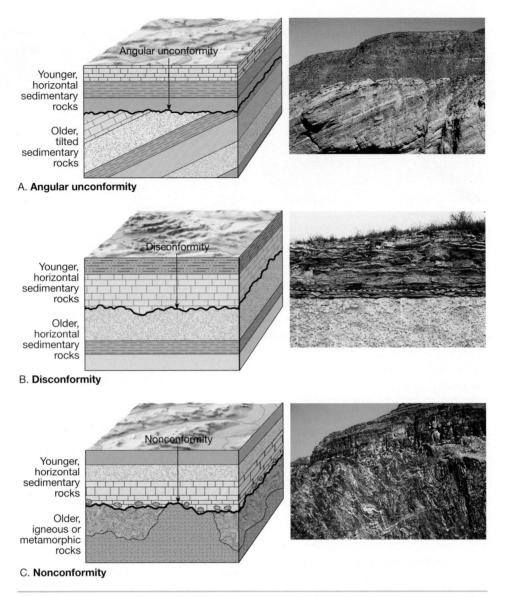

Figure 6.7 Three common types of unconformities. On the diagrams, wavy dashed lines mark the unconformity.

Figure 6.8 This basalt dike (black) is younger than the sandstone layers that it cuts through. (Photo by E. J. Tarbuck)

called a *sill*, C, is more recent than both of the rock layers B and D and older than the igneous intrusion H?

Fossils and the Principle of Fossil Succession

Fossils (Figure 6.10) are among the most important tools used to interpret Earth's history. They are used to define the ages of rocks, correlate one rock unit with another, and determine past environments on Earth.

Earth has been inhabited by different assemblages of plants and animals at different times. As rocks form, they often incorporate the preserved remains of these organisms as fossils. According to the principle of **fossil succession**, fossil organisms succeed each other in a definite and determinable order. Therefore, the time that a rock originated can frequently be determined by noting the kinds of fossils that are found within it.

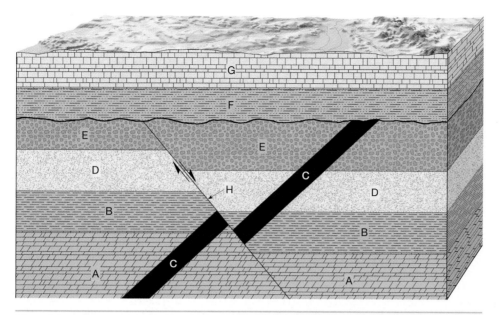

Figure 6.9 Geologic block diagram of a hypothetical area showing an igneous intrusion (C), a fault (H), and sedimentary rocks.

Figure 6.10 Various types of fossilization. In photo **A.** the mineral quartz now occupies the internal spaces of what was once wood. **B.** is the replica of fish after the carbonized remains were removed. In photo **C.** mineral matter occupies the hollow space where a shell was once located. **D.** is a track left by a dinosaur in formerly soft sediment. (Photos by E. J. Tarbuck)

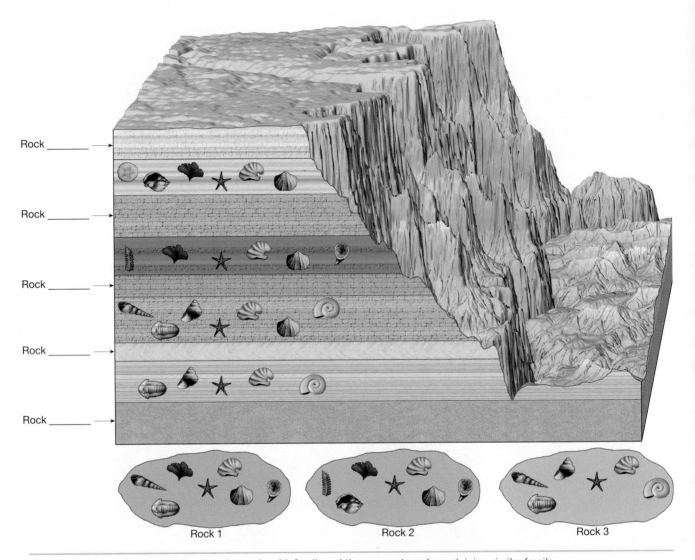

Rock _____

Rock _____

Rock _____

Rock _____

Rock _____

Rock 1   Rock 2   Rock 3

Figure 6.11 Layered sequence of sedimentary rocks with fossils and three separate rocks containing similar fossils.

Using the materials supplied by your instructor, answer questions 14 and 15.

14. At the discretion of your instructor, there may be several stations with fossils and questions set up in the laboratory. Following the specific directions of your instructor, proceed to the stations.

15. What are the conditions that would favor the preservation of an organism as a fossil?

_____

_____

16. Refer to Figure 6.10. Select the photo, A, B, C, or D, that best illustrates each of the following methods of fossilization or fossil evidence.

    Petrification: The small internal cavities and pores of the original organism are filled with precipitated mineral matter. Photo: _____

Cast: The space once occupied by a dissolved shell or other structure is subsequently filled with mineral matter. Photo: _____

Impression: A replica of a former fossil left in fine-grained sediment after the fossilizing material, often carbon, is removed. Photo: _____

Indirect evidence: Traces of prehistoric life, but not the organism itself. Photo: _____

Figure 6.11 shows a sequence of undeformed sedimentary rocks. Each layer of rock contains the fossils illustrated within it. The three rocks, Rocks 1, 2, and 3, illustrated below the layered sequence were found nearby and each rock contains the fossils indicated. Answer question 17 using Figure 6.11.

17. Applying the principle of fossil succession, indicate the proper position of each of the three rocks relative to the rock layers by writing the words Rock 1, Rock 2, or Rock 3 at the appropriate position in the sequence.

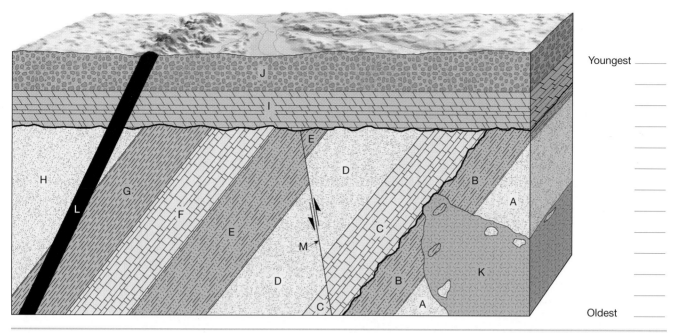

Figure 6.12 Geologic block diagram of a hypothetical area showing igneous intrusive features (K and L), a fault (M), and sedimentary rocks.

Applying Relative Dating Techniques

Geologists often apply several of the techniques of relative dating when investigating the geologic history of an area.

Figure 6.12 is a geologic cross section of a hypothetical area. Letters K and L are igneous rocks. Letter M is a fault. All the remaining letters represent sedimentary rocks. Using Figure 6.12 to complete questions 18–24 will provide insight into how the relative geologic history of an area is determined.

18. Identify and label the unconformities indicated in the cross section.

19. Rock layer I is (older, younger) than layer J. Circle your answer. What law or principle have you applied to determine your answer?

20. The fault is (older, younger) than rock layer I. Circle your answer. What law or principle have you applied to determine your answer?

21. The igneous intrusion K is (older, younger) than layers A and B. Circle your answer. What two laws or principles have you applied to determine your answer?

_____ and _____

22. The age of the igneous intrusion L is (older, younger) than layers J, I, H, G, and F.

23. List the entire sequence of events, in order from oldest to youngest, by writing the appropriate letter in the space provided on the figure.

24. Explain why it was difficult to place the fault, letter M, in a specific position among the sequence of events in Figure 6.12.

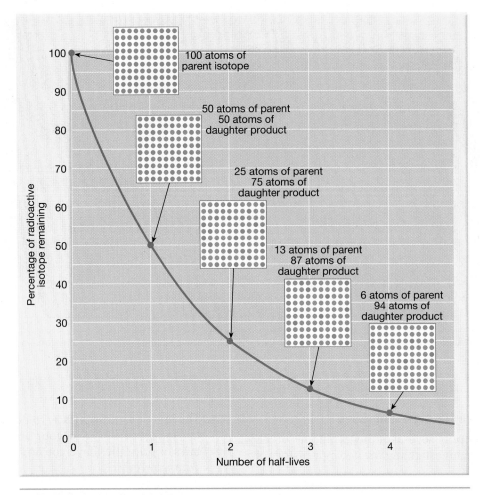

Figure 6.13 Radioactive decay curve.

Radiometric Dating

The discovery of radioactivity and its subsequent understanding has provided a reliable means for calculating the *numerical age* in years of many Earth materials. Radioactive atoms, such as the isotope uranium-238, emit particles from their nuclei that we detect as radiation. Ultimately, this process of decay produces an atom that is stable and no longer radioactive. For example, eventually the stable atom lead-206 is produced from the radioactive decay of uranium-238.

Determining Radiometric Ages

The radioactive isotope used to determine a **radiometric date** is referred to as the *parent isotope*. The amount of time it takes for half of the radioactive nuclei in a sample to change to their stable end product is referred to as the **half-life** of the isotope. The isotopes resulting from the decay of the parent are termed the *daughter products*. For example, if we begin with one gram of radioactive material, half a gram would decay and become a daughter product after one half-life. After the second half-life, half

of the remaining radioactive isotope, 0.25 g or $\frac{1}{4}$ of the original amount ($\frac{1}{2}$ of $\frac{1}{2}$), would still exist. *With each successive half-life, the remaining parent isotope would be reduced by half.*

Figure 6.13 graphically illustrates how the ratio of a parent isotope to its stable daughter product continually changes with time. Use Figure 6.13 to help answer questions 25–29.

25. What percentage of the original parent isotope remains after each of the following half-lives has elapsed?

**FRACTION OF
PARENT ISOTOPE REMAINING**

One half-life: _____

Two half-lives: _____

Three half-lives: _____

Four half-lives: _____

26. Assume you begin with 10.0 g of a radioactive parent isotope. How many grams of parent isotope will be present in the sample after each of the following half-lives?

REMAINING PARENT ISOTOPE

One half-life: _____ g

Four half-lives: _____ g

27. If a radioactive isotope has a half-life of 400 million years, how long will it take for 50 percent of the material to change to the daughter product?

_____ years

28. A sample is brought to the laboratory and the chemist determines that the percentage of the parent isotope remaining is 13 percent of the total amount that was originally present. If the half-life of the material is 600 million years, how old is the sample?

_____ years old

29. Determine the numerical ages of rock samples that contain a parent isotope with a half-life of 100 million years and have the following percentages of original parent isotope.

50 percent: Age = _____

25 percent: Age = _____

6 percent: Age = _____

Applying Radiometric Dates

When used in conjunction with relative dates, radiometric dates help Earth scientists refine their interpretation of the geologic history of an area. Completing questions 30–35 will aid in understanding how both types of dates are often used together.

Previously in the exercise you determined the geologic history of the area represented in Figure 6.12 using relative dating techniques. Assume that the rock layers H and I in Figure 6.12 each contain radioactive materials with known half-lives.

30. An analysis of a sample of rock from layer H indicates an equal proportion of parent isotope and daughter produced from the parent. The half-life of the parent is known to be 425 million years.

   a. (Fifty, Twenty-five, Thirteen) percent of the original parent has decayed to the daughter product. Circle your answer.

   b. How many half-lives of the parent isotope have elapsed since rock H formed?

   _____

   c. What is the numerical age of rock layer H? Write your answer below and at rock layer H on Figure 6.12.

   Age of rock layer H = _____ years

31. The analysis of a sample of rock from layer I indicates its age to be 400 million years. Write the numerical age of layer I on Figure 6.12.

Refer to the relative and numerical ages you determined for the rocks in Figure 6.12 to answer the following questions.

32. How many years long is the interval of time represented by the unconformity that separates rock layer H from layer I? Explain how you arrived at your answer.

The unconformity represents an interval of time that was _____ million years long.

Explanation: _____

_____

_____

_____

33. The age of fault M is (older, younger) than 400 million years. Circle your answer. Explain how you arrived at your answer.

Explanation: _____

_____

_____

_____

34. What is the approximate maximum numerical age of the igneous intrusion L?

The igneous intrusion L formed more recently than _____ million years ago.

35. Complete the following general statement describing the numerical ages of rock layers G, F, and E.

All of the rock layers are (younger, older) than _____ million years.

The Geologic Time Scale

Applying the techniques of geologic dating, the history of Earth has been subdivided into several different units which provide a meaningful time frame within which the events of the geologic past are arranged. Since the span of a human life is but a "blink of an eye" compared to the age of Earth, it is often difficult to comprehend the magnitude of geologic time. By completing questions 36–41, you will be better able to grasp the great age of Earth and appreciate the sequence of events that have brought it to this point in time.

36. Obtain a piece of adding machine paper slightly longer than 5 meters and a meterstick or metric measuring tape from your instructor. Draw a line at one end of the paper and label it "PRESENT." Using the following scale, construct a time line by completing the indicated steps.

SCALE

1 meter = 1 billion years

10 centimeters = 100 million years

1 centimeter = 10 million years

1 millimeter = 1 million years

Step 1. Using the geologic time scale, Figure 6.14, as a reference, divide your time line into the **eons** and **eras** of geologic time. Label each division with its name and indicate its absolute age.

Step 2. Using the scale, plot and label the plant and animal events listed in Figure 6.14 on your time line.

After completing your time line, answer questions 37–41.

37. What fraction or percent of geologic time is represented by the Precambrian eon?

Approximately _____ of geologic time.

38. Suggest a reason(s) why approximately 540 million years ago was selected to mark the end of the Precambrian eon and the beginning of the Phanerozoic eon.

_____

_____

39. Suggest a reason(s) why the periods of the Cenozoic era have been further subdivided into several epochs with reasonably reliable accuracy.

_____

_____

_____

_____

40. How many times longer is the whole of geologic time than the time represented by recorded history, about 5000 years?

Geologic time is _____ times longer than recorded history.

41. For what fraction or percent of geologic time have land plants been present on Earth?

Approximately _____ of geologic time.

Geologic Time on the Internet

Apply what you have learned in this exercise to write a geologic interpretation of a rock outcrop and to explore the fossil record by completing the corresponding online activities on the *Applications & Investigations in Earth Science* website at http://prenhall.com/earthsciencelab

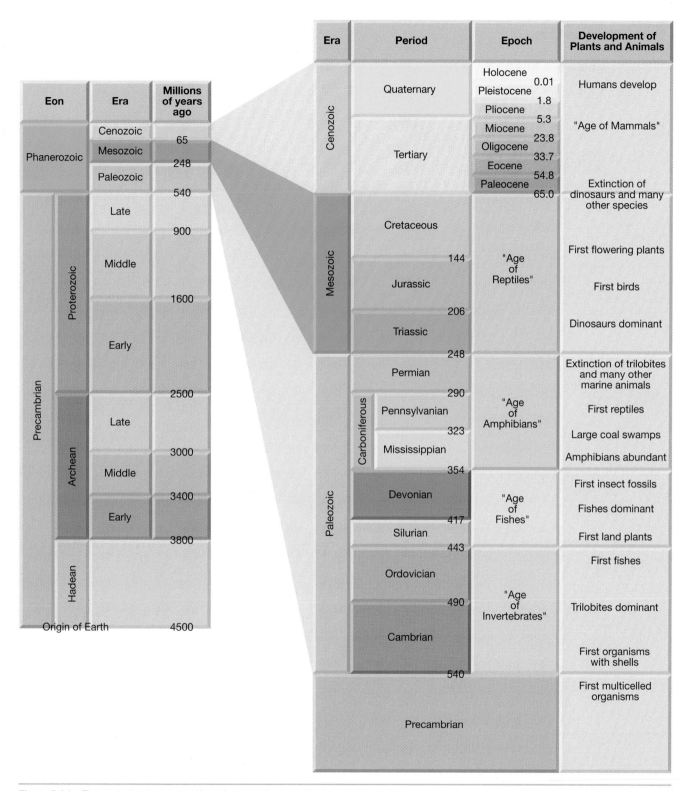

Figure 6.14 The geologic time scale. (Data from the Geologic Society of America)

Notes and calculations.

Determining Geologic Ages

Date Due: _____

Name: _____

Date: _____

Class: _____

After you have finished Exercise 6, complete the following questions. You may have to refer to the exercise for assistance or to locate specific answers. Be prepared to submit this summary/report to your instructor at the designated time.

1. Determine the sequence of geologic events that have occurred at the hypothetical area illustrated in Figure 6.15. List your answers from oldest to youngest in the space provided by the figure. Letters M and N are faults, J, K, and L are igneous intrusions, and all other layers are sedimentary rocks.

2. The following questions refer to Figure 6.15.

   a. What type of unconformity separates layer G from layer F?

   _____

   b. Which law, principle, or doctrine of relative dating did you apply to determine that rock layer H is older than layer I?

   _____

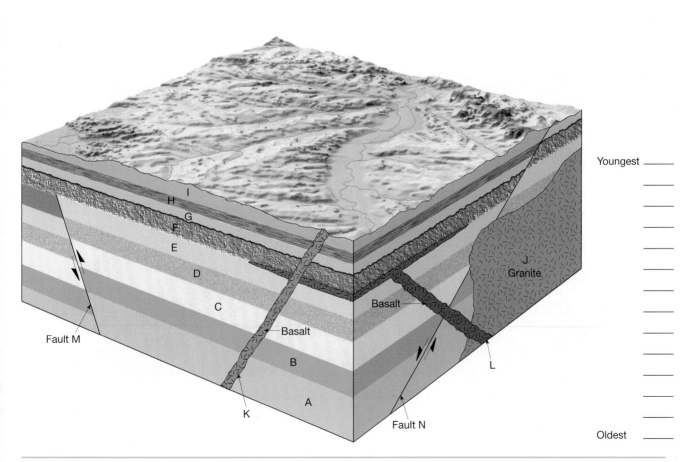

Figure 6.15 Geologic block diagram of a hypothetical region.

c. Which law, principle, or doctrine of relative dating did you apply to determine that fault M is older than rock layer F?

_____

d. Explain why you know that fault N is older than the igneous intrusion J.

_____

e. If rock layer F is 150 million years old and layer E is 160 million years old, what is the approximate age of fault M?

_____ million years

f. The analysis of samples from layers G and F indicates the following proportions of parent isotope to the daughter product produced from it. If the half-life of the parent is known to be 75 million years, what are the ages of the two layers?

	PARENT	DAUGHTER	AGE
Layer G:	50%	50%	_____
Layer F:	25%	75%	_____

g. What absolute time interval is represented by the unconformity at the base of rock layer G?

From _____ to _____ million years

3. List the sequence of geologic events that you determined took place in the area represented by Figure 6.12, question 23, in the exercise.

Oldest _____ Youngest

4. What fraction of time is represented by each of the following geologic eons?

Phanerozoic eon: _____ Precambrian eon: _____

5. How many meters long would the time line you constructed in the exercise, question 36, have been if you had used a scale of 1 millimeter equals 1000 years?

_____

6. Examine the photograph in Figure 6.16 closely. Applying the principles of relative dating, describe as accurately as possible the relative geologic history of the area.

_____

_____

_____

_____

Figure 6.16 Photo of sedimentary beds to be used with question 6.

Geologic Maps and Structures

Earth is a dynamic planet. Its internal forces result in crustal movements that are continuously causing rocks to deform (Figure 7.1). If the magnitude of the force exceeds the strength of the rocks, the rocks will yield and folding or breaking may occur. In this exercise, you will investigate some of the common rock structures that are produced during crustal deformation and learn how Earth scientists map and analyze these geologic features to gain insights into the nature of the evolving Earth.

Objectives

After you have completed this exercise, you should be able to:

1. Explain how geologists describe the orientation of folded rocks and faults using the measurements strike and dip.
2. Draw and interpret a simple geologic block diagram.

Figure 7.1 Deformed sedimentary strata. (Photo by E. J. Tarbuck)

3. Describe the various types of folds and how they form.

4. Recognize and diagram anticlines, synclines, domes, and basins in both geologic map and cross-sectional views.

5. Discuss the formation and types of dip-slip and strike-slip faults.

6. Recognize and diagram the various types of faults in both geologic map and cross-sectional views.

7. Interpret a simplified geologic map and use it to construct a geologic cross section.

Materials

ruler   hand lens
colored pencils   protractor

Terms

geologic map   joint   basin
stress   strike   dip-slip fault
compressional stress dip   strike-slip fault
tensional stress   cross section   normal fault
shear stress   anticline   reverse fault
plastic deformation syncline   right-lateral
fold   axial plane   fault
fault   dome   left-lateral fault

Introduction

Geologists are continuously attempting to understand the nature of the forces that cause rocks to deform. They often begin their study by preparing a **geologic map** that shows the types, ages, distribution, and orientation of rocks on the surface of Earth. Working from these maps, geologists can interpret the nature of the rocks below the surface and assess any forces that may have caused their deformation.

Stress is the term used to describe the force that acts on a rock unit to change its shape or volume. The forces that act to shorten a rock body are known as **compressional stresses**, whereas those that elongate or pull apart a rock unit are called **tensional stresses**. **Shear stresses** tend to bend or break a rock unit (Figure 7.2).

If the stress applied to a rock unit exceeds its strength, the rock may undergo plastic deformation or fracture. **Plastic deformation** takes place at high temperatures and pressures within Earth and results in permanent changes in the rock unit. A rock's size and shape may be altered through folding or flowing. Plastic deformation often produces wavelike undulations called **folds** (see Figure 7.1) in formerly flat-lying rocks. Under the lower temperature and pressure conditions found near the surface, most rocks behave like a brittle solid and fracture or break when stressed beyond their

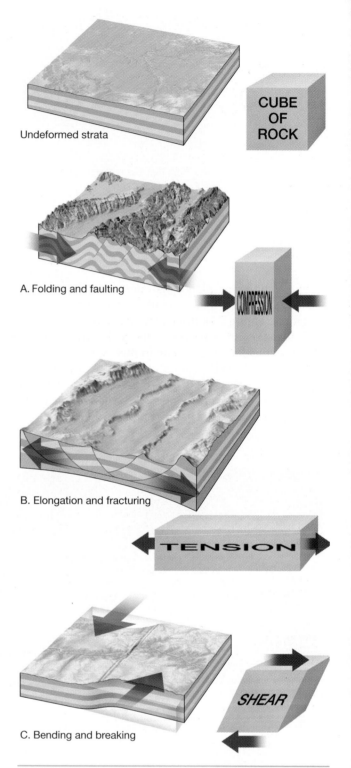

Figure 7.2 Simplified diagram showing the deformation of rock layers. **A.** Compressional stresses tend to shorten a rock body, often by folding. **B.** Tensional stresses act to elongate, or pull apart, a rock unit. **C.** Shear stresses act to bend or break a rock unit.

limit. If the rocks on either side of the fracture move, the geologic feature is called a **fault** (see Figure 7.1). A **joint** is a fracture or break in a rock along which there has been no displacement (Figure 7.3).

Figure 7.3 Joints are fractures along which no appreciable displacement has occurred. (Photo by E. J. Tarbuck)

Strike and Dip

Geologists use measurements called **strike** (trend) and **dip** (inclination) to help define the orientation or attitude of a rock layer or fault (Figure 7.4). By knowing the strike and dip of rocks at the surface, geologists can predict the nature and structure of rock units and faults that are hidden beneath the surface, beyond their view.

Strike is the compass direction of the line produced by the intersection of an inclined rock layer or fault with a horizontal plane at the surface (see Figure 7.4). The strike, or compass bearing, is generally expressed as an angle relative to north. For example, "north 10° east" (N10°E) means the line of strike is ten degrees to the east of north. The strike of the rock units illustrated in Figure 7.4 is approximately north 60° east (N60°E).

Dip is the angle of inclination of the surface of the rock unit or fault from the horizontal plane. Dip includes both an angle of inclination and a direction toward which the rock is inclined. In Figure 7.4 the dip angle of the rock layer is 30°. A good way to visualize dip is to imagine that a water line will always run down the rock surface parallel to the dip. The direction of dip will always be at a 90° angle to the strike. (To illustrate this fact, hold your closed textbook at an angle to the tabletop. The upper edge of your text represents the strike. Regardless of the way you point the text, the direction of dip of the book is always at 90°, or a right angle, to the strike.)

Typically, the strike and dip of rock units are shown on geologic maps. The standard map symbol for strike and dip is ‾‾20°‾ . The long line shows the strike direction and the short line points in the direction of the dip. The number written at the end of the short dip line is the angle of dip. In Figure 7.4, the strike-dip symbol indicates that the rocks are dipping toward the southeast at a 30° angle from the horizontal plane (30°SE).

1. Figure 7.5 illustrates geologic map views (views from directly overhead) of two hypothetical areas showing a strike-dip symbol for a rock layer in each area. Complete the information requested below each map and draw a single large arrow on each map illustrating the direction of dip of the rock layer.

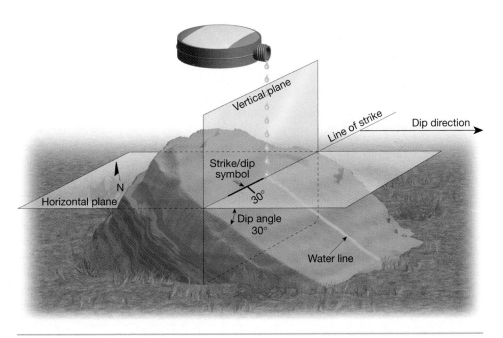

Figure 7.4 Determining the strike and dip of a rock layer.

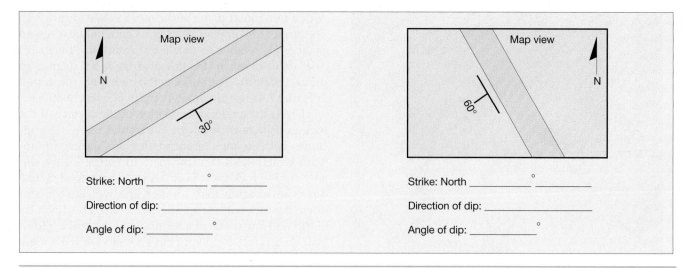

Strike: North _____°_____

Direction of dip: _____

Angle of dip: _____°

Strike: North _____°_____

Direction of dip: _____

Angle of dip: _____°

Figure 7.5   Geologic map views of two hypothetical areas.

Block Diagrams

Block diagrams allow geologists to illustrate both the geologic map view and geologic **cross section** (view from the side or beneath the surface) of an area (Figure 7.6).

Use the block diagram as a guide while you answer questions 2, 3, and 4.

2. Use the information presented on the map view to complete the cross sections on the front and side of the block diagram illustrated in Figure 7.7.

3. Use the information supplied on the geologic map views to complete each of the block diagrams illustrated in Figures 7.8 and 7.9.

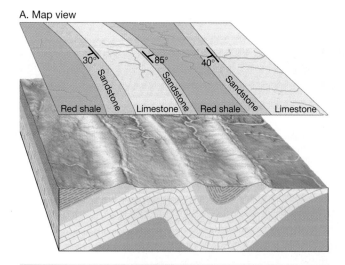

Figure 7.6   Block diagram illustrating inclined rock layers as they would appear in a geologic map view (top of block) and cross sections (front and side of block). By establishing the strike and dip of outcropping sedimentary beds on a map, geologists can infer the orientation of the structure below ground.

4. On Figure 7.10, use the following guidelines to sketch both a geologic map view and block diagram of an area.

Guidelines:

a. Four sedimentary layers of equal thickness exposed at the surface.

b. The strike of each layer is N45°W.

c. The direction of dip of each layer is to the northeast.

d. The angle of dip of each layer is 60°.

5. Show strike-dip symbols for each rock layer on both illustrations in Figure 7.10.

Types of Folds

During the process of mountain building, formerly flat-lying rocks are often compressed into a series of waves called *folds*. **Anticlines** and **synclines** are the two most common types of folds (Figure 7.11). Rock layers that fold upward forming an arch are called anticlines (Figure 7.12). Often associated with anticlines are downfolds, or troughs, called synclines (Figure 7.13).

The **axial plane** is an imaginary plane drawn through the long axis of a fold that divides it as equally as possible into two halves called *limbs* (Figure 7.14). In a *symmetrical fold*, the limbs are mirror images of each other and diverge at the same angle (see Figures 7.11 and 7.13). In an *asymmetrical fold* the limbs each have different angles of dip (see Figure 7.12). A fold where one limb is tilted beyond the vertical is referred to as an *overturned fold* (see Figure 7.11).

Folds do not continue forever. Where folds "die out" and end, the axis is no longer horizontal and the fold is said to be plunging (Figure 7.14B and Figure 7.15).

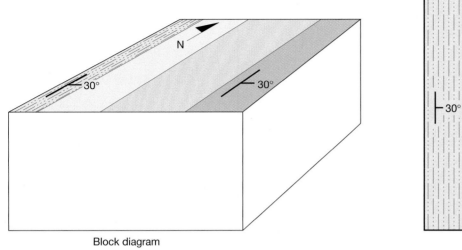

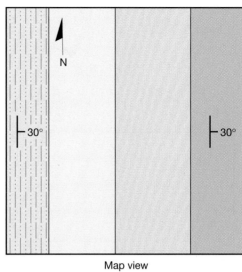

Block diagram

Map view

Figure 7.7 Block diagram and map view of a hypothetical area.

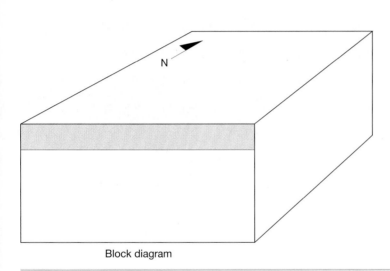

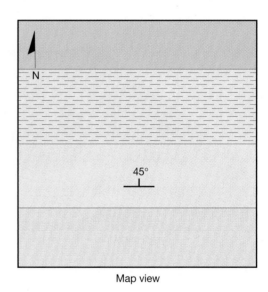

Block diagram

Map view

Figure 7.8 Block diagram and map view of a hypothetical area.

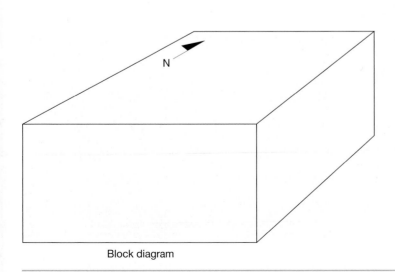

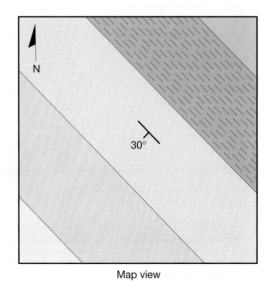

Block diagram

Map view

Figure 7.9 Block diagram and map view of a hypothetical area.

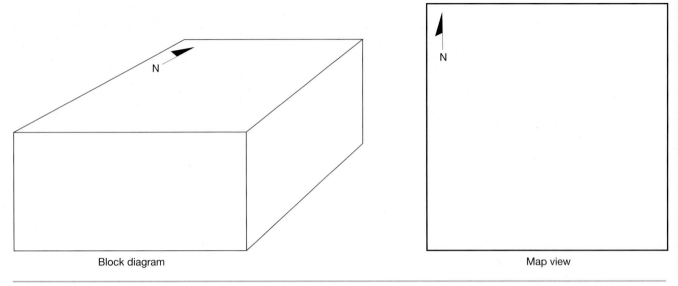

Figure 7.10 Block diagram and map view of a hypothetical area.

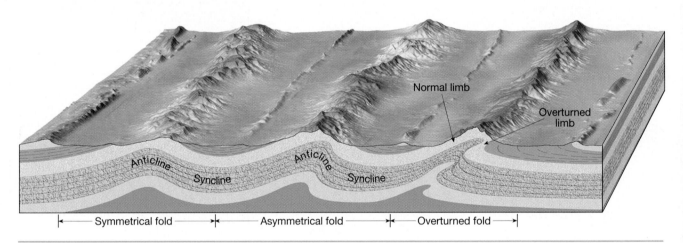

Figure 7.11 Block diagram illustrating the principal types of folded geologic structures.

Figure 7.12 An asymmetrical anticline. (Photo by E. J. Tarbuck)

Figure 7.13 A nearly symmetrical syncline. (Photo by E. J. Tarbuck)

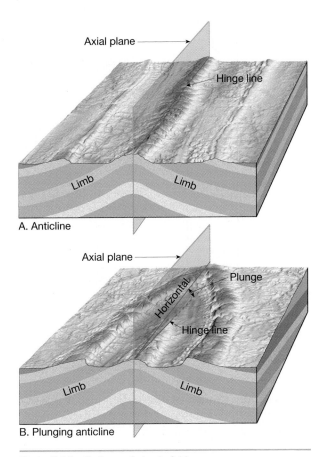

Figure 7.14   Features of simple folds.

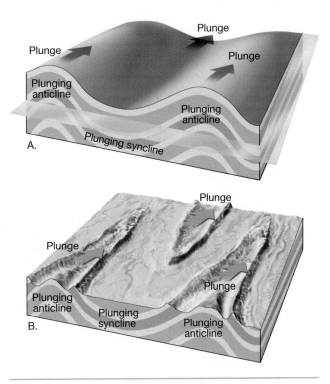

Figure 7.15   Plunging folds. **A.** Idealized view. **B.** View after extensive erosion.

In an anticline, rock layers dip away from the axial plane. If erosion levels an anticline, then the formerly deepest, and hence oldest, rock layers will be exposed on the surface at the axial plane (see Figures 7.11 and 7.15). In an eroded syncline, the layers dip toward the axial plane where the youngest, most recently deposited rock layer occurs (see Figures 7.11 and 7.15).

Figure 7.16 illustrates an eroded anticline and an eroded syncline. Use the figure to answer questions 6–15.

6. On each block diagram, label the type of fold, anticline or syncline, illustrated.

7. The *law of superposition* (see Exercise 6) states that, in most situations concerning layered rocks, the oldest rocks are at the bottom. With this in mind, list by number the rock layers in each block diagram from oldest to youngest.

Anticline:

Oldest ————————————— Youngest

Syncline:

Oldest ————————————— Youngest

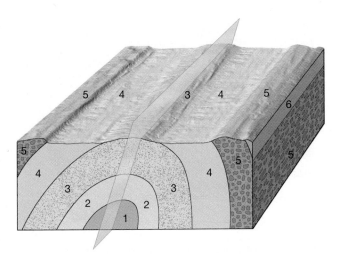

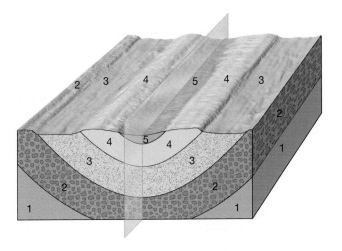

Figure 7.16   Idealized eroded folds.

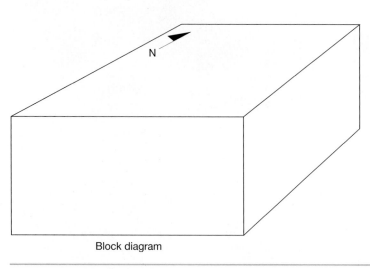

Block diagram   Map view

Figure 7.17 Block diagram and geologic map of a hypothetical area.

8. The near-vertical plane through each block diagram represents each fold's _____ _____ .

9. Draw appropriate strike and dip symbols on the surface of each block diagram for a rock layer on both sides of the axial plane.

10. In the anticline, all the rock layers are dipping (toward, away from) the axial plane. Circle your answer.

11. In the syncline, all the rock layers are dipping (toward, away from) the axial plane. Circle your answer.

12. The anticline is (symmetrical, asymmetrical), and the syncline is (symmetrical, asymmetrical). Circle your answers.

13. Both folds are (plunging, nonplunging) folds. Circle your answer.

14. Using the word "old," label the area(s) of the oldest rocks exposed on the surface of the map view portion of both block diagrams.

15. By circling the correct response, complete the following statements that describe what happens to the ages of the surface rocks as you walk away from the axial plane of each of the following structures.

    a. On an *eroded anticline* the surface rocks get (older, younger) as you walk away from the axial plane.

    b. On an *eroded syncline* the surface rocks get (older, younger) as you walk away from the axial plane.

16. On Figure 7.17, complete the block diagram using the information provided on the map view.

Rocks illustrated with the same pattern are part of the same rock layer.

17. Write the names of the two types of geologic structures illustrated at the appropriate place on the block diagram in Figure 7.17.

Anticlines and synclines are linear features caused by compressional forces. Two other types of folds, **domes** (Figure 7.18) and **basins**, are often nearly circular features that result from vertical displacement. Upwarping of sedimentary rocks produces a dome, whereas a basin is a downwarped structure.

18. Draw several appropriate strike and dip symbols on Figure 7.18.

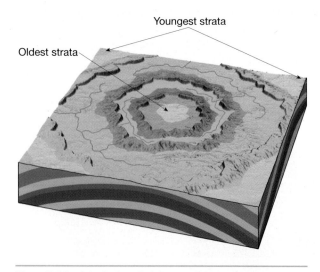

Figure 7.18 A typical eroded dome results in an outcrop pattern that is roughly circular with the rocks dipping away from the center.

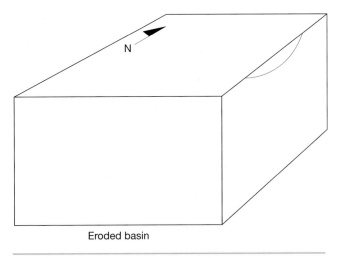

Figure 7.19 Block diagram of an eroded basin.

19. On Figure 7.19, complete the geologic block diagram for the indicated feature. Draw a minimum of four rock layers with appropriate strike and dip symbols on the diagram and label the oldest and youngest rocks.

20. Describe the directions of dip and map view locations of the oldest and youngest surface rocks that comprise the feature.

_____

_____

_____

_____

Types of Faults

Faults are fractures or breaks in rocks along which movement has occurred. As with folds, geologists also use strike and dip to describe the attitude of faults. Faults in which the relative movement of rock units is primarily vertical (along the dip of the fault plane) are called **dip-slip faults** (Figure 7.20). When the dominant motion of rock units is horizontal (in the direction of the strike), the faults are called **strike-slip faults**. Although most faults exhibit both vertical and horizontal displacement, this investigation will only consider the two fundamental motions.

Dip-slip Faults

Since very few faults have a vertical dip, geologists describe the different types of dip-slip faults by noting the relative motion of the blocks on top and below the fault. If the hanging wall (the top of the fault, where miners who may be extracting minerals deposited in

Figure 7.20 A typical dip-slip fault where the relative movement of the rock units is primarily vertical. (Photo by E. J. Tarbuck)

the fault would hang their tools) moves down relative to the footwall (the bottom of the fault, where the miner stands), the fault is classified as a **normal fault** (Figure 7.21A). Normal faults result from tensional forces. Compressional forces will often result in a **reverse fault**, where the hanging wall moves up relative to the footwall (see Figure 7.21B).

Use Figure 7.21 to answer questions 21 and 22.

21. Draw the appropriate strike-dip symbol for each fault on the surface of the large block diagrams in Figure 7.21.

22. On each large block diagram in Figure 7.21, write the word "older" or "younger" to indicate the relative ages of the surface rocks on both sides of the fault trace. Then complete the following statements by circling the correct response.

   a. On an eroded normal fault, the rocks at the surface on the direction of dip side of the fault are the (youngest, oldest) surface rocks.

   b. On an eroded reverse fault, the rocks at the surface on the direction of dip side of the fault are the (youngest, oldest) surface rocks.

23. Complete Figure 7.22 by illustrating an eroded reverse fault on the right-hand diagram. On diagram B, draw an appropriate strike-dip symbol

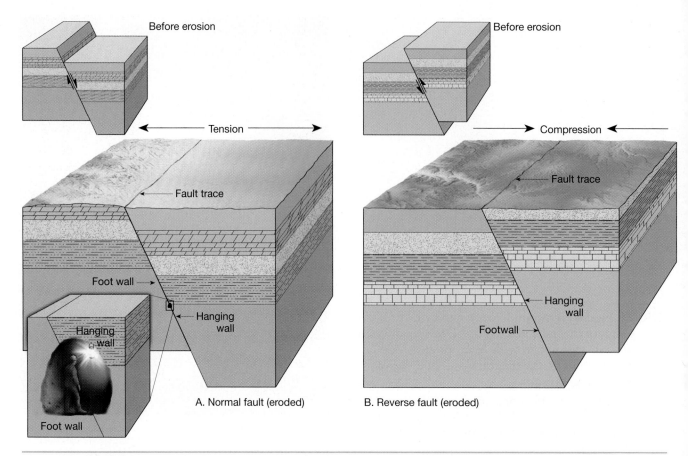

Figure 7.21 Block diagrams illustrating simplified eroded versions of the two common types of dip-slip faults. **A.** Normal fault. **B.** Reverse fault. The arrows by each fault show the relative motions of the blocks on both sides of the fault.

for the fault and arrows on both sides of the fault to illustrate the relative motion of the blocks.

Strike-slip Faults

Strike-slip faults result from horizontal displacement of rock units along the strike or trend of a fault (Figure 7.23).

Many strike-slip faults, such as the famous San Andreas fault system in California, are near-vertical faults that exhibit tens or hundreds of kilometers of displacement along their boundaries.

A strike-slip fault is described as being either a **right-lateral fault** or a **left-lateral fault** depending on the relative motion of the blocks on either side of the

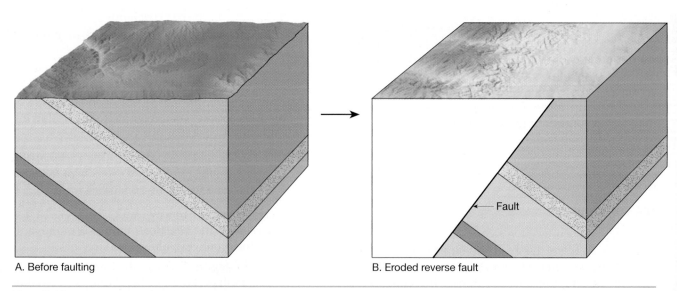

Figure 7.22 Block diagram of an eroded reverse fault. **A.** Before faulting, **B.** after faulting and erosion.

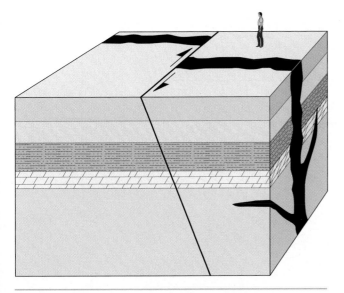

Figure 7.23 Block diagram illustrating a simplified eroded version of a strike-slip fault. The arrows on the surface show the relative motions of the blocks on both sides of the fault.

fault. If you are standing on one side of the fault, looking across the fault to the other side, and the opposite side has been displaced to your right, the fault is called a right-lateral fault. On the other hand, if the side opposite the fault has moved to the left, the fault is a left-lateral fault.

24. The fault illustrated in Figure 7.23 is a (right, left)-lateral fault. Circle your answer.

25. On Figure 7.24, complete the right-hand block diagram by illustrating a left-lateral fault. Draw arrows on both sides of the fault on the surface to illustrate the relative motion of each block.

Examining a Geologic Map

Figure 7.25 is a portion of the Devils Fence, Montana, geologic map with an abbreviated "Explanation" for the map shown in Figure 7.26. Examine the map and explanation, then answer questions 26–34. (*Note:* You may find the geologic time scale, Figure 6.14 in Exercise 6, helpful.)

26. What is the scale of the Devils Fence map?

Scale: _____ : _____, which means one inch on the map equals _____ inches, or approximately _____ mile(s).

27. What are the names and approximate age in years of the youngest and oldest *sedimentary* rock units shown in the map "Explanation"?

Youngest sedimentary rock unit shown in the "Explanation":

Name: _____

Age in years: _____ million

Oldest sedimentary rock unit shown in the "Explanation":

Name: _____

Age in years: _____ million

28. On the geologic structure shown on the eastern half of the map, write the word "oldest" where the oldest sedimentary rock unit is exposed at the surface and the word "youngest" where the youngest sedimentary rocks occur.

29. The igneous intrusive rocks near the center of the structure shown on the eastern half of the map are (younger, older) than the youngest sedimentary rocks. Circle your answer.

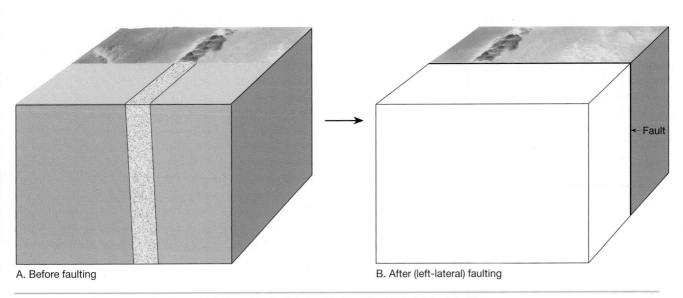

A. Before faulting

B. After (left-lateral) faulting

Figure 7.24 Block diagram of an eroded left-lateral fault. **A**. before faulting, **B**. after left-lateral faulting.

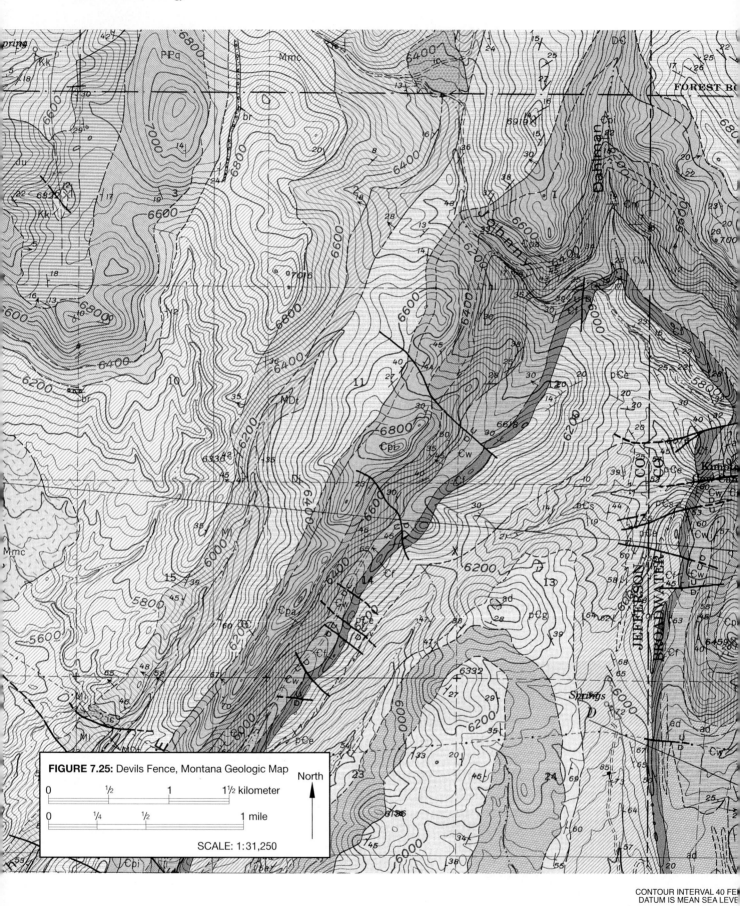

FIGURE 7.25: Devils Fence, Montana Geologic Map

North

| 0 | ½ | 1 | 1½ kilometer |

| 0 | ¼ | ½ | 1 mile |

SCALE: 1:31,250

CONTOUR INTERVAL 40 FEET
DATUM IS MEAN SEA LEVEL

Figure 7.25   Geologic Map—Devils Fence, Montana. (Source: U.S. Geologic Survey, Professional Paper 292)

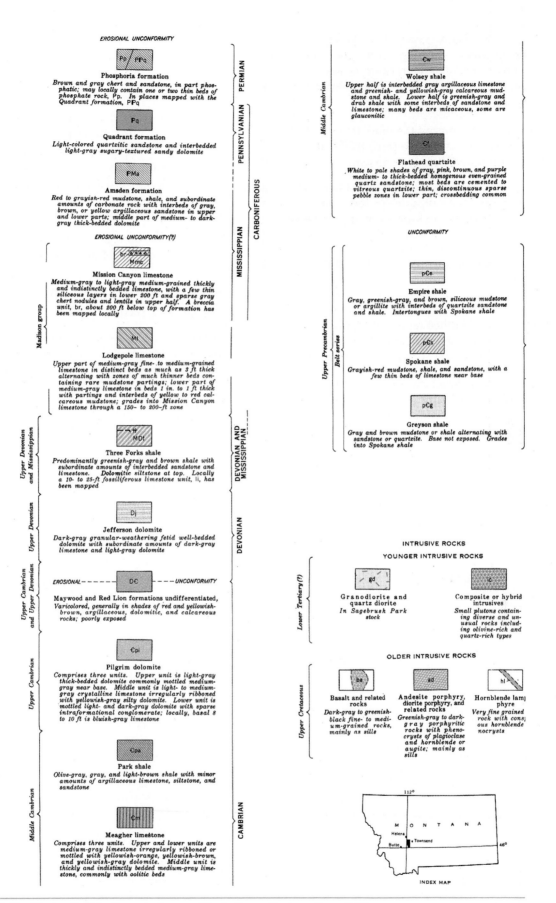

Figure 7.26 "Explanation" for the geologic map of Devils Fence, Montana, Figure 7.25.

30. Examine the strike and dip of the rock units on the Devils Fence geologic map. Draw several large arrows on the map pointing in the direction of dip of several of the rock units. Make sure you examine the entire map. Then answer questions 30a and 30b.

    a. Near the center of the map, in sections 11 and 14, the rocks are dipping to the (northwest, southeast). Circle your answer.

    b. The same rocks that occur in the southeast of section 14 are dipping to the (south, east) in the southeast area of the map, in the eastern halves of sections 13 and 24.

31. What is the approximate angle of dip of the rock units in section 15?

    Angle of dip = _____ °

32. Draw a dashed line representing the hinge line of the large geologic structure illustrated in the eastern half of the map. Label the line "hinge line."

33. The geologic structure present over the majority of the eastern half of the map is a plunging (anticline, syncline). Circle your answer and give two lines of evidence that support your choice.

    _____

    _____

Examine the faults marked with bold black lines that occur in section 14 on the northwest limb of the fold. The relative motion of the faults is shown with a "U" on the side of the fault that has moved up and a "D" on the downward side. Although strike and dip of the faults is not given on the map, assume the direction of dip is to the southwest.

34. Label an example of a normal fault and a reverse fault in section 14 on the geologic map.

Geologic Maps and Structures on the Internet

Use the concepts presented in this exercise to investigate geologic structures and maps by completing the corresponding online activity on the *Applications & Investigations in Earth Science* website at http://prenhall.com/earthsciencelab

Geologic Maps and Structures

Date Due: _____

Name: _____

Date: _____

Class: _____

After you have finished Exercise 7, complete the following questions. You may have to refer to the exercise for assistance or to locate specific answers. Be prepared to submit this summary/report to your instructor at the designated time.

1. Figure 7.27 is a simplified geologic map. In the spaces provided, describe the strike and dip of the igneous intrusion illustrated on the map.

   Strike: North _____ ° _____

   Angle of dip: _____ °

   Direction of dip: _____

2. In the center and upper-left of Figure 7.27, draw a map view of an eroded plunging syncline. Show at least three different sedimentary beds

and indicate the dip and strike of each bed with the proper symbol.

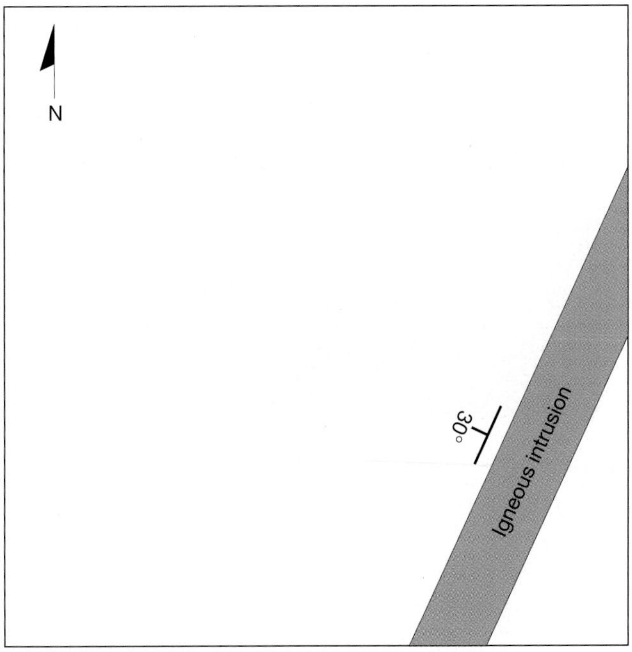

3. What type of geologic structure is illustrated in the eastern half of the Devils Fence geologic map, Figure 7.25? List two lines of evidence in support of your choice.

   _____

   _____

4. Describe the geologic structures that are illustrated in Figure 7.6. What force(s) caused their formation?

   _____

   _____

   _____

5. (Compression, Tension) was the type of stress most likely responsible for producing the geologic structures in Figure 7.25. Circle your answer.

6. Define the following terms.

   Strike: _____

   _____

   Dip: _____

   _____

   Symmetrical anticline: _____

   _____

   Normal fault: _____

   _____

   Left-lateral fault: _____

   _____

Figure 7.27 Simplified geologic map.

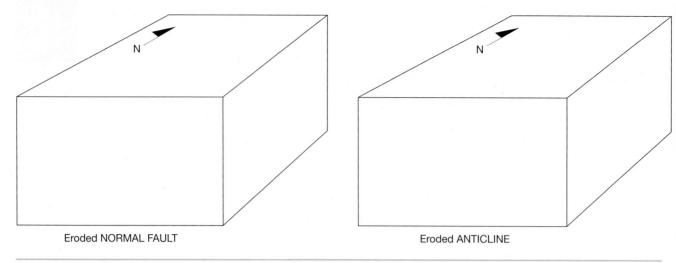

Eroded NORMAL FAULT   Eroded ANTICLINE

Figure 7.28   Block diagrams of an eroded normal fault and an eroded anticline.

7. On Figure 7.28, complete each of the block diagrams by illustrating the indicated geologic structure. Show a minimum of four rock layers and draw the appropriate strike-dip symbols on the surface of each block diagram.

8. With reference to the axial plane, where are the youngest rocks located in an eroded syncline?

_____

9. Label the hanging wall and footwall on each of the two faults illustrated in Figure 7.29. On each

photo, draw arrows showing the relative movement on both sides of the fault. Name the type of fault illustrated on each photo and describe the forces that most likely produced it.

A. _____

_____

B. _____

_____

A.

B.

Figure 7.29   Photographs of two faults to be used with question 9. (Photos by Marli Miller)

Earthquakes and Earth's Interior

Almost all of Earth lies beneath us, yet its accessibility to direct examination is limited. Therefore, one of the most difficult problems faced by Earth scientists is determining the physical properties of Earth's interior. The branch of Earth science called **seismology** combines mathematics and physics to explain the nature of earthquakes and how they can be used to gather information about Earth beyond our view. This exercise introduces some of the techniques that are used by seismologists to determine the location of an earthquake and to investigate the structure of Earth's interior.

Objectives

After you have completed this exercise, you should be able to:

1. Examine an earthquake seismogram and recognize the P waves, S waves, and surface waves.

2. Use a seismogram and travel-time graph to determine how far a seismic station is from the epicenter of an earthquake.

3. Determine the actual time that an earthquake occurred using a seismogram and travel-time graph.

4. Locate the epicenter of an earthquake by plotting seismic data from three seismic stations.

5. Explain how earthquakes are used to determine the structure of Earth's interior.

6. List the name, depth, composition, and state of matter of each of Earth's interior zones.

7. Describe the temperature gradient of the upper Earth.

8. Explain why Earth scientists think that the asthenosphere consists of partly melted, plastic material at a depth of about 100 kilometers.

9. Explain how earthquakes and Earth's temperature gradient have been used to explain the fact that large, rigid slabs of the lithosphere are descending into the mantle at various locations on Earth.

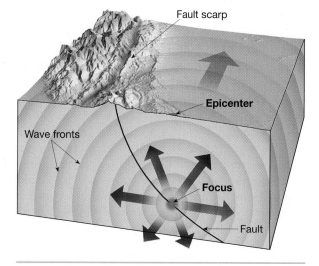

Figure 8.1 Earthquake focus and epicenter. The focus is the zone within Earth where the initial displacement occurs. The epicenter is the surface location directly above the focus.

Materials

calculator	colored pencils
drawing compass	ruler

Materials Supplied by Your Instructor

atlas or wall map

Terms

seismology	P wave	asthenosphere
lithosphere	S wave	mantle
focus	surface wave	outer core
seismic wave	amplitude	inner core
seismograph	period	geothermal gradient
seismogram	epicenter	

Earthquakes

Earthquakes are vibrations of Earth that occur when the rigid materials of the **lithosphere** are strained beyond their limit, yield, and "spring back" to their original shape, rapidly releasing stored energy. This energy radiates in all directions from the source of the earthquake, called the **focus,** in the form of **seismic waves.** (Figure 8.1). **Seismograph** instruments (Figure 8.2) located throughout the world amplify and record the ground motions produced by passing seismic waves on **seismograms** (Figure 8.3). The seismograms are then used to determine the time of occurrence and location of an earthquake, as well as to define the internal structure of Earth.

Examining Seismograms

The three basic types of seismic waves generated by an earthquake at its focus are **P waves, S waves,** and **surface waves** (Figure 8.4). P and S waves travel through Earth while surface waves are transmitted along the outer layer. Of the three wave types, P waves have the greatest velocity and, therefore, reach the seismograph station first. Surface waves arrive at the seismograph station last. P waves also have smaller **amplitudes** (range from the mean, or average, to the extreme) (Figure 8.3) and shorter **periods** (time interval between the arrival of successive wave crests) than S and surface waves.

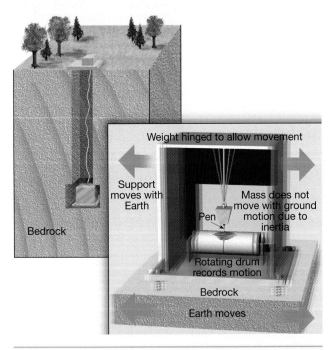

Figure 8.2 Principle of the seismograph. The inertia of the suspended mass tends to keep it motionless, while the recording drum, which is anchored to bedrock, vibrates in response to seismic waves.

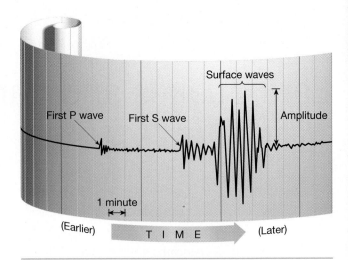

Figure 8.3 Typical earthquake seismogram.

On Figure 8.3, a typical earthquake recording on a seismogram, each vertical line marks a one-minute time interval. Answer questions 1–6 by referring to Figure 8.3.

1. It took approximately (5, 7, 11) minutes to record the entire earthquake from the first recording of the P waves to the end of the surface waves. Circle your answer.

2. (Five, Seven) minutes elapsed between the arrival of the first P wave and the arrival of the first S wave on the seismogram. Circle your answer.

3. (Five, Seven) minutes elapsed between the arrival of the first P wave and the first surface wave.

4. The maximum amplitude of the surface waves is approximately (5, 9) times greater than the maximum amplitude of the P waves.

5. The approximate period of the surface waves is (10, 30, 60) seconds.

6. The period of the surface waves is (greater, less) than the period for P waves.

Locating an Earthquake

The focus of an earthquake is the actual place within Earth where the earthquake originates. Earthquake foci have been recorded to depths as great as 600 km. When locating an earthquake on a map, seismologists plot the **epicenter,** the point on Earth's surface directly above the focus (Figure 8.1).

The difference in the velocities of P and S waves provides a method for locating the epicenter of an earthquake. Both P and S waves leave the earthquake focus at the same instant. Since the P wave has a greater velocity, the further away the recording instrument is from the focus, the greater will be the difference in the arrival times of the first P wave compared to the first S wave.

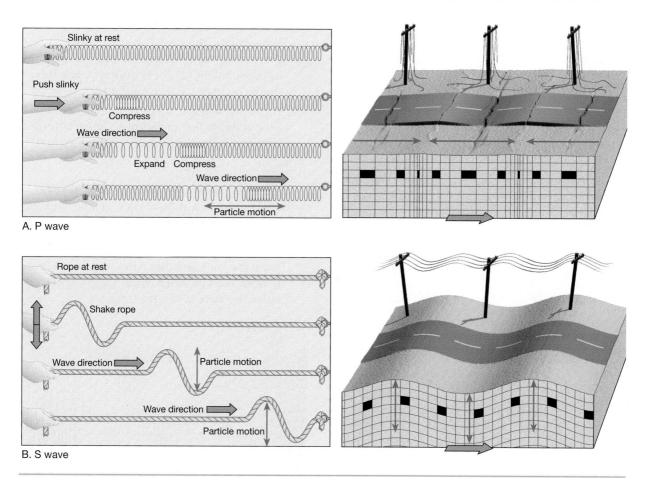

Figure 8.4 Types of seismic waves and their characteristic motion. (Note that during a strong earthquake, ground shaking consists of a combination of various kinds of seismic waves.) **A.** As illustrated by a slinky, P waves are compressional waves that alternately compress and expand the material through which they pass. The back-and-forth motion produced as compressional waves travel along the surface can cause the ground to buckle and fracture, and may cause power lines to break **B.** S waves cause material to oscillate at right angles to the direction of wave motion. Because S waves can travel in any plane, they produce up-and-down and sideways shaking of the ground.

To determine the distance between a recording station and an earthquake epicenter, find the place on the travel-time graph, Figure 8.5, where the vertical separation between the P and S curves is equal to the number of minutes difference in the arrival times between the first P and first S waves on the seismogram. From this position, a vertical line is drawn that extends to the top or bottom of the graph and the distance to the epicenter is read.

To accurately locate an earthquake epicenter, records from three different seismograph stations are needed. First, the distance that each station is from the epicenter is determined using Figure 8.5. Then, for each station, a circle centered on the station with a radius equal to the station's distance from the epicenter is drawn. The geographic point where all three circles, one for each station, intersect is the earthquake epicenter (Figure 8.6).

Answering questions 7–16 will help you understand the process used to determine an earthquake epicenter.

7. An examination of Figure 8.5 shows that the difference in the arrival times of the first P and the first S waves on a seismogram (increases, decreases), the farther a station is from the epicenter. Circle your answer.

8. Use Figure 8.5 to determine the difference in arrival times (in minutes) between the first P wave and first S wave for stations that are the following distances from an epicenter.

1000 miles: _____ minutes difference

2400 km: _____ minutes difference

3000 miles: _____ minutes difference

On the seismogram in Figure 8.3, you determined the difference in the arrival times between the first P and the first S waves to be five minutes.

9. Refer to the travel-time graph. What is the distance from the epicenter to the station that recorded the earthquake in Figure 8.3?

_____ miles

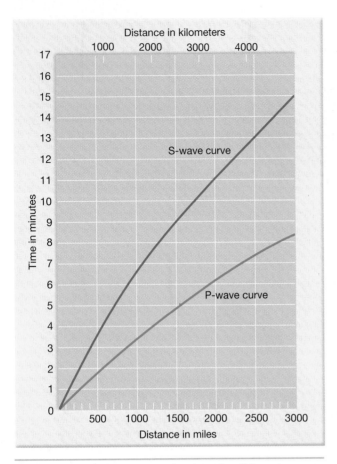

Figure 8.5 Travel-time graph used to determine distance to an earthquake epicenter.

Figure 8.6 An earthquake epicenter is located using the distances obtained from three or more seismic stations.

10. From the travel-time (minutes) axis of the travel-time graph, the first P waves from the seismogram in Figure 8.3 arrived at the recording station approximately (3, 7, 14) minutes after the earthquake occurred. Circle your answer.

11. If the first P wave was recorded at 10:39 P.M. local time at the station in Figure 8.3, what was the local time when the earthquake actually occurred?

_____ P.M. local time

Figure 8.7 illustrates seismograms for the same earthquake recorded at New York, Seattle, and Mexico City. Use this information to answer questions 12–16.

12. Use the travel-time graph, Figure 8.5, to determine the distance that each station in Figure 8.7 is from the epicenter. Write your answers in the epicenter data table, Table 8.1.

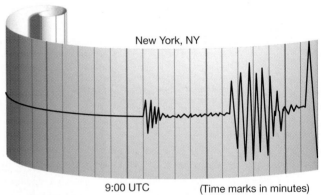

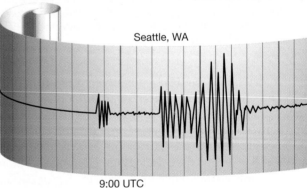

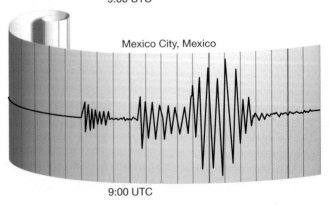

Figure 8.7 Three seismograms of the same earthquake recorded at three different locations.

Table 8.1 **Epicenter data table**

	NEW YORK	SEATTLE	MEXICO CITY
Elapsed time between first P and first S waves			
Distance from epicenter in miles			

13. After referring to an atlas or wall map, accurately place a small dot showing the location of each of the three stations on the map provided in Figure 8.8. Also label each of the three cities.

14. On Figure 8.8, use a drawing compass to draw a circle around each of the three stations with a radius, in miles, equal to its distance from the epicenter. (*Note:* Use the distance scale provided on the map to set the distance on the drawing compass for each station.)

15. What is the approximate latitude and longitude of the epicenter of the earthquake that was recorded by the three stations?

    _____ latitude and _____ longitude

16. Note on the seismograms that the first P wave was recorded in New York at 9:01 Coordinated Universal Time (UTC, the international standard on which most nations base their civil time). At what time (UTC) did the earthquake actually occur?

    _____ UTC

Global Distribution of Earthquakes

Earth scientists have determined that the global distribution of earthquakes is not random but follows a few relatively narrow belts that wind around Earth. Figure 8.9 illustrates the world distribution of earthquakes for a period of several years. Use the figure to answer questions 17 and 18.

17. With what Earth feature is each of the following earthquake belts associated?

    Western and southern Pacific Ocean basin: _____

    _____

    Western South America: _____

    _____

    Mid-Atlantic Ocean basin: _____

    _____

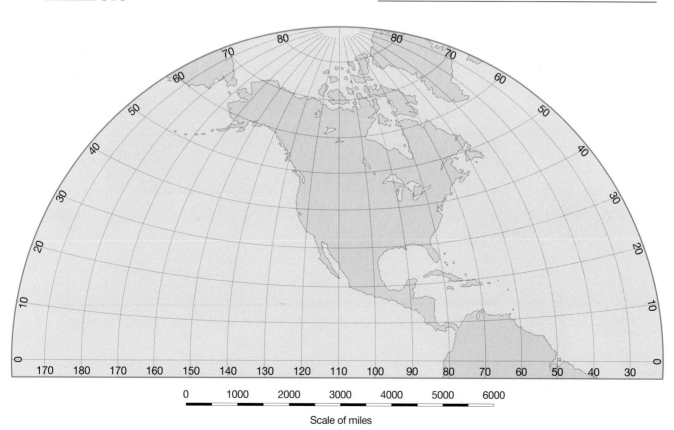

Figure 8.8 Map for locating an earthquake epicenter.

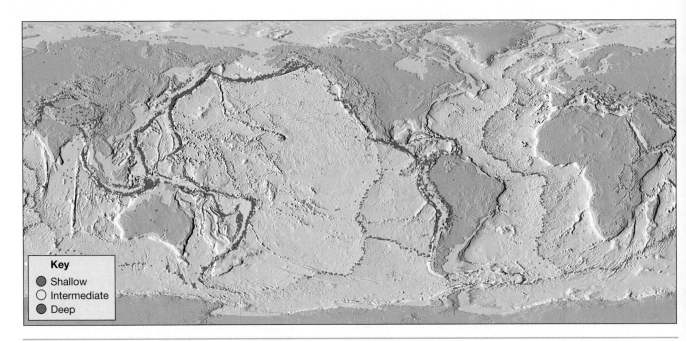

Figure 8.9 World distribution of shallow-, intermediate-, and deep-focus earthquakes. (Data from NOAA)

18. The belts of earthquake activity follow closely the boundaries of what Earth phenomenon?

_____

The Earth Beyond Our View

The study of earthquakes has contributed greatly to Earth scientists' understanding of the internal structure of Earth. Variations in the travel times of P and S waves as they journey through Earth provide scientists with an indication of changes in rock properties. Also, since S waves cannot travel through fluids, the fact that they are not present in seismic waves that penetrate deep into Earth suggests a fluid zone near Earth's center.

In addition to the lithosphere, the other major zones of Earth's interior include the **asthenosphere, mantle, outer core,** and **inner core.** After you have reviewed these zones and the general structure of the Earth's interior, use Figure 8.10 to answer questions 19–24.

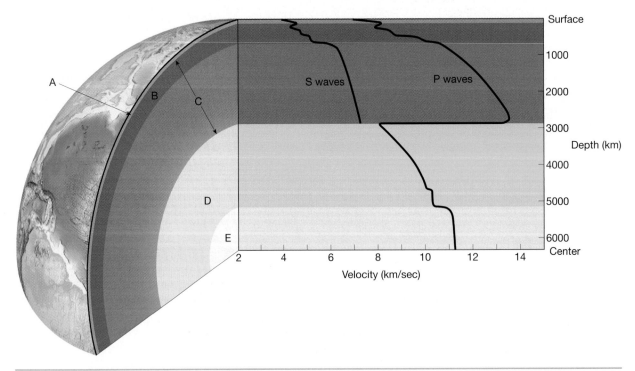

Figure 8.10 Earth's interior with variations in P and S wave velocities. (Data from Bruce A. Bolt)

19. The layer labeled A on Figure 8.10 is the solid, rigid, upper zone of Earth that extends from the surface to a depth of about (100, 500, 1000) kilometers. Circle your answer.

    a. Zone A is called the (core, mantle, lithosphere).

    b. What are the approximate velocities of P and S waves in zone A?

    P wave velocity: _____ km/sec

    S wave velocity: _____ km/sec

    c. The velocity of both P and S waves (increases, decreases) with increased depth in zone A. Circle your answer.

    d. List the two parts of Earth's *crust* that are included in zone A and briefly describe the composition of each.

    1) _____ : _____

    _____

    2) _____ : _____

    _____

20. Zone B is the part of Earth's upper mantle that extends from the base of zone A to a depth of up to about (180, 660, 2250) kilometers in some regions of Earth. Circle your answer.

    a. Zone B is called the (crust, asthenosphere, core).

    b. The velocity of P and S waves (increases, decreases) immediately below zone A in the upper part of zone B.

    c. The change in velocity of the S waves in zone B indicates that it (is, is not) similar to zone A.

21. Zone C (which includes the lower part of zone A and zone B) extends to a depth of

    _____ kilometers.

    a. Zone C is called Earth's _____ .

    b. What fact concerning S waves indicates that zone C is not liquid?

    _____

    _____

    c. What is the probable composition of zone C?

    _____

22. Zone D extends from 2885 km to about (5100, 6100) kilometers.

    a. Zone D is Earth's _____ _____ .

    b. What happens to S waves when they reach zone D, and what does this indicate about the zone?

    _____

    _____

    _____

    _____

    c. The velocity of P waves (increases, decreases) as they enter zone D. Circle your answer.

23. Zone E is Earth's _____ _____ .

    a. Zone E extends from a depth of _____ km to the _____ of Earth.

    b. What change in velocity do P waves exhibit at the top of zone E, and what does this suggest about the zone?

    _____

    _____

    _____

    _____

    c. What is the probable composition of Earth's core?

    _____

    _____

24. Label Figure 8.10 by writing the name of each interior zone at the appropriate letter.

Earth's Internal Temperature

Measurements of temperatures in wells and mines have shown that Earth's temperatures increase with depth. The rate of temperature increase is called the **geothermal gradient.** Although the geothermal gradient varies from place to place, it is possible to calculate an average. Table 8.2 shows an idealized average temperature gradient for the upper Earth compiled from many different sources. Use the information in Table 8.2 to answer questions 25–29.

Table 8.2   Idealized internal temperatures of Earth compiled from several sources

DEPTH (KILOMETERS)	TEMPERATURE (°C)
0	20°
25	600°
50	1000°
75	1250°
100	1400°
150	1700°
200	1800°

Table 8.3   Melting temperatures of granite (with water) and basalt at various depths within Earth

GRANITE (WITH WATER)		BASALT	
DEPTH (KM)	MELTING TEMP. (°C)	DEPTH (KM)	MELTING TEMP. (°C)
0	950°	0	1100°
5	700°	25	1160°
10	660°	50	1250°
20	625°	100	1400°
40	600°	150	1600°

25. Plot the temperature values from Table 8.2 on the graph in Figure 8.11. Then draw a single line that fits the pattern of points from the surface to 200 km. Label the line "temperature gradient."

26. Refer to the graph in Figure 8.11. The rate of increase of Earth's internal temperature (is constant, changes) with increasing depth. Circle your answer.

27. The rate of temperature increase from the surface to 100 km is (greater, less) than the rate of increase below 100 km.

28. The temperature at the base of the lithosphere, which is about 100 kilometers below the surface, is approximately (600, 1400, 1800) degrees Celsius.

29. Use the data and graph to calculate the average temperature gradient (temperature change per unit of depth) for the upper 100 km of Earth in °C/100 km and °C/km.

    °C/100 km: _____, °C/km: _____

Melting Temperatures of Rocks

Geologists have always been concerned with the conditions required for pockets of molten rock (magma) to form near the surface, as well as at what depth within Earth a general melting of rock may occur. The melting temperature of a rock changes as pressure increases deeper within Earth. The approximate melting points of the igneous rocks, granite and basalt, under various pressures (depths) have been determined in the laboratory and are shown in Table 8.3. Granite and basalt have been selected because they are the common materials of the upper Earth. Use the data in Table 8.3 to answer questions 30–35.

30. Plot the melting temperature data from Table 8.3 on the Earth's internal temperature graph you have prepared in Figure 8.11. Draw a different colored line for each set of points and label them "melting curve for wet granite" and "melting curve for basalt."

Use the graphs you have drawn in Figure 8.11 to help answer questions 31–33.

31. Assume your Earth temperature gradient is accurate. At approximately what depth within Earth would wet granite reach its melting temperature and form granitic magma?

    _____ km within Earth

32. Evidence suggests that the oceanic crust and the remaining lithosphere down to a depth of about 100 km are similar in composition to basalt. The melting curve for basalt indicates that the lithosphere above approximately 100 km (has, has not) reached the melting temperature for basalt and therefore should be (solid, molten). Circle your answers.

33. Figure 8.11 indicates that basalt reaches its melting temperature within Earth at a depth of approximately _____ km. (Solid, Partly melted) basaltic material would be expected to occur below this depth. Circle your answer.

34. Referring to Figure 8.10, what is the name of the zone within Earth that begins at a depth of about 100 km and may extend to approximately 700 km?

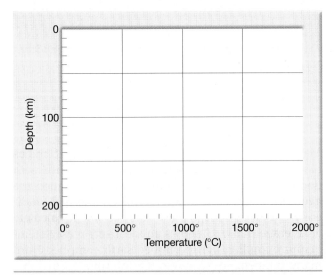

Figure 8.11   Graph for plotting temperature curves.

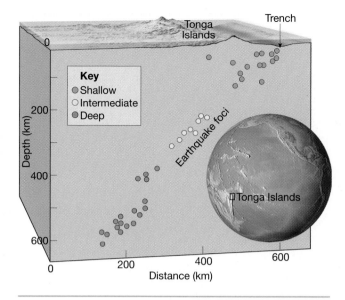

Figure 8.12 Distribution of earthquake foci in 1965 in the vicinity of the Tonga Islands. (Data from B. Isacks, J. Oliver, and L. R. Sykes)

35. Why do scientists believe that the zone in question 34 is capable of "flowing"?

_____

_____

Earthquakes and Earth Temperatures—A Practical Application

The study of earthquakes and Earth's internal temperature has contributed greatly to the understanding of plate tectonics. One part of the plate tectonics theory is that large, rigid slabs of the lithosphere are descending into the mantle where they generate deep focus earthquakes. Using earthquakes and Earth temperatures, Earth scientists have confirmed that this major Earth process is currently taking place near the Tonga Islands in the South Pacific and elsewhere.

Figure 8.12 illustrates the distribution of earthquake foci during a one-year period in the vicinity of the Tonga Islands. Use the figure to answer questions 36–40.

36. At approximately what depth do the deepest earthquakes occur in the area represented on Figure 8.12?

_____ kilometers

37. The earthquake foci in the area are distributed (in a random manner, nearly along a line). Circle your answer.

38. On the figure, outline the area of earthquakes within Earth.

39. Using previous information from this exercise, draw a line on Figure 8.12 at the proper depth that indicates the top of the *asthenosphere*—the zone of partly melted or plastic Earth material. Label the line "top of asthenosphere."

40. Recall the cause and mechanism of earthquakes. Why have Earth scientists been drawn to the conclusion of a descending slab of solid lithosphere being consumed into the mantle near Tonga?

_____

_____

_____

_____

Earthquakes on the Internet

Continue your exploration of earthquakes by completing the corresponding online activity on the *Applications & Investigations in Earth Science* website at http://prenhall.com/earthsciencelab

Notes and calculations.

Earthquakes and Earth's Interior

Date Due: _____

Name: _____

Date: _____

Class: _____

After you have finished Exercise 8, complete the following questions. You may have to refer to the exercise for assistance or to locate specific answers. Be prepared to submit this summary/report to your instructor at the designated time.

1. Use the minute marks provided below to sketch a typical seismogram where the first P wave arrives three minutes ahead of the first S wave. Label each type of wave.

. . . .

(minute marks)

2. How far from the earthquake epicenter is the seismic station that recorded the seismogram in question 1 of this Summary/Report page?

_____ miles

3. Use a diagram to explain how the epicenter of an earthquake is located.

Explanation: _____

_____

_____

_____

_____

_____

Epicenter Diagram

4. What are three Earth features associated with earthquakes?

_____

_____

5. The change in velocity of S waves at the top of the asthenosphere suggests that it is (similar to, different from) the lithosphere. Circle your answer.

6. Why don't S waves make it through Earth's outer core?

_____

7. List the depths of the following interior zones of Earth.

   Crust: depth (km) from _____ to _____

   Mantle: depth (km) from _____ to _____

   Outer core: depth (km) from _____ to _____

   Inner core: depth (km) from _____ to _____

8. On the internal temperature graph you constructed in Figure 8.11, at what depth did you determine granitic magma should form?

   _____ kilometers

9. Why do Earth scientists think that rigid slabs of the lithosphere are descending into the mantle near the Tonga Islands?

   _____

   _____

10. Define the following terms:

    Earthquake focus: _____

    _____

    Earthquake epicenter: _____

    _____

    Seismogram: _____

    _____

    Asthenosphere: _____

    _____

    Geothermal gradient: _____

    _____

    Lithosphere: _____

    _____

11. Identify, label, and describe each of Earth's interior zones on Figure 8.13.

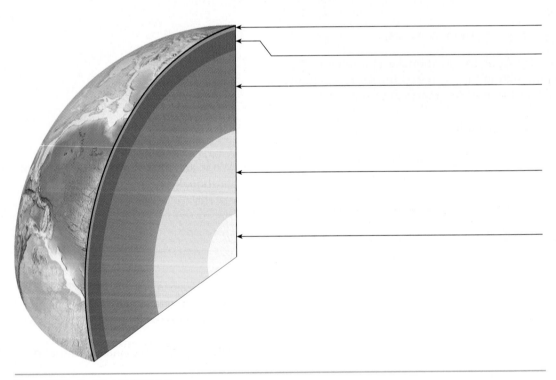

Figure 8.13 Earth's interior zones.

Source: Larry Ulrich/DRK Photo

PART

Oceanography

133

9

Introduction to Oceanography

The global ocean covers nearly three quarters of Earth's surface and **oceanography** is an important focus of Earth science studies. This exercise investigates some of the physical characteristics of the oceans. To establish a foundation for reference, the extent, depths, and distribution of the world's oceans are the first topics examined. Salinity and temperature, two of the most important variables of seawater, are studied to ascertain how they influence the density of water and the deep ocean circulation (Figure 9.1).

Objectives

After you have completed this exercise, you should be able to:

1. Locate and name the major water bodies on Earth.

2. Discuss the distribution of land and water in each hemisphere.

3. Locate and describe the general features of ocean basins.

4. Explain the relation between salinity and the density of seawater.

5. Describe how seawater salinity varies with latitude and depth in the oceans.

6. Explain the relation between temperature and the density of seawater.

7. Describe how seawater temperature varies with latitude and depth in the oceans.

Materials

colored pencils   ruler

Materials Supplied by Your Instructor

measuring cylinder	world wall map,	test tubes
(100 ml, clear,	globe, or atlas	dye
Pyrex or plastic)	ice	salt
salt solutions	beaker	rubber band

Figure 9.1 The deep-diving submersible *Alvin* is 7.6 meters long, weighs 16 tons, has a cruising speed of 1 knot, and can reach depths as great as 4000 meters. A pilot and two scientific observers are along during a normal 6- to 10-hour dive. (Courtesy of Rod Catanach/Woods Hole Oceanographic Institution)

Terms

oceanography	deep-ocean trench	submarine
continental shelf	mid-ocean ridge	canyons
continental slope	density	turbidity
abyssal plain	density current	currents
seamount	salinity	

Extent of the Oceans

1. Refer to a globe, wall map of the world, or world map in an atlas and identify each of the oceans

and major water bodies listed below. Locate and label each on the world map, Figure 9.2.

Oceans	Other Major Water Bodies	
A. Pacific	1. Caribbean Sea	11. Arabian Sea
B. Atlantic	2. North Sea	12. Weddell Sea
C. Indian	3. Coral Sea	13. Bering Sea
D. Arctic	4. Sea of Japan	14. Red Sea
	5. Sea of Okhotsk	15. Bay of Bengal
	6. Gulf of Mexico	16. Caspian Sea
	7. Persian Gulf	
	8. Mediterranean Sea	
	9. Black Sea	
	10. Baltic Sea	

Area

The area of Earth is about 510 million square kilometers (197 million square miles). Of this, approximately 360 million square kilometers (140 million square miles) are covered by oceans and marginal seas.

2. What percentage of Earth's surface is covered by oceans and marginal seas?

$$\frac{\text{Area of oceans and marginal seas}}{\text{Area of Earth}} \times 100$$

$$= \underline{\hspace{2cm}} \text{ \% oceans}$$

3. What percentage of Earth's surface is land?

_____ % land

Distribution of Land and Water by Hemisphere

Answer questions 4–7 by examining either a globe, wall map of the world, world map in an atlas, or Figure 9.2.

4. a. Which hemisphere, Northern or Southern, could be called the "water" hemisphere and which the "land" hemisphere?

   "Water" hemisphere: _____

   "Land" hemisphere: _____

   b. The oceans become (wider, more narrow) as you go from the equator to the pole in the Northern Hemisphere. Circle your answer.

   c. In the Southern Hemisphere the width of the oceans (increases, decreases) from the equator to the pole.

5. Follow a line around a globe, world map, and Figure 9.3 at the latitudes listed on the following page and estimate what percentage of Earth's surface is ocean at each latitude.

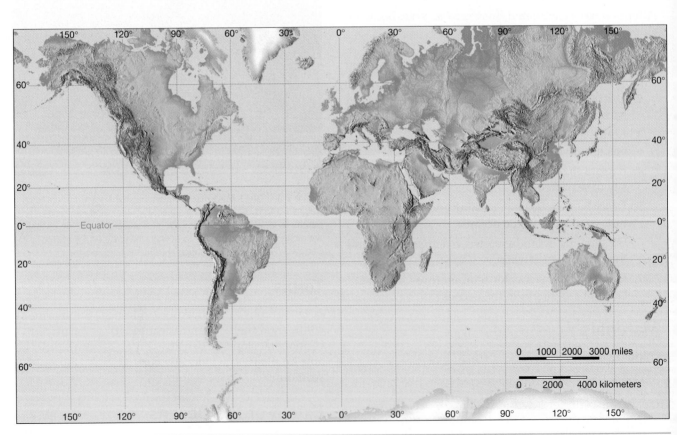

Figure 9.2 World map.

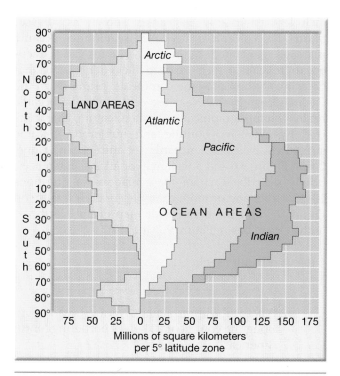

Figure 9.3   Distribution of land and water in each 5° latitude belt. (After M. Grant Gross, *Oceanography: A View of the Earth*, 2nd ed., Englewood Cliffs, N.J.: Prentice-Hall 1977)

NORTHERN HEMISPHERE	SOUTHERN HEMISPHERE
40°: _____ % ocean	_____ % ocean
60°: _____ % ocean	_____ % ocean

6. Which ocean covers the greatest area?

_____

7. Which ocean is almost entirely in the Southern Hemisphere?

_____

Measuring Ocean Depths

Charting the shape or topography of the ocean floor is a fundamental task of oceanographers. In the 1920s a technological breakthrough for determining ocean depths occurred with the invention of electronic depth-sounding equipment. The **echo sounder** (also referred to as *sonar*, an acronym for *so*und *nav*igation *a*nd *r*anging) works by measuring the precise time that a sound wave, traveling at about 1500 meters per second in water, takes to reach the ocean floor and return to the instrument (Figure 9.4). Today, in addition to using sophisticated echo sounders such as *multibeam sonar*,

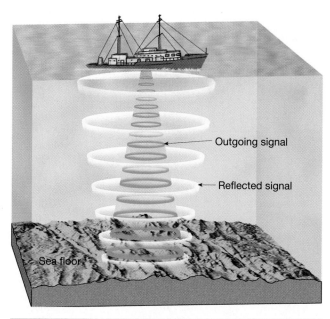

Figure 9.4   An echo sounder determines the water depth by measuring the time interval required for an acoustic wave to travel from a ship to the seafloor and back. The speed of sound in water is 1500 m/sec. Therefore, depth = 1/2(1500 m/sec × echo travel time).

oceanographers are also using satellites to map the ocean floor.

8. Using the formula in Figure 9.4, calculate the depth of the ocean for each of the following echo soundings.

   5.2 seconds: _____

   6.0 seconds: _____

   2.8 seconds: _____

   Ships generally don't make single depth soundings. Rather, as the ship makes a traverse from one location to another, it is continually sending out sound impulses and recording the echoes. In this way, oceanographers obtain many depth recordings from which a *profile* (side view) of the ocean floor can be drawn.

   The data in Table 9.1 were gathered by a ship equipped with an echo sounder as it traveled the North Atlantic Ocean eastward from Cape Cod, MA. The depths were calculated using the same technique used in question 8.

9. Use the data in Table 9.1 to construct a generalized profile of the ocean floor in the North Atlantic on Figure 9.5. Begin by plotting each point at its proper distance from Cape Cod, at the indicated depth. Complete the profile by connecting the depth points.

Table 9.1 **Echo sounder depths eastward from Cape Cod, MA.**

POINT	DISTANCE (KM)	DEPTH (M)
1	0	0
2	180	200
3	270	2700
4	420	3300
5	600	4000
6	830	4800
7	1100	4750
8	1130	2500
9	1160	4800
10	1490	4750
11	1770	4800
12	1800	500
13	1830	4850
14	2120	4800
15	2320	4000
16	2650	3000
17	2900	1500
18	2950	1000
19	2960	2700
20	3000	2700
21	3050	1000
22	3130	1900

Ocean Basin Topography

Various features are located along the continental margins and on the ocean basin floor (Figure 9.6). **Continental shelves**, flooded extensions of the continents, are gently sloping submerged surfaces extending from the shoreline toward the ocean basin. The seaward edge of the continental shelf is marked by the **continental slope**, a relatively steep structure (as compared with the shelf) that marks the boundary between continental crust and oceanic crust. Deep, steep-sided valleys known as **submarine canyons**, eroded in part by the periodic downslope movements of dense, sediment-laden water called **turbidity currents**, are often cut into the continental slope. The ocean basin floor, which constitutes almost 30 percent of Earth's surface,

includes remarkably flat areas known as **abyssal plains**, tall volcanic peaks called **seamounts**, **oceanic plateaus** generated by mantle plumes, and **deep-ocean trenches**, which are deep linear depressions that occasionally border some continents, primarily in the Pacific Ocean basin. Near the center of most oceanic basins is a topographically elevated feature, characterized by extensive faulting and numerous volcanic structures, called the **oceanic** (or **mid-ocean**) **ridge**. Using Figure 9.6 and a wall map or atlas as references, briefly describe each of these features in questions 10–15. Label one or more examples of each feature on Figure 9.5 and the ocean floor map of the North Atlantic Ocean basin, Figure 9.7. (p. 140)

10. Continental shelf: _____

_____

a. What is the approximate average ocean depth along the continental shelves bordering North America?

_____

b. Write a brief statement comparing the width of the continental shelf along the east coast, west coast, and gulf coast of North America.

_____

_____

11. Continental slope: _____

_____

a. Briefly describe the origin of submarine canyons and label at least one on Figure 9.7.

_____

_____

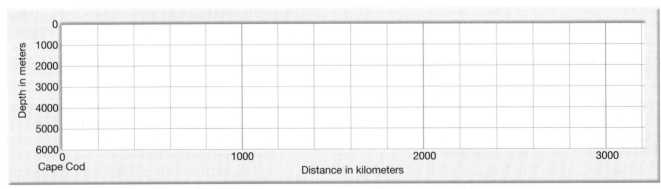

Figure 9.5 North Atlantic Ocean floor profile (exaggerated).

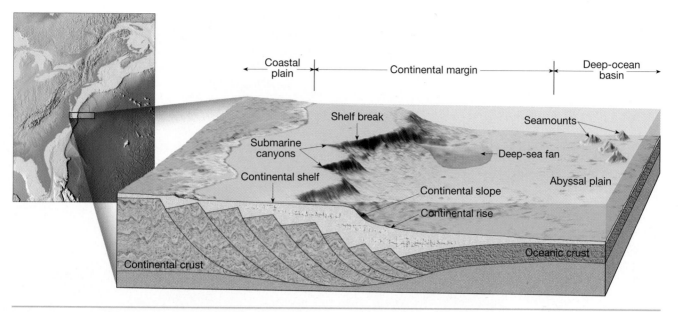

Figure 9.6 Generalized continental margin. Note that the slopes shown for the continental shelf and continental slope are greatly exaggerated. The continental shelf has an average slope of one tenth of 1 degree, while the continental slope has an average of about 5 degrees.

---

12. Abyssal plain: _____

_____

   a. The general topography of abyssal plains is (flat, irregular). Circle your answer.

   b. How do abyssal plains form and what is their composition?

_____

_____

13. Seamount: _____

_____

14. Deep-ocean trench (Not shown on Figure 9.5):

_____

_____

_____

   a. Approximately how deep is the Puerto Rico trench?

   _____ meters

   b. Use a map or globe to locate three deep-ocean trenches in the western Pacific Ocean. Give the name, location, and depth of each.

Trench 1: _____

_____

Trench 2: _____

_____

Trench 3: _____

_____

15. Mid-ocean ridge: _____

_____

   a. Examine the mid-ocean ridge system on a world map. Follow the ridge eastward from the Atlantic Ocean into the Indian Ocean and then into the Pacific. Describe what happens to the ridge along the southwest coast of North America.

_____

   b. Approximately how high above the adjacent ocean floor does the Mid-Atlantic Ridge rise?

   _____ meters

16. Note that Figures 9.5 and 9.7 illustrate only the western side of the North Atlantic floor. Using a globe or map, write a brief statement comparing

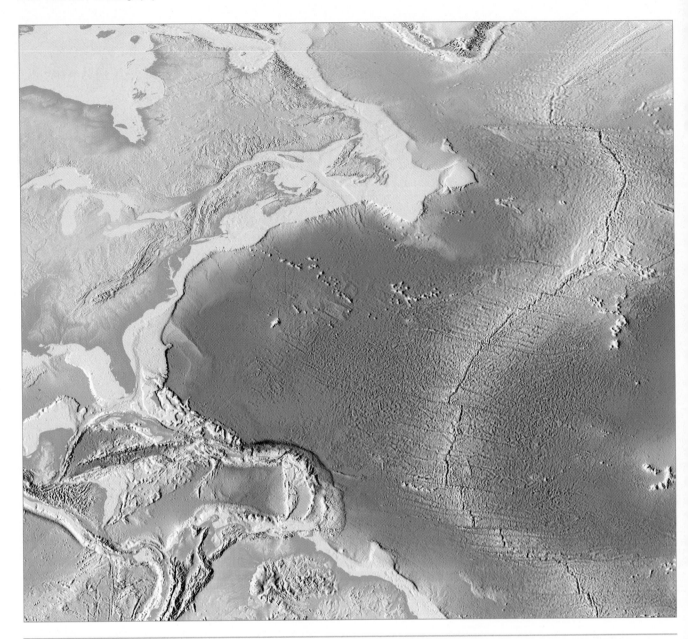

Figure 9.7 North Atlantic basin.

the topography of the North Atlantic Ocean floor east of the mid-ocean ridge to that on the west side.

_____

_____

_____

_____

Characteristics of Ocean Water

Ocean circulation has two primary components: surface ocean currents and deep-ocean circulation. Both are examined with greater detail in Exercise Eleven,

"Waves, Currents, and Tides." While surface currents like the famous Gulf Stream are driven primarily by the prevailing world winds, the deep-ocean circulation is largely the result of differences in ocean water **density** (mass per unit volume of a substance). A **density current** is the movement (flow) of one body of water over, under, or through another caused by density differences and gravity. Variations in **salinity** and temperature are the two most important factors in creating the density differences that result in the deep-ocean circulation.

Salinity

Salinity is the amount of dissolved solid material in water, expressed as parts per thousand parts of water. The symbol for parts per thousand is 0/00. Although

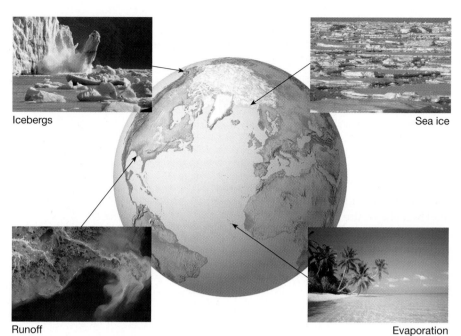

Icebergs

Sea ice

Runoff

Evaporation

Figure 9.8 Processes affecting seawater salinity. Processes that *decrease* seawater salinity include precipitation, runoff, icebergs melting, and sea ice melting. Processes that *increase* seawater salinity include formation of sea ice and evaporation. Source: (upper left) Tom Bean/Tom and Susan Bean, Inc., (upper right) Wolfgang Kaehler Photography, (lower left) NASA Headquarters, (lower right) Paul Steel/Corbis/Stock Market.

there are many dissolved salts in seawater, sodium chloride (common table salt) is the most abundant.

Variations in the salinity of seawater are primarily a consequence of changes in the water content of the solution. In regions where evaporation is high, the proportionate amount of dissolved material in seawater is increased by removing the water and leaving behind the salts. On the other hand, in areas of high precipitation and high runoff, the additional water dilutes seawater and lowers the salinity. Since the factors that determine the concentration of salts in seawater are not constant from the equator to the poles, the salinity of seawater also varies with latitude and depth (Figure 9.8).

Salinity-Density Experiment

To gain a better understanding of how salinity affects the density of water, examine the equipment in the lab (see Figure 9.9) and conduct the following experiment by completing each of the indicated steps.

Step 1. Fill the measuring cylinder with cool tap water up to the rubber band or other marker near the top of the cylinder.

Step 2. Fill a test tube about half full of solution A (saltwater) and pour it slowly into the cylinder. Observe and describe what happens.

Observations: _____

_____

Step 3. Repeat steps 1 and 2 two additional times and measure the time required for the front edge of the saltwater to travel from the rubber band to the bottom of the cylinder. Record the times

for each test in the data table, Table 9.2. *Make certain* that you drain the cylinder after each trial and refill it with fresh water and use the same amount of solution with each trial.

Step 4. Determine the travel time two times for solution B exactly as you did with solution A and enter your measurements in Table 9.2.

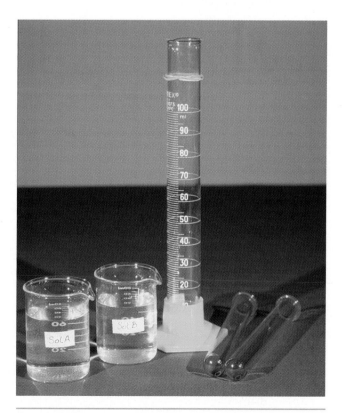

Figure 9.9 Lab setup for salinity-density experiment.

Table 9.2 **Salinity-density experiment data table**

SOLUTION	TIMED TRIAL #1	TIMED TRIAL #2	AVERAGE OF BOTH TRIALS
A			
B			
Solution B plus salt		XXXX	XXXX

Step 5. Fill a test tube about half full of solution B and add to it some additional salt. Then shake the test tube vigorously. Determine the travel time of this solution and enter your results in Table 9.2.

Step 6. Clean all your glassware.

17. Questions 17a and 17b refer to the salinity-density experiment.

  a. Write a brief summary of the results of your salinity-density experiment.

  _____

  _____

  b. Since the solution that traveled fastest has the greatest density, solution (A, B) is most dense. Circle your answer.

Table 9.3 lists the approximate surface water salinity at various latitudes in the Atlantic and Pacific Oceans. Using the data, construct a salinity curve for each ocean on the graph, Figure 9.10. *Use a different colored pencil for each ocean.* Then answer questions 18–22.

Table 9.3 **Ocean surface water salinity in parts per thousand (0/00) at various latitudes in the Atlantic and Pacific Oceans**

LATITUDE	ATLANTIC OCEAN	PACIFIC OCEAN
60°N	33.0 0/00	31.0 0/00
50°	33.7	32.5
40°	34.8	33.2
30°	36.7	34.2
20°	36.8	34.2
10°	36.0	34.4
0°	35.0	34.3
10°	35.9	35.2
20°	36.7	35.6
30°	36.2	35.7
40°	35.3	35.0
50°	34.3	34.4
60°S	33.9	34.0

18. At which latitudes are the highest surface salinities located?

  _____

19. What are two factors that control the concentration of salts in seawater?

  _____ and _____

20. Refer to the factors listed in question 19. What is the cause of the difference in surface water salinity between equatorial and subtropical regions in the Atlantic Ocean?

  _____

  _____

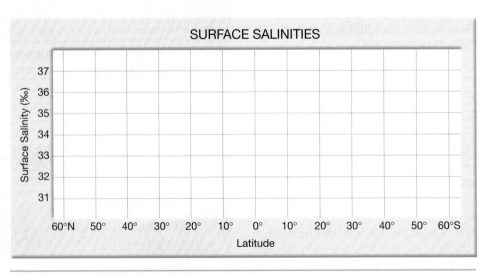

Figure 9.10 Graph for plotting surface salinities.

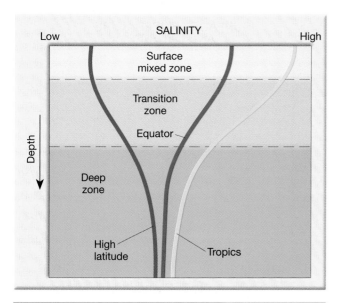

Figure 9.11 Ocean water salinity changes with depth at high latitudes, equatorial regions, and the tropics.

21. Of the two oceans, the (Atlantic, Pacific) Ocean has higher average surface salinities. Circle your answer.

22. Suggest a reason(s) for the difference in average surface salinities between the oceans.

_____

_____

_____

Figure 9.11 shows how ocean water salinity varies with depth at different latitudes. Use the figure to answer questions 23–26.

23. In general, salinity (increases, decreases) with depth in the equatorial and tropical regions and (increases, decreases) with depth at high latitudes. Circle your answers.

24. Why are the surface salinities higher than the deepwater salinities in the lower latitudes?

_____

The *halocline* (*halo*-salt, *cline*-slope) is a layer of ocean water where there is a rapid change in salinity with depth.

25. Label the halocline on Figure 9.11. Where does it occur?

_____

26. Below the halocline the salinity of ocean water (increases rapidly, remains fairly constant, decreases rapidly). Circle your answer.

Ocean Water Temperatures

Seawater temperature is the most extensively determined variable of the oceans because it is easily measured and has an important influence on marine life. Like salinity, ocean water temperatures vary from the equator to poles and also changes with depth.

Temperature, like salinity, also affects the density of seawater. However, the density of seawater is more sensitive to temperature fluctuations than salinity.

Temperature-Density Experiment

To illustrate the effects of temperature on the density of water, examine the equipment in the lab (see Figure 9.12) and then conduct the following experiment by completing each of the indicated steps.

Step 1. Fill a measuring cylinder with *cold* tap water up to the rubber band.

Step 2. Put 2–3 drops of dye in a test tube and fill it half full with *hot* tap water.

Step 3. Pour the contents of the test tube *slowly* into the cylinder and then record your observations.

Observations: _____

_____

_____

Figure 9.12 Lab setup for temperature-density experiment.

Step 4. Empty the cylinder and refill it with *hot* water.

Step 5. Add a test tube full of cold water and 2–3 drops of dye to some ice in a beaker. Stir the solution for a few seconds. Fill the test tube three-fourths full with some liquid (no ice) from your beaker. Pour this cold liquid *slowly* into the cylinder. Then record your observations.

Observations: _____

_____

_____

Step 6. Clean the glassware and return it along with the other materials to your instructor.

27. Questions 27a and 27b refer to the temperature-density experiment.

    a. Write a brief summary of your temperature-density experiment.

    _____

    _____

    _____

    b. Given equal salinities, (cold, warm) seawater would have the greatest density. Circle your answer.

Table 9.4 shows the average surface temperature and density of seawater at various latitudes. Using the data, plot a line on the graph in Figure 9.13 for temperature and a separate line for density using a different color. Then answer questions 28–30.

28. (Warm, Cool) surface temperatures and (high, low) surface densities occur in the equatorial regions. While at high latitudes, (warm, cool) surface temperatures and (high, low) surface densities are found. Circle your answers.

Table 9.4 **Idealized ocean surface water temperatures and densities at various latitudes.**

LATITUDE	SURFACE TEMPERATURE (C°)	SURFACE DENSITY (g/cm³)
60°N	5	1.0258
40°	13	1.0259
20°	24	1.0237
0°	27	1.0238
20°	24	1.0241
40°	15	1.0261
60°S	2	1.0272

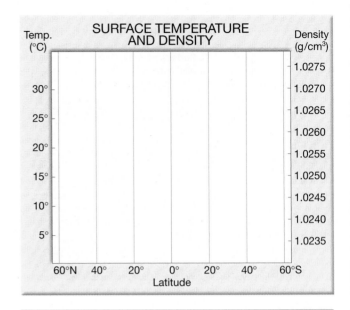

Figure 9.13 Graph for plotting surface temperatures and densities.

29. What is the reason for the fact that higher average surface densities are found in the Southern Hemisphere?

    _____

In question 18 you concluded that surface salinities were greatest at about latitudes 30°N and 30°S.

30. Refer to the density curve in Figure 9.13. What evidence supports the fact that the temperature of seawater is more of a controlling factor of density than salinity?

    _____

    _____

Figure 9.14 shows how ocean water temperature varies with depth at different latitudes. Use the figure to answer questions 31–33.

31. Temperature decreases most rapidly with depth at (high, low) latitudes. Circle your answer and give the reason that the decrease with depth is most rapid at these latitudes.

    _____

The layer of water where there is a rapid change of temperature with depth is called the *thermocline* (*thermo* = heat, *cline* = slope). The thermocline is a very important structure in the ocean because it creates a vertical barrier to many types of marine life.

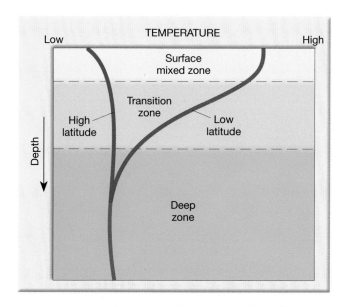

Figure 9.14 Ocean water temperature changes with depth at high and low latitudes.

32. Label the thermocline on Figure 9.14. Where does it occur?

   _____

33. Below the thermocline the temperature of ocean water (increases rapidly, remains fairly constant, decreases rapidly). Circle your answer.

Oceanography on the Internet

Continue your exploration of the oceans by applying the concepts in this exercise to investigate real-time ocean water characteristics on the *Applications & Investigations in Earth Science* website at http://prenhall.com/earthsciencelab

Notes and calculations.

Introduction to Oceanography

Date Due: _____

Name: _____

Date: _____

Class: _____

After you have finished Exercise 9, complete the following questions. You may have to refer to the exercise for assistance or to locate specific answers. Be prepared to submit this summary/report to your instructor at the designated time.

1. Give the approximate latitude and longitude of the centers of each of the following water bodies.

   Mediterranean Sea: _____

   Sea of Japan: _____

   Indian Ocean: _____

2. Write a brief statement comparing the distribution of water and land in the Northern Hemisphere to the distribution in the Southern Hemisphere.

   _____

   _____ .

3. On the ocean basin profile in Figure 9.15, label the continental shelf, continental slope, abyssal plain, seamounts, mid-ocean ridge, and deep-ocean trench.

4. List the names and depths of two Pacific Ocean trenches.

NAME	DEPTH
_____	_____
_____	_____

5. Explain how an echo sounder is used to determine the shape or topography of the ocean floor.

   _____

   _____

   _____

   _____

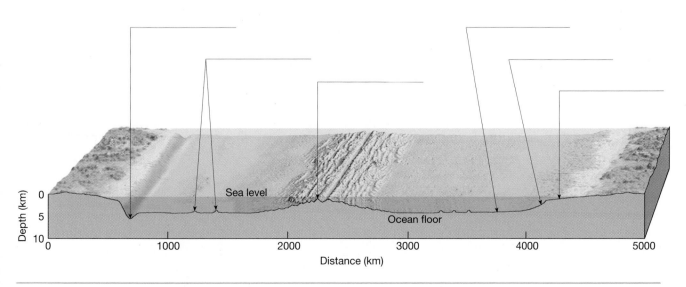

Figure 9.15 Hypothetical ocean basin.

6. The following are some short statements. Circle the most appropriate response.

   a. The higher the salinity of seawater, the (lower, higher) the density.

   b. The lower the temperature of seawater, the (lower, higher) the density.

   c. Surface salinity is greatest in (polar, subtropical, equatorial) regions.

   d. (Temperature, Salinity) has the greatest influence on the density of seawater.

   e. (Warm, Cold) seawater with (high, low) salinity would have the greatest density.

   f. Vertical movements of ocean water are most likely to begin in (equatorial, subtropical, polar) regions, because the surface water there is (most, least) dense.

7. Why is the surface salinity of an ocean higher in the subtropics than in the equatorial regions?

   _____

   _____

   _____

8. Refer to the salinity-density experiment you conducted. Solution (A, B) had the greatest density. Circle your answer.

9. Describe the change in salinity *and* temperature with depth that occurs at low latitudes.

   Salinity: _____

   _____

   Temperature: _____

   _____

10. Are the following statements true or false? Circle your response.

   T F a. The Atlantic Ocean covers the greatest area of all the world oceans.

   T F b. Continental shelves are part of the deep-ocean floor.

   T F c. Deep-ocean trenches are located in the middle of ocean basins.

   T F d. High evaporation rates in the subtropics cause the surface ocean water to have a lower than average salinity.

The Dynamic Ocean Floor

One of the most significant scientific revelations of the 20th century was the realization that the ocean basins are young, ephemeral features. Based upon this discovery, a revolutionary theory called *plate tectonics* evolved that helps to explain and interrelate earthquakes, mountain building, the origins of ocean basins, and other geologic events and processes. This exercise examines some of the lines of evidence that have been used to verify this comprehensive model of the way Earth scientists view the restless Earth.

Objectives

After you have completed this exercise, you should be able to:

1. List and explain the lines of evidence that support the theory of plate tectonics.
2. Locate and describe the mid-ocean ridge system and deep-ocean trenches.
3. Describe the relation between earthquakes and plate boundaries.
4. Describe the magnetic polarity reversals that have taken place on Earth in the last four million years.
5. Determine the rate of seafloor spreading that occurs along a mid-ocean ridge by using paleomagnetic evidence.
6. Use the rate of seafloor spreading for an ocean basin to determine its age.
7. Describe the three types of plate boundaries and the motion of the plates that occurs along each boundary.

Materials

calculator	ruler	colored pencils

Materials Supplied by Your Instructor

atlas, globe, or world wall map

Terms

plate tectonics	rift valleys	paleomagnetism
lithosphere	deep-ocean trench	seafloor
plates	Pangaea	spreading
mid-ocean ridge	continental drift	

Introduction

Since the early days of sailing vessels, the investigation of the ocean floor has been both a challenge and a source of new understanding about the mechanisms of the dynamic Earth. Using the information gathered over the years concerning the topography, age, and method of evolution of the ocean basins, Earth scientists have developed an important theory called **plate tectonics**. Plate tectonics is the foundation used by modern geology to help explain the origin of mountains and continents, the occurrence of earthquakes, the evolution of ocean basins, the development and distribution of plants and animals, as well as many other geologic processes.

Plate tectonics postulates that the **lithosphere** of Earth is broken into several large, rigid slabs called **plates**. These plates and the continents on them are moving. Where the plates are separating along the **mid-ocean ridges**, new ocean floor crust is forming. Further, along the axis of some ridge segments are deep down-faulted structures called **rift valleys**. Along the plate margins, earthquakes are generated as plates slide past each other, collide to form mountains, or override each other causing **deep-ocean trenches**.

The idea that our present-day continents at one time were parts of a single supercontinent was stated in 1912 by Alfred Wegener. His hypothesis, called **continental drift**, proposed that long ago a supercontinent called

Pangaea broke apart into the present continents (Figure 10.1). Over a period of millions of years, each of the continents drifted across Earth's surface to its current position. Among the lines of evidence used to support this hypothesis were (1) the geometric fit of the continents, (2) the fit of geologic structures (mountains, etc.) and rock ages across oceans, (3) the global distribution of fossils (*paleontology*), and (4) ancient climates (*paleoclimatology*).

In the 1960s great advances resulting from new technologies in oceanographic research, such as the ocean research vessel *Glomar Challenger,* brought new evidence in support of many of Wegener's ideas. Research concerning rock magnetism, the cause and distribution of earthquakes, and the age of ocean sediments led to the development of a much broader theory than continental drift. The expanded theory is known as *plate tectonics.*

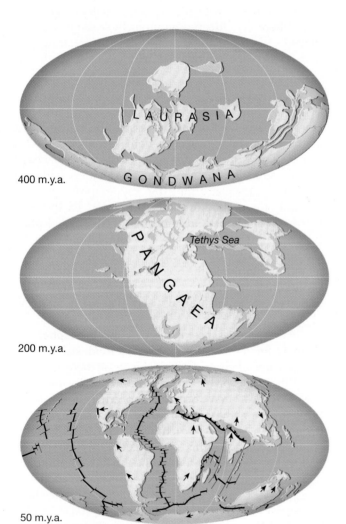

400 m.y.a.

200 m.y.a.

50 m.y.a.

Figure 10.1 Pangaea, as it is thought to have appeared 200 million years ago. (After R. S. Dietz and J. C. Holden, *Journal of Geophysical Research* 75: 4943.)

The Evidence for Plate Tectonics

Prior to the plate tectonics theory, the origin of many ocean floor features was uncertain. However, today Earth scientists have a much better understanding of these features, especially when viewed as parts of a dynamic, evolving lithosphere. Two of these features, mid-ocean ridges and deep-ocean trenches, are of major importance in understanding plate tectonics.

1. In the following spaces, write brief statements that describe each of the following ocean-floor features.

   Mid-ocean ridge: _____

   _____

   _____

   Mid-ocean ridge rift valley: _____

   _____

   _____

   Deep-ocean trench: _____

   _____

   _____

2. Using an atlas, and/or world wall map as a reference, accurately draw the global mid-ocean ridge system on the world map, Figure 10.2.

3. Use an atlas and/or world wall map to locate and label each of the following deep-ocean trenches on Figure 10.2. Draw a line to represent the trench. To conserve space, write only the letter of each trench on the map.

a. Puerto Rico	**f.** Japan
b. Cayman	**g.** Mariana
c. Peru-Chile	**h.** Tonga
d. Aleutian	**i.** Kermadec
e. Kuril	**j.** Java

Fit of the Continents

Examine the east coast of South America and the west coast of Africa on the world map, Figure 10.2. Then answer questions 4–7.

4. The shape of the east coast of South America conforms to the shape of the (east, west) side of the mid-ocean ridge in the central South Atlantic Ocean. Circle your answer.

5. The shape of the west coast of Africa conforms to the shape of the (east, west) side of the mid-ocean ridge in the South Atlantic Ocean. Circle your answer.

6. If South America and Africa were moved to the mid-ocean ridge, their shapes (would, would not) fit together along the ridge. Circle your answer.

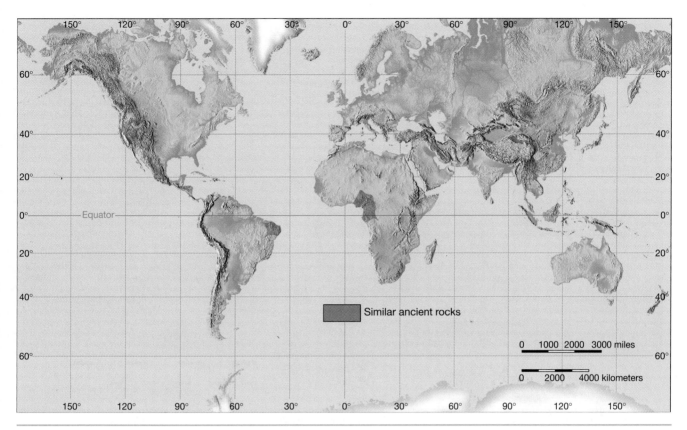

Figure 10.2 World map.

Notice on Figure 10.2 that the ages of the rocks in eastern South America and western Africa are similar. The rocks that comprise the ocean floor separating the two continents are much younger.

7. If the South American and African continents are brought together at the mid-ocean ridge, do the areas of ancient rocks match? Does the match support the idea that these continents were once joined?

_____

_____

(*Note:* Actually, the fit of the continents is more exact on a globe because flat maps have distortions. Also, using the seaward edge of the continental shelf rather than the coastline would be more accurate.)

8. In Figure 10.1, identify each of the present-day continents that comprised Pangaea by writing their names at the proper locations on the center figure.

Earthquakes

The distribution and depths of earthquakes provide evidence for the mechanics of plate tectonics.

Examine the world map of plates, Figure 10.3, and compare it to the earthquake distribution map,

Figure 8.9, in Exercise 8. Use the maps and Figure 10.2 to answer questions 9–11.

9. Most deep-focus earthquakes are associated with (mid-ocean ridges, deep-ocean trenches). Circle your answer.

10. Most mid-ocean, shallow-focus earthquakes are associated with (mid-ocean ridges, deep-ocean trenches).

Compare the earthquake distribution map to the map showing the plates. Notice that plate boundaries are outlined by earthquake zones.

11. On the world map, Figure 10.2, outline and label, by name, the major plates.

Paleomagnetism

Some minerals in igneous rocks develop a slight magnetism in alignment with Earth's magnetic field at the time of their formation. Also, scientists have discovered that the polarity of Earth's magnetic field has periodically reversed and the North Magnetic Pole becomes the South Magnetic Pole, while the South Magnetic Pole becomes the North Magnetic Pole. Putting these facts together provides additional support for plate tectonics.

The ancient magnetism, called **paleomagnetism**, present in rocks on the ocean floor can be used to determine the rate at which the plates are separating and,

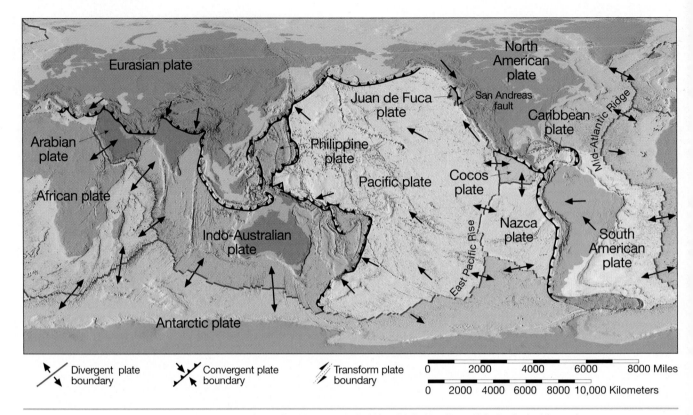

Figure 10.3 The mosaic of rigid plates that constitutes Earth's outer shell. (After W. B. Hamilton, U.S. Geological Survey)

consequently, the time when they began to separate. Where plates separate along the mid-ocean ridge, magma from the mantle rises to the surface and creates new ocean floor. As the magma cools, the minerals assume a magnetism equal to the prevailing magnetic field. As the plates continue to separate and Earth's magnetic field reverses polarity, new material forming at the ridge is magnetized in an opposite (or *reversed*) direction.

Scientists have reconstructed Earth's magnetic polarity reversals over the past several million years. A generalized record of these polarity reversals is shown in graphic form in Figure 10.4. The periods of normal polarity, when a compass needle would have pointed North as it does today, are shown in color and labeled a–f for reference. Use Figure 10.4 to answer questions 12–17.

12. The magnetic field of Earth has had (3, 5, 7) intervals of reversed polarity during the past 4 million years. Circle your answer.

13. Approximately how long ago did the current normal polarity begin?

    _____ years ago

14. One and a half million years ago the indicator on a compass needle would have pointed to the (North, South). Circle your answer.

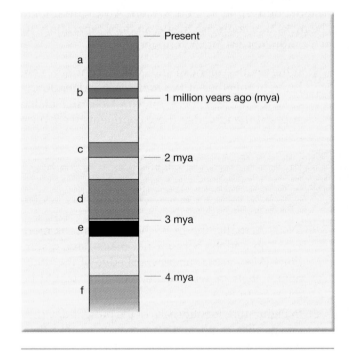

Figure 10.4 Chronology of magnetic polarity reversals on Earth during the last 4 million years. Periods of normal polarity, when a compass would have pointed North as it does today, are shown in color. Periods of reverse polarity are shown in white. (Data from Allan Cox and G. Dalrymple)

15. The period of normal polarity, c, began (1, 2, 3) million years ago.

16. During the past 4 million years, each interval of reverse polarity has lasted (more, less) than 1 million years. Circle your answer.

17. Based upon the pattern of magnetic changes exhibited in Figure 10.4, does it appear as though Earth is due for another magnetic polarity reversal in the near future?

Paleomagnetism and the Ocean Floor

The records of the magnetic polarity reversals that have been determined from the oceanic crust across sections of the mid-ocean ridges in the Pacific, South Atlantic, and North Atlantic oceans are illustrated in Figure 10.5. As new ocean crust forms along the mid-ocean ridge, it spreads out equally on both sides of the ridge. Therefore, a record of the reversals is repeated (mirrored). Notice that the general pattern of polarity reversals presented in Figure 10.4 can be matched with the polarity of the rocks on either side of the ridge for each ocean basin in Figure 10.5.

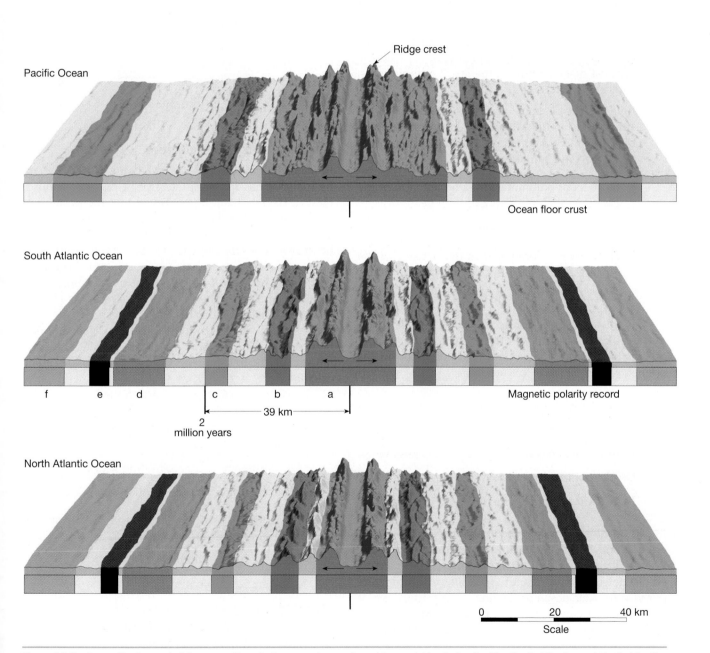

Figure 10.5   Generalized record of the magnetic polarity reversals near the mid-ocean ridge in the Pacific, South Atlantic, and North Atlantic oceans. Periods of normal polarity are shown in color and correspond to those illustrated in Figure 10.4.

Use Figures 10.4 and 10.5 to answer questions 18–23.

18. On Figure 10.5, identify and mark the periods of normal polarity with the letters a–f (as in Figure 10.4) along each ocean floor. Begin at the ridge crest and label along both sides of each ridge. (*Note:* The left side of the South Atlantic has already been done and can act as a guide. Also, only the periods of normal polarity through c are illustrated in the Pacific basin.)

19. Using the South Atlantic as an example, label the beginning of the normal polarity period c, "2 million years ago," on the left sides of the Pacific and North Atlantic diagrams.

20. Using the distance scale in Figure 10.5, the (Pacific, South Atlantic, North Atlantic) has spread the greatest distance during the last 2 million years. Circle your answer.

21. Refer to the distance scale. Notice that the left side of the South Atlantic basin has spread approximately 39 kilometers from the center of the ridge crest in 2 million years.

    a. How many kilometers has the left side of the Pacific basin spread in 2 million years?

       _____ kilometers

    b. How many kilometers has the left side of the North Atlantic basin spread in 2 million years?

       _____ kilometers

    The distances in question 21 are for only one side of the ridge. Assuming that the ridge spreads equally on both sides, the actual distance each ocean basin has opened would be twice this amount.

22. How many kilometers has each ocean basin opened in the past 2 million years?

    Pacific Ocean basin: _____ kilometers

    South Atlantic Ocean basin: _____ kilometers

    North Atlantic Ocean basin: _____ kilometers

    If both the *distance* that each ocean basin has opened and the *time* it took to open that distance are known, the rate of **seafloor spreading** can be calculated. To determine the rate of spreading in centimeters per year for each ocean basin, first convert the distance the basin has opened (see question 22) from kilometers to centimeters and then divide this distance by the time, 2 million years.

23. Determine the *rate of seafloor spreading* for the Pacific and North Atlantic Ocean basins. As an example, the South Atlantic has already been done.

a. South Atlantic: distance =

   78 km × 100,000 cm/km = 7,800,000 cm

   $$\text{Rate of spreading} = \frac{7{,}800{,}000 \text{ cm}}{2{,}000{,}000 \text{ yr}} = 3.9 \text{ cm/yr}$$

b. Pacific: distance = _____ km × 100,000 cm/km

   = _____ cm

   $$\text{Rate of spreading} = \frac{\text{cm}}{\text{yr}}$$

   = _____ cm/yr

c. North Atlantic: distance = _____ km

   × 100,000 cm/km = _____ cm

   $$\text{Rate of spreading} = \frac{\text{cm}}{\text{yr}}$$

   = _____ cm/yr

Hot Spots and Plate Velocities

Researchers have proposed that a rising plume of mantle material has formed a *hot spot* below the island of Hawaii (Figure 10.6). It is believed that the position of this hot spot within Earth has remained constant during a very long period of time. In the past, as the Pacific plate has moved over the hot spot, the successive volcanic islands of the Hawaiian chain have been built. Today the island of Hawaii is forming over this mantle plume.

Radiometric dating of the volcanoes in the Hawaiian chain has revealed that they increase in age with increasing distance from the island of Hawaii (see Figure 10.6). Knowing the age of an island and its distance from the hot spot, the velocity of the plate can be calculated.

Use Figure 10.6 to answer questions 24–26.

24. What are the minimum and maximum ages of the island of Kauai?

    Minimum age: _____ million years

    Maximum age: _____ million years

25. What is the distance of Kauai from the hot spot in both kilometers and centimeters?

    _____ kilometers

    _____ centimeters

26. Using the data in questions 24 and 25, calculate the approximate maximum and minimum velocities, in centimeters per year (cm/yr), of the Pacific plate as it moved over the Hawaiian hot spot.

    Maximum velocity: _____ cm/yr

    Minimum velocity: _____ cm/yr

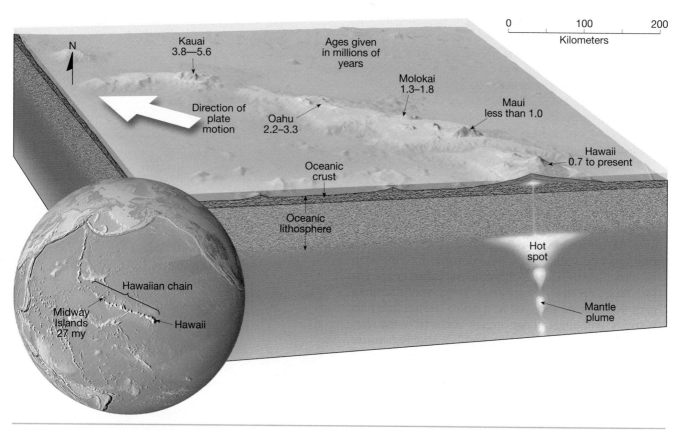

Figure 10.6 Movement of the Pacific plate over a stationary hot spot and the corresponding radiometric ages of the Hawaiian Islands in millions of years.

Ocean Basin Ages

To estimate how many millions of years ago the North Atlantic and South Atlantic Ocean basins began forming, complete questions 27–29.

27. On a large wall map or globe, accurately measure the distance between the seaward edges of the continental shelves from eastern North America at North Carolina to northwestern Africa at Mauritania (20°N latitude). Determine the distance in kilometers, then convert to centimeters. (*Note:* If you have completed Exercise 21, "Location and Distance on Earth," you may find it easier to determine the great circle distance between the two.)

    Distance = _____ kilometers

    = _____ centimeters

28. Divide the distance in centimeters separating the continents that you measured in question 27 by the rate of seafloor spreading for the North Atlantic basin, calculated in question 23c. Your answer is the approximate age in years of the North Atlantic Ocean basin.

    Distance/Rate of seafloor spreading =

    _____ =

    Age of the North Atlantic Basin: _____ years

Use the same procedure you used in question 28 to determine the age of the South Atlantic basin. Measure the distance between the continents along an east-west line from the eastern edge of Brazil directly east to Africa. (*Note:* You may want to determine the total number of degrees of longitude separating the two and then use the Table 21.1, "Longitude as distance," in Exercise 21, to calculate the distance.) Use the rate of seafloor spreading for the South Atlantic in your calculation.

29. How many years ago did South America and Africa begin to separate?

    Age of the South Atlantic Basin: _____ years

30. Write your calculated ages of the North and South Atlantic Ocean basins on the map in Figure 10.2. Also on Figure 10.2, use the world map of plates in Figure 10.3 as a reference and draw arrows on all the major plates showing their directions of movement.

Plate Boundaries

Earth scientists recognize three distinct types of plate boundaries, with each distinguished by the movement of the plates associated with it. Along *transform fault boundaries*, plates slide horizontally past one another without the production or destruction of lithosphere. Transform faults were first identified where they join offset segments of an ocean ridge. Although most are located within the ocean basins, a few cut through continental crust. One classic example of such a boundary is the earthquake-prone San Andreas Fault of California. Constituting a second type of boundary, *divergent boundaries* are located along the crests of oceanic ridges and can be thought of as constructive plate margins since this is where new oceanic lithosphere is generated. Although new lithosphere is constantly being produced at the oceanic ridges, our planet is not growing larger—its total surface area remains the same. To balance the addition of newly created lithosphere, older portions of oceanic lithosphere descend into the mantle along the third type of boundary, called *convergent boundaries*. Because lithosphere is "destroyed" at convergent boundaries, they are also called destructive plate margins. Although all convergent zones have the same basic characteristics, they are highly variable features. Each is controlled by the type of crustal material involved and the tectonic setting. Convergent boundaries can form between two oceanic plates, one oceanic plate and one continental plate, or two continental plates (Figure 10.3).

The three diagrams in Figure 10.7 illustrate each type of plate boundary. Use the figure to answer questions 31–33.

31. Figure 10.7A represents a (convergent, divergent, transform) plate boundary. Circle your answer.

    a. The plates along the boundary are (spreading, colliding). Circle your answer.

    b. This type of plate boundary occurs at (deep-ocean trenches, mid-ocean ridges). Circle your answer.

    c. This type of plate boundary results in (construction, destruction) of lithospheric material. Circle your answer.

32. Figure 10.7B represents a (convergent, divergent, transform) plate boundary. Circle your answer.

    a. The plates along this type of boundary are (spreading, colliding). Circle your answer.

    b. Briefly describe each of the following types of plate convergence. Using Figure 10.3 as a general reference, give a specific example of a physical feature on Earth that has formed as a result of each type.

    Oceanic-continental convergence: _____

    Feature: _____

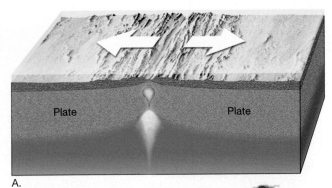

A.

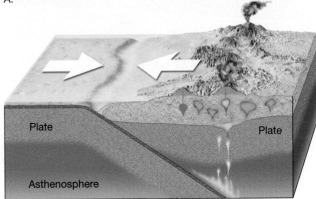

B.

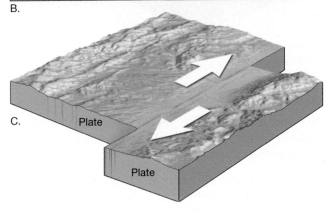

C.

Figure 10.7 Schematic illustrations of the three types of plate boundaries.

Oceanic-oceanic convergence: _____

Feature: _____

Continental-continental convergence: _____

Feature: _____

33. Figure 10.7C represents a (convergent, divergent, transform) plate boundary. Circle your answer.

    a. Lithospheric material (is being created, is being destroyed, remains unchanged) along this type of boundary. Circle your answer.

    b. What type of faults are likely to parallel the direction of plate movement along this type of boundary? Write a brief description of the fault.

_____ faults; _____

_____

Igneous Rocks and Plate Boundaries

The theory of plate tectonics provides a framework for explaining the occurrence of many igneous rocks and the features they compose. The formation of magma is most frequently associated with weaknesses along the boundaries between lithospheric plates.

Figure 10.8 illustrates a few generalized geologic events that are happening along some plate boundaries. Examine the figure closely. Then, use Figure 10.8 to complete questions 34–39.

34. On Figure 10.8, indicate the direction of movement of the plates on both sides of the mid-ocean ridge by drawing arrows on the ocean floor.

35. The ocean floor is composed of the igneous rock (granite, basalt). Circle your answer.

36. The eruption of basaltic magma from the part of the mantle called the *asthenosphere* is associated with what ocean features?

_____ and _____

37. Volcanic islands in the ocean, such as the Hawaiian Islands, form over (trenches, hot spots) in the athenosphere and are made of the igneous rock (granite, basalt). Circle your answers.

38. (Deep-ocean trenches, Mid-ocean ridges) form on the ocean floor where oceanic plates are descending beneath continents.

39. As oceanic plates descend beneath the continents, the plates (remain solid, partially melt) and produce pockets of magma in the continents. These magma chambers are the sources of lava that form (ridges, volcanoes) on the continents. Circle your answers.

The Ocean Floor on the Internet

Continue your exploration of the topics presented in this exercise by applying the concepts you have learned to the corresponding online activity on the *Applications & Investigations in Earth Science* website at http://prenhall.com/earthsciencelab

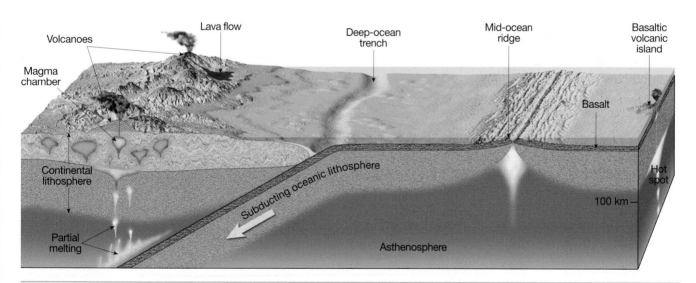

Figure 10.8 Ocean-floor features and the occurrence of magma along plate boundaries.

Notes and calculations.

The Dynamic Ocean Floor

Date Due: _____

Name: _____

Date: _____

Class: _____

After you have finished Exercise 10, complete the following questions. You may have to refer to the exercise for assistance or to locate specific answers. Be prepared to submit this summary/report to your instructor at the designated time.

1. In the following space, draw a general profile of an ocean floor between two continents illustrating a mid-ocean ridge and a deep-ocean trench. Label each of the features and draw arrows showing plate motion.

Ocean Floor Profile

2. Deep-focus earthquakes are associated with what ocean-floor features?

_____

3. Shallow-focus earthquakes are associated with what ocean-floor features?

_____

4. Earthquake foci can be used to mark the boundaries of what Earth features?

_____

5. Describe how paleomagnetism is used to calculate the rate of seafloor spreading.

_____

_____

6. From question 23 in the exercise what were your calculated rates of seafloor spreading for the following ocean basins?

Pacific Ocean basin: _____ cm/yr

North Atlantic Ocean basin: _____ cm/yr

7. From question 26 in the exercise, what was your calculated maximum velocity for the Pacific plate near the Hawaiian Islands?

_____

8. From questions 28 and 29 in the exercise, what were your calculated ages for the North and South Atlantic Ocean basins?

North Atlantic: _____ million years old

South Atlantic: _____ million years old

9. In the following space, draw a profile of a typical divergent plate boundary. Show the motion of the plates with arrows. Explain what will happen along the boundary between the plates.

Profile of a Divergent Plate Boundary

Explanation: _____

_____

_____

159

10. Describe the origin of volcanoes on the ocean floor and along continental margins bordering ocean trenches.

    Ocean-floor volcanoes: _____

    _____

    _____

    Volcanoes along continental margins bordering ocean trenches:

    _____

    _____

    _____

11. Figure 10.9 illustrates a generalized cross-section of the plate boundary along the western edge of South America. Use Figure 10.9 to answer the following questions.

    a. Label each of the following features on Figure 10.9.

       asthenosphere   continental crust
       deep-ocean trench   oceanic crust
       continental   continental lithosphere
          volcanic arc   oceanic lithosphere

b. What type of plate boundary is illustrated by the figure? Write a general description of this type of boundary.

   _____

   _____

   _____

   _____

c. Referring to Figure 10.3, identify by name the plates illustrated in the block diagram in Figure 10.9. Write the name of each plate at the appropriate location on the block diagram.

   _____

12. In your own words, describe the theory of plate tectonics. List at least two lines of evidence used to support the theory.

   _____

   _____

   _____

   _____

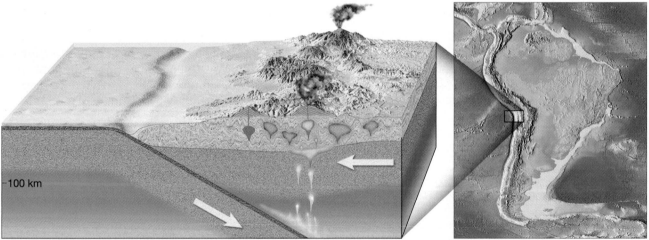

Figure 10.9 Generalized cross-section of the plate boundary along the western edge of South America.

-100 km

Waves, Currents, and Tides

The world's ocean waters are in constant motion via waves, currents, and tides. The immediate cause of each varies; however, the ultimate source of energy is the Sun. Investigating the causes, mechanics, and results of these ocean-water movements will provide a greater understanding of some important systems that operate over 70 percent of Earth's surface—the world oceans (Figure 11.1).

Objectives

After you have completed this exercise, you should be able to:

1. Explain how waves and currents are generated in the ocean.
2. Name the parts of a wave and describe the motion of water particles in a deepwater and shallow-water wave.
3. Use a formula to calculate wavelength, wave velocity, and wave period.
4. Explain why waves are refracted and what causes them to break and form surf.
5. Locate each of the major surface ocean currents.
6. List the names and characteristics of the principal deepwater masses.
7. Identify the features of erosion and deposition that occur along shorelines and explain how each is formed.
8. Explain the cause of tides and identify the different types of tides.

Materials

colored pencils   hand lens   calculator

Materials Supplied by Your Instructor

atlas or world wall map

Figure 11.1 Cliff undercut by wave erosion along the Oregon coast. (Photo by E. J. Tarbuck)

Terms

wave crest	surface current	estuary
wave trough	Coriolis effect	beach
wave height	density current	spit
wavelength	longshore current	tombolo
wave period	tidal current	baymouth bar
surf zone	emergent coast	diurnal tide
tsunami	wave-cut cliff	semidiurnal tide
refraction	platform	mixed tide
headland	submergent coast	

Waves

Most waves are set in motion when friction with wind begins rotating water particles in circular orbits (Figure 11.2). If you were to watch a ball floating on the surface, you would notice that while the wave form moves forward, the ball, and hence the water, does not. On the surface, as water particles reach the highest point in their circular orbits, a **wave crest** is formed, while particles at their lowest orbital points form **wave troughs**. **Wave height** is the vertical distance between the crest and trough of a wave.

Beneath the surface in deep water the circular orbits of water particles become progressively smaller with depth. At a depth equal to about half the **wavelength** (the horizontal distance separating two successive wave crests), the circular motion of water particles becomes negligible.

1. From Figure 11.2, select the letter that identifies each of the following.

	LETTER		**LETTER**
wave crest	_____	wavelength	_____
wave trough	_____	depth of negligible water particle motion	
wave height	_____		_____

2. Below what depth would a submarine have to submerge so that it would not be swayed by surface waves with a wavelength of 24 meters?

Below _____ meters

Wave Mechanics

In deep water, where the depth is greater than half the wavelength, the velocity (V) of a wave depends upon the **wave period** (T) (the time interval between successive wave crests, measured from a stationary point) and the wavelength (L). The mathematical equation that expresses the relation between these variables is velocity = wavelength divided by wave period ($V = L/T$).

As a wave approaches the shore and the depth of water becomes less than half the deepwater wavelength, the ocean bottom begins to interfere with the orbital motion of water particles, and the wave begins to "feel bottom" (Figure 11.2). Interference between the bottom and water particle motion causes changes to occur in the wave. At a depth of water equal to about one-twentieth of the deepwater wavelength [$(\frac{1}{20})L$ or $0.05L$], the top of the wave begins to fall forward and the wave breaks. In the **surf zone**, where waves are breaking and releasing energy, a significant amount of water is transported toward the shoreline.

3. What are three wind factors that determine the height, length, and period of waves?

Factor 1: _____

Factor 2: _____

Factor 3: _____

Refer to Figure 11.2 to answer questions 4–7.

4. The shape of the orbits of surface water particles in deepwater waves is (circular, elliptical). Near the shore in shallow water, the shapes become (circular, elliptical). Circle your answers.

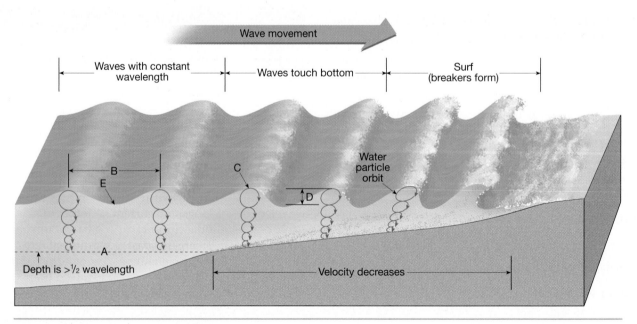

Figure 11.2 Deep and shallow water waves.

5. In shallow water, water particles in the wave crest are (ahead of, behind) those at the bottom of the wave.

6. As waves approach the shore in shallow water, their heights (increase, decrease) and the wavelength becomes (longer, shorter).

7. In the surf zone, water particles in the crest of a wave are (falling forward, standing still).

8. What would be the velocity of deepwater waves with a wavelength of 40 meters and a wave period of 6.3 seconds?

$$\text{Velocity} = \frac{\text{wavelength}(L)}{\text{wave period}(T)} = \frac{40 \text{ m}}{6.3 \text{ sec}}$$

$$= \underline{\hspace{3cm}} \text{ m/sec}$$

9. What would be the wavelength of deepwater waves that have a period of 8 seconds and a velocity of 2 meters/sec? (HINT: $V \times T = L$)

Wavelength(L) = _____ meters

   a. What would be the *wave base* (depth below which water particle motion in the wave ceases) for the waves in question 9?

   Wave base = _____ meters

   b. The waves in question 9 will begin to break at a water depth of about (1, 3, 5) meter(s). Circle your answer.

10. What factor(s) determine the distance between where waves begin to break and the shoreline?

   _____

11. Imagine that you are standing on the shore considering walking out into the surf zone where the waves are beginning to break, but you cannot swim. You estimate the wavelength of the incoming deepwater waves to be 80 meters. Would it be safe to walk out to where the waves are breaking? Explain how you arrived at your answer.

   _____

   _____

12. Along some shorelines, why does the water simply rise and fall rather than forming a surf zone?

   _____

13. What effect will *breakwaters* (walls of concrete or rock built offshore and parallel to the beach) have on waves?

   _____

Tsunamis (ocean waves produced by a submarine earthquake, sometimes mistakenly called "tidal waves") can have a wavelength of 125 miles and a wave period of 20 minutes.

14. If a tsunami had a wavelength of 125 miles and a period of 20 minutes, what would be its velocity?

   Velocity = _____ miles per hour

Wave Refraction

Waves that approach the shore at an angle are **refracted** (bent) because that part of the wave that touches bottom first is slowed down, while the remaining part of the wave continues to move forward. Refraction causes most waves to reach the shore approximately parallel to the shoreline.

Figure 11.3 illustrates a map-view of a **headland** along a coastline with water depths shown by contour lines. Assume that waves, with a wavelength of 80 feet, are approaching the shoreline from the lower margin of the figure.

Use Figure 11.3 to answer questions 15–21.

15. The approaching waves will begin to touch bottom and slow down at a water depth of about (10, 20, 30, 40) feet. Circle your answer.

16. At a water depth of approximately (4, 8, 12, 16) feet, the waves will begin to break.

17. Indicate where the waves will begin to break with a dashed line. Write the words "surf zone" along the line.

18. Beginning with the wave shown, sketch a succession of lines to illustrate the wave refraction that will take place as the waves approach shore.

19. Use arrows to indicate where most of the wave energy will be concentrated as the waves are refracted and impact the shore.

20. Erosion by waves will be most severe (on the headland, in the bay). Circle your answer.

21. What effect will the concentrated energy from wave impact eventually have on the shape of the coastline?

   _____

   _____

Currents

Moving masses of water on the surface or within the ocean are called *currents*. The primary generating force for surface currents is wind, whereas deep-ocean circulation is a response to density differences among water masses.

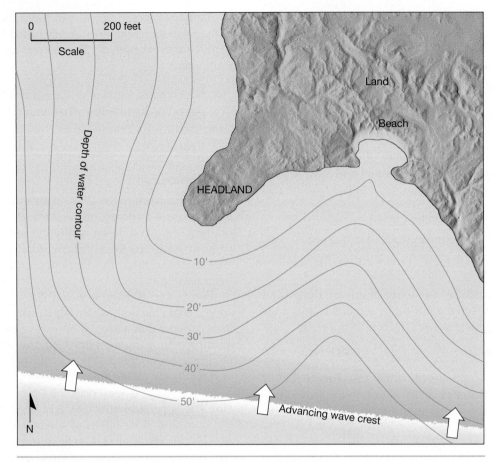

Figure 11.3 Coastline with depth of water contours and an approaching wave.

Surface Currents

Surface currents develop when friction between the moving atmosphere and the water causes the surface layer of the ocean to move as a single, large mass. Once set in motion, surface currents are influenced by the **Coriolis effect**, which deflects the path of the moving water to the right in the Northern Hemisphere and to the left in the Southern Hemisphere. *Warm currents* carry equatorial water toward the poles, while *cold currents* move water from higher latitudes toward the equator.

22. Many surface ocean currents flow with great persistence. On the world map, Figure 11.4, draw arrows representing each of the following principal surface ocean currents. Use an atlas or, if available, a large wall map that depicts surface currents as a reference. *Show warm currents with red arrows and cold currents with blue arrows.* To conserve space on the map, indicate the name of each current by writing the number that has been assigned to it.

PRINCIPAL SURFACE OCEAN CURRENTS

1. Equatorial   4. Canaries
2. Gulf Stream   5. Brazil
3. California   6. Benguela
7. Kuro Siwo   10. North Atlantic Drift
8. West Wind Drift 11. North Pacific Drift
9. Labrador   12. Peruvian

Using Figure 11.4, or a world map of surface ocean currents, answer questions 23–26.

23. Which surface ocean current travels completely around the globe, west to east, without interruption?

_____

24. Which surface ocean current flows along the eastern coast of the United States? The current is a (warm, cold) current. Circle your answer.

_____

25. What is the name of the surface ocean current located along the western coast of the United States? The current is a (warm, cold) current. Circle your answer.

_____

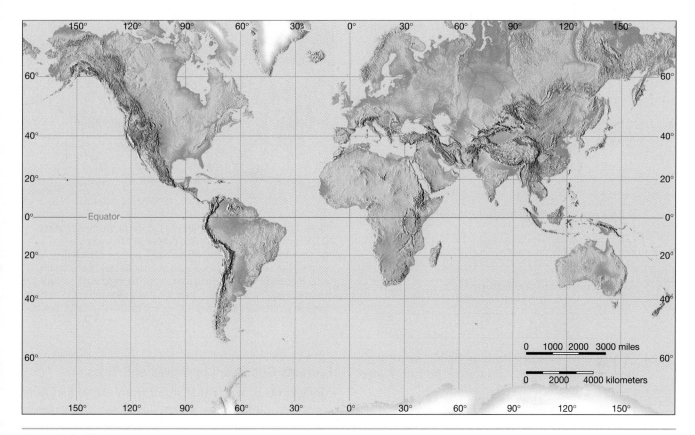

Figure 11.4 World map.

26. The general circulation of the surface currents in the North Atlantic Ocean is (clockwise, counterclockwise). In the South Atlantic, circulation is (clockwise, counterclockwise). Circle your answers.

Density Currents

Density currents result when water of greater density flows under or through water of a lower density. At any given depth, the density of water is influenced by its temperature and salinity—factors which you may have investigated in Exercise 9, "Introduction to Oceanography."

Figure 11.5 is a cross section of the Atlantic Ocean illustrating the deep (thermohaline) circulation. Use the figure to answer questions 27 and 28.

27. After you examine their latitude of origin, describe the probable temperature and/or

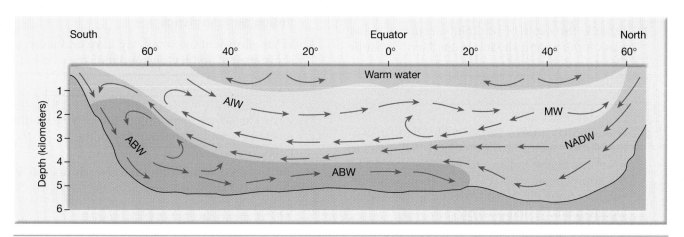

Figure 11.5 Cross section of the deep circulation of the Atlantic Ocean. (After Gerhard Neumann and Willard J. Pierson, Jr., *Principles of Physical Oceanography*, 1966. Reprinted by permission of Gerhard Neumann)

salinity characteristics and general movements of each of the following water masses.

ABW (Antarctic Bottom Water): _____

_____

_____

NADW (North Atlantic Deep Water): _____

_____

_____

AIW (Antarctic Intermediate Water): _____

_____

_____

MW (Mediterranean Water): _____

_____

_____

28. What is the mechanism responsible for causing the very high density of Antarctic Bottom Water?

_____

_____

Deep-ocean circulation begins in high latitudes where water becomes cold and its salinity increases as sea ice forms. When this surface water becomes dense enough, it sinks and moves throughout the ocean basins in sluggish currents. Oceanographers estimate that after sinking from the surface of the ocean, deep waters will not reappear at the surface for an average of 500 to 2000 years. A simplified model of deep-ocean circulation is similar to a conveyor belt that travels from the Atlantic Ocean through the Indian and Pacific oceans and back again (Figure 11.6). Use Figure 11.6 to answer questions 29 and 30.

29. What is the name of the cold subsurface water mass forming and sinking in the North Atlantic Ocean?

_____

30. Assume that it takes surface water that sinks in the North Atlantic near Greenland 1000 years to resurface in the Indian Ocean. What would be the approximate velocity of the deep-ocean circulation from the North Atlantic to the Indian Ocean in km/yr and cms/hr?

_____ km(s)/yr

_____ cm(s)/hr

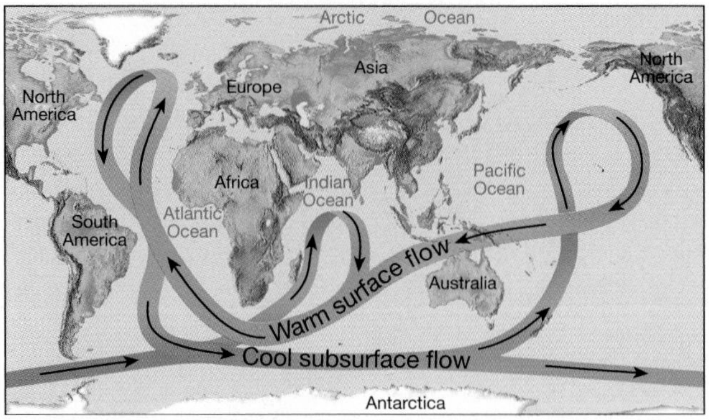

Figure 11.6 Idealized "conveyor belt" model of ocean circulation, which is initiated in the North Atlantic Ocean when warm water transfers its heat to the atmosphere, cools, and sinks below the surface. This water moves southward as a subsurface flow and joins water that encircles Antarctica. From here, this deep water spreads into the Indian and Pacific oceans, where it slowly rises and completes the conveyer as it travels along the surface into the North Atlantic Ocean.

Currents Generated by Waves and Tides

Movements of water that result from waves and tides constitute a third class of currents. Whenever waves or tides push water against a shore, currents form that transport the water along the coast and return it seaward. Two of these currents are longshore currents and tidal currents.

Longshore currents form when waves strike the coast at an angle and the water moves in a zigzag pattern parallel to the shore in the surf zone. These currents transport tremendous amounts of sediment which, when deposited, form many types of coastal features.

Tidal currents, which reverse their direction of flow with each tide, submerge and then expose low-lying coastal zones.

In Figure 11.3 you completed a diagram illustrating wave refraction. Answer questions 31–34 by referring to Figure 11.3.

31. Indicate with arrows the probable directions of the longshore currents.

32. What effect will the small bay have on the longshore current and its transportation of sediment?

_____

33. Write the word "deposition" where you would most likely find sediment being deposited by the longshore current.

34. Explain the cause of the sandy beach deposit at the head of the small bay.

_____

_____

Shoreline Features

The nature of shorelines varies considerably from place to place. One way that geologists classify coasts is based upon changes that have occurred with respect to sea level. This very general classification divides coasts into two types, emergent and submergent.

Emergent coasts have been raised above the sea as a result of rising land or falling sea level and are characterized by **wave-cut cliffs** or **platforms**.

Submergent coasts, resulting from a rising sea level or subsiding land, are often irregular due to the fact that many river mouths are flooded and become **estuaries**.

Nevertheless, whether along the rugged New England coast or the steep coastlines of California, the effects of wave erosion and sediment deposition by currents produce many similar features. Some of the more common depositional features include **beaches, spits, tombolos**, and **baymouth bars**.

Features of Emergent and Submergent Coasts

Figure 11.7 illustrates several erosional and depositional features of emergent and submergent coastlines. Using Figure 11.7, complete questions 35–37.

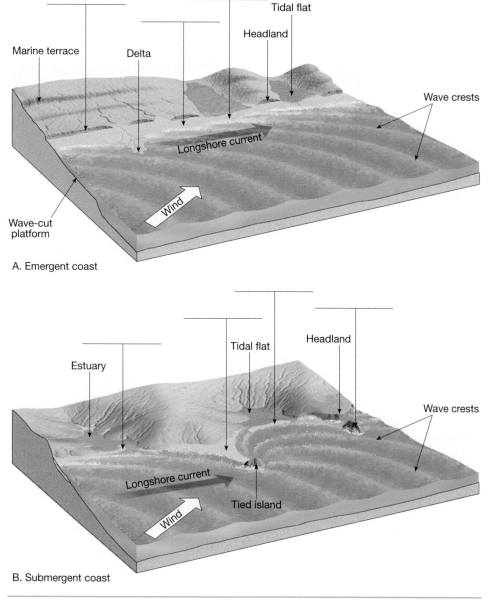

A. Emergent coast

B. Submergent coast

Figure 11.7   Hypothetical illustrations showing general features of **A.** emergent and **B.** submergent coastlines.

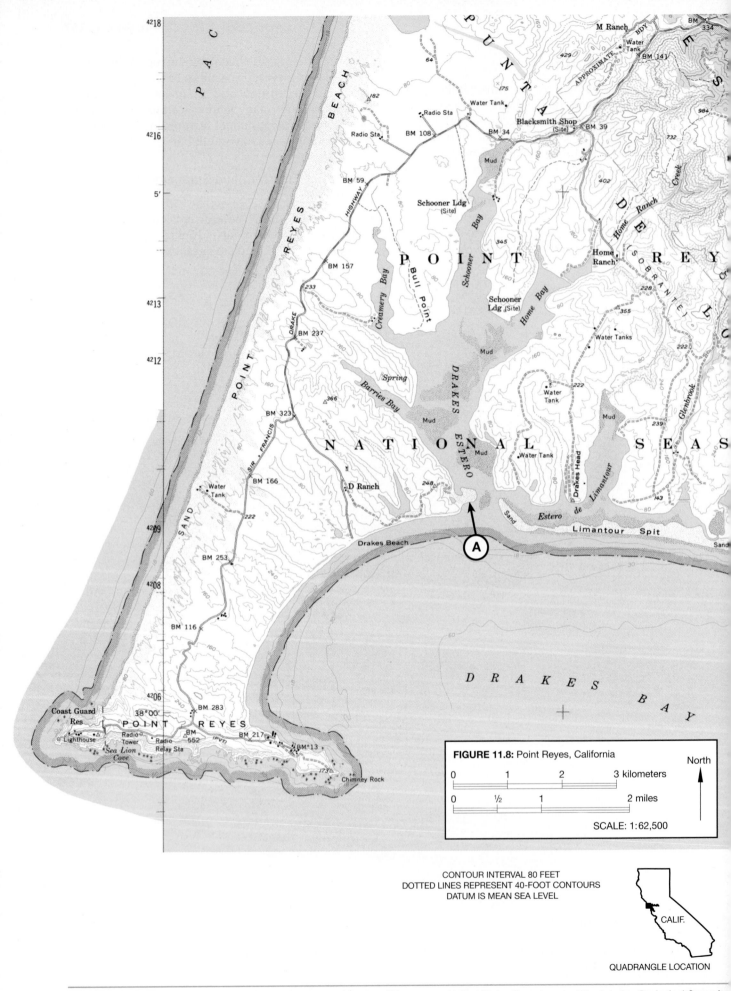

Figure 11.8 Portion of the Point Reyes, California, topographic map. (Map source: United States Department of the Interior, Geological Survey)

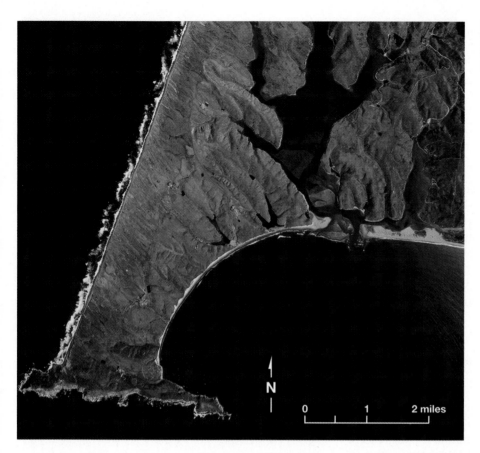

Figure 11.9 High-altitude false-color image of the Point Reyes area north of San Francisco, California. (Courtesy of USDA-ASCS)

N

0   1   2 miles

35. On Figure 11.7 identify each of the following coastal features by writing their name above the appropriate vertical line.

FEATURES CAUSED BY EROSION	FEATURES PRODUCED BY DEPOSITION	
wave-cut cliff	beach	baymouth bar
sea stack	spit	barrier island
	tombolo	

36. What is the purpose for constructing each of the following artificial features along a coast?

a *groin:* _____

a pair of *jetties:* _____

37. Draw and label a pair of jetties at the most appropriate location along the emergent coast illustrated in Figure 11.7A.

Identifying Coastal Features on a Topographic Map

Figure 11.8 is a portion of the Point Reyes, CA, topographic map. Compare the map with the high-altitude image of the same area in Figure 11.9. Then use Figure 11.8, Figure 11.9, and Figure 11.7 to answer questions 38–44.

38. The features along the shoreline of Drakes Bay suggest that the coast is (emergent, submergent). Circle your answer.

39. Drakes Estero and other bays shown on the map are (estuaries, headlands).

40. Point Reyes, a typical headland, is undergoing severe wave erosion. What type of feature is Chimney Rock and the other rocks located off the shore of Point Reyes? How have they formed?

_____

_____

41. Several depositional features in Drakes Bay are related to the movement of sediment by longshore currents. The feature labeled A on the map is one of these features, called a (spit, tombolo).

42. Using a large arrow, indicate the direction of the current in the vicinity of Limantour Spit.

43. Assume a groin is constructed by the word "Limantour" on Limantour Spit. On which side of the groin, east or west, will sand accumulate? What will be the effect on the opposite side of the groin?

_____

_____

44. What is the probable origin of the "U-shaped" lake east of D Ranch?

_____

_____

Tides

Tides are the cyclical rise and fall of sea level caused by the gravitational attraction of the Moon and, to a lesser extent, by the Sun. Gravitational pull creates a bulge in the ocean on the side of Earth nearest the Moon and on the opposite side of Earth from the Moon. Tides develop as the rotating Earth moves through these bulges causing periods of high and low water. Using tidal information from many sources, tides are classified into three types:

- **Diurnal** (*diurnal* = daily) **tides** are characterized by a single high tide and a single low tide each tidal day.
- **Semidiurnal** (*semi* = twice, *diurnal* = daily) **tides** exhibit two high tides and two low tides each tidal day.
- **Mixed tides** are similar to semidiurnal tides except that they are characterized by large inequalities in high water heights, low water heights, or both (Figure 11.10).

Identifying Types of Tides

Tidal curves for the month of September at several locations are illustrated in Figure 11.11. Use the figure to answer questions 45–47.

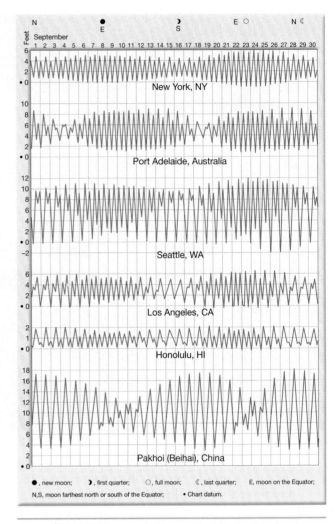

Figure 11.11 Tidal curves for the month of September at various locations. (Source: U.S. Navy Hydrograph Office, *Oceanography*, U.S. Government Printing Office, 1966)

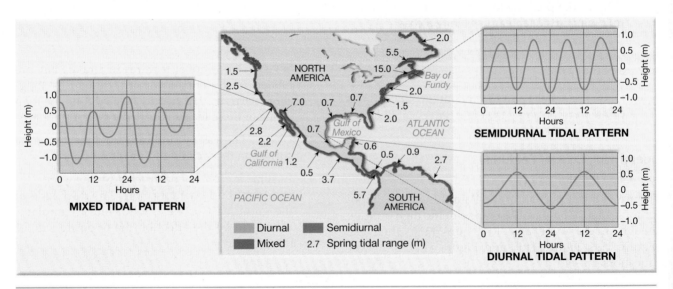

Figure 11.10 Tidal patterns and their occurrence along North and Central American coasts. A diurnal tidal pattern (*lower right*) shows one high and low tide each tidal day. A semidiurnal pattern (*upper right*) shows two highs and lows of approximately equal heights during each tidal day. A mixed tidal pattern (left) shows two highs and lows of unequal heights during each tidal day.

45. Classify each of the tidal curves shown in Figure 11.11 as to the most appropriate type.

Diurnal tides occur at: _____

_____

Semidiurnal tides: _____

_____

Mixed tides: _____

_____

46. Of the locations you classified as having a *mixed* tide, (Port Adelaide, Seattle, Los Angeles) had the greatest inequality between successive low water heights on September 5. Circle your answer.

47. Write a general statement comparing the type of tide that occurs along the Pacific coast of the United States to the type found along the Atlantic coast.

_____

_____

Tidal Variations

In Figure 11.11, notice that during the month of September, at any given location, the heights of the tides were not constant. Two important factors that influence this variation are (1) the alignment of the Sun, Earth, and Moon, and (2) the distance between Earth and the Moon. Although these two controls are significant, they alone cannot be used to predict the height or time of actual tides at a particular place. Other factors, such as the shape of the coastline and the configuration of ocean basins, are also important. Consequently, tides at various locations respond differently to the tide-producing forces.

Use Figure 11.11 to answer questions 48–53.

48. As shown in Figure 11.11, the *tidal range* (difference in height between high tide and the following low tide) at any one location (changes, remains the same) throughout September. Circle your answer.

The lunar phases for the month are shown at the top of the figure (new moon on the 8th and full moon on the 23rd of the month).

49. What general relation seems to exist between the phases of the Moon and the tidal ranges at New York?

_____

_____

50. Explain the cause of *spring tides* and *neap tides.* Sketch a diagram showing the relative positions of the Earth, Moon, and Sun, viewed from above, that would cause each situation.

Spring tide

Spring tides: _____

_____

_____

Neap tide

Neap tides: _____

_____

_____

51. On Figure 11.11, label the times of spring tide and times of neap tide for New York.

52. Do the tides at Pakhoi, China, have the same relations to the lunar phases as those that occur at New York? What are some other factors that may be influencing the tides at Pakhoi?

_____

_____

Table 11.1 Tidal Data for Long Beach, New York, January 2003

| | TIMES ARE LISTED IN LOCAL STANDARD TIME (LST) — ALL HEIGHTS ARE IN FEET | | | | | | | |
DAY	TIME	HEIGHT	TIME	HEIGHT	TIME	HEIGHT	TIME	HEIGHT
1	05:45 A.M.	5.5	12:16 P.M.	−0.7	06:12 P.M.	4.4	—	—
2	12:18 A.M.	−0.5	06:35 A.M.	5.6	01:07 P.M.	−0.8	07:03 P.M.	4.4
3	01:10 A.M.	−0.5	07:23 A.M.	5.5	01:56 P.M.	−0.8	07:53 P.M.	4.4
4	01:59 A.M.	−0.4	08:11 A.M.	5.4	02:42 P.M.	−0.7	08:42 P.M.	4.3
5	02:45 A.M.	−0.2	08:59 A.M.	5.1	03:25 P.M.	−0.5	09:32 P.M.	4.2
6	03:30 A.M.	0.0	09:47 A.M.	4.8	04:07 P.M.	−0.3	10:23 P.M.	4.0
7	04:14 A.M.	0.3	10:35 A.M.	4.6	04:49 P.M.	−0.1	11:12 P.M.	3.9
8	05:01 A.M.	0.6	11:22 A.M.	4.3	05:32 P.M.	0.2	11:59 P.M.	3.9
9	05:54 A.M.	0.8	12:09 P.M.	4.0	06:18 P.M.	0.4	—	—
10	12:45 A.M.	3.9	06:56 A.M.	0.9	12:57 P.M.	3.7	07:10 P.M.	0.5
11	01:31 A.M.	3.9	07:59 A.M.	0.9	01:47 P.M.	3.5	08:02 P.M.	0.5
12	02:19 A.M.	4.0	08:57 A.M.	0.8	02:41 P.M.	3.4	08:53 P.M.	0.5
13	03:10 A.M.	4.1	09:50 A.M.	0.6	03:39 P.M.	3.5	09:41 P.M.	0.4
14	04:02 A.M.	4.3	10:38 A.M.	0.3	04:34 P.M.	3.6	10:28 P.M.	0.2
15	04:51 A.M.	4.6	11:26 A.M.	0.1	05:23 P.M.	3.7	11:15 P.M.	0.1
16	05:36 A.M.	4.8	12:12 P.M.	−0.1	06:08 P.M.	3.9	—	—
17	12:02 A.M.	−0.1	06:17 A.M.	5.0	12:57 P.M.	−0.3	06:51 P.M.	4.1
18	12:49 A.M.	−0.2	06:58 A.M.	5.1	01:40 P.M.	−0.5	07:32 P.M.	4.2
19	01:35 A.M.	−0.4	07:38 A.M.	5.2	02:22 P.M.	−0.6	08:15 P.M.	4.3
20	02:20 A.M.	−0.4	08:21 A.M.	5.2	03:30 P.M.	−0.7	09:01 P.M.	4.4

(Source: Center for Operational Oceanographic Products and Services, National Oceanographic and Atmospheric Association, National Ocean Service.)

At some locations, tidal power is being considered as a means of generating electricity.

53. Suggest two criteria that a bay must meet before its tidal energy can be economically harnessed.

   Criterion 1: _____

   Criterion 2: _____

Examining Tidal Data

Table 11.1 presents January 2003 tidal data for Long Beach, NY. Accurately plot the data on the graph in Figure 11.12. After you have plotted the data, answer questions 54–58.

54. What type of tide occurs at Long Beach, NY? What fact(s) support your conclusion?

   _____

   _____

   _____

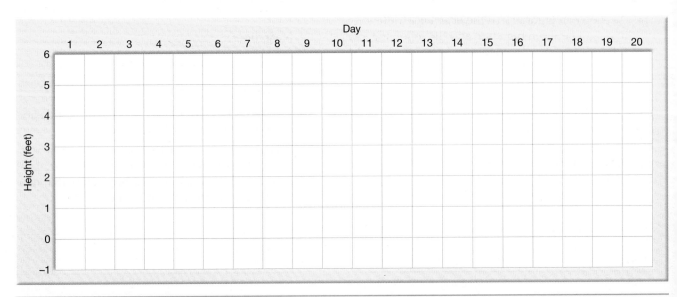

Figure 11.12 Tidal curve for Long Beach, NY.

55. The greatest tidal range occurs on day _____,
while the smallest range is on day _____ .

56. Selecting from the tidal curves in Figure 11.11,
the tides at Long Beach are most like those at:

_____

Examine closely the association between the tides
shown for the city you selected in question 56 and the
phases of the Moon depicted at the top of Figure 11.11.

57. Using Figure 11.11 as a reference, label the most
likely lunar phases associated with the tidal curve
above the appropriate days on Figure 11.12.

58. Using Table 11.1 and Figure 11.12, assume that at
9:00 A.M. on January 5, a boat was anchored near a
beach in 4 feet of water. When the owner returned
at 3:30 P.M., the boat was resting on sand. What
had happened? Approximately how long did

the owner have to wait to sail the boat away
from the area?

_____

_____

_____

Waves, Currents, and Tides on the Internet

Continue your exploration of waves and tides by completing the corresponding online activity on the *Applications & Investigations in Earth Science* website at
http://prenhall.com/earthsciencelab

Notes and calculations.

Waves, Currents, and Tides

Date Due: _____

Name: _____

Date: _____

Class: _____

After you have finished Exercise 11, complete the following questions. You may have to refer to the exercise for assistance or to locate specific answers. Be prepared to submit this summary/report to your instructor at the designated time.

1. On Figure 11.13 sketch a profile view of deep- and shallow-water waves approaching a shore. Label all parts and measurements of a typical wave. Also, illustrate the motion of several water particles at increasing depths in both a deep- and shallow-water wave.

2. What will happen to the shapes of waves as they approach a headland that is surrounded by shallow water?

_____

3. Describe the formation and appearance of each of the following features:

   Spit: _____

   _____

Stack: _____

_____

Tombolo: _____

_____

Estuary: _____

_____

4. Refer to Figure 11.8. What types of coastal features are Point Reyes, Drakes Estero, and Chimney Rock?

   Point Reyes: _____

   Drakes Estero: _____

   Chimney Rock: _____

5. The circulation of the surface currents in the South Atlantic Ocean is (clockwise, counterclockwise). Circle your answer.

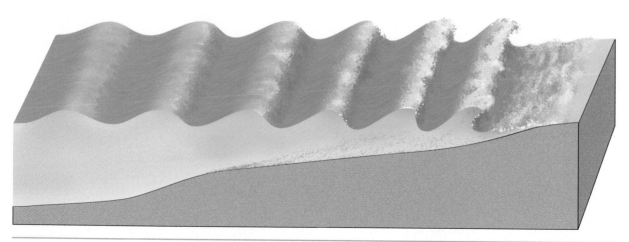

Figure 11.13 Deep water waves approaching a shallow coast.

175

6. What are the names of the surface currents that are located along the east and west coasts of the United States? Is each a warm or a cold current?

   East coast: ———————————————

   West coast: ———————————————

7. List the characteristics and describe the movement of the following deep-ocean water masses in the Atlantic Ocean.

   NADW: ———————————————

   ———————————————

   ABW: ———————————————

   ———————————————

8. Spring tides are most likely to occur during which lunar phase(s)?

   ———————————————

9. Which of the tidal curves illustrated in Figure 11.11 exhibits the greatest tidal range?

   ———————————————

10. Refer to Figure 11.11. Of the three types of tides, name the type that occurs at each of the following locations.

    Pakhoi: ———————————————

    Honolulu: ———————————————

    New York: ———————————————

11. Referring to question 54 of the exercise, what type of tide occurs at Long Beach, NY? Describe this type of tide.

    ———————————————

    ———————————————

    ———————————————

Source: Warren Faidley/DRK Photo

Meteorology

3

Earth–Sun Relations

To life on this planet, the relations between Earth and the Sun are perhaps the most important of all astronomical phenomena. The variations in solar energy striking Earth as it rotates and revolves around the Sun cause the seasons and therefore are an appropriate starting point for studying weather and climate.

In this exercise you will investigate the reasons why the amount of solar radiation intercepted by Earth varies for different latitudes and changes throughout the year at a particular place (Figure 12.1). The next exercise, Exercise 13, examines how the atmosphere is warmed by this radiation.

Objectives

After you have completed this exercise, you should be able to:

1. Describe the effect that Sun angle has on the amount of solar radiation a place receives.

2. Explain why the intensity and duration of solar radiation varies with latitude.

3. Explain why the intensity and duration of solar radiation varies at any one place throughout the year.

4. Describe the significance of these special parallels of latitude: Tropic of Cancer, Tropic of Capricorn, Arctic Circle, Antarctic Circle, and equator.

5. Diagram the relation between Earth and the Sun on the dates of the solstices and equinoxes.

6. Determine the latitude where the overhead Sun is located on any day of the year.

7. Calculate the noon Sun angle for any place on Earth on any day.

8. Calculate the latitude of a place using the noon Sun angle.

Materials

metric ruler   colored pencils
protractor   calculator

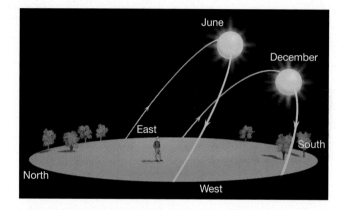

Figure 12.1   Daily paths of the Sun for June and December for an observer in the middle latitudes in the Northern Hemisphere. Notice that the angle of the Sun above the horizon is much greater in the summer than in the winter.

Materials Supplied by Your Instructor

globe
large rubber band or string

Terms

weather	solar constant	solstice
weather element	equator	equinox
weather control	Tropic of Cancer	analemma
solar intensity	Tropic of Capricorn	noon Sun
solar duration	Arctic Circle	angle
langley	Antarctic Circle	
calorie		

Introduction

Weather is the state of the atmosphere at a particular place for a short period of time. The condition of the atmosphere at any location and time is described by measuring the four basic **elements** of weather: temperature, moisture, air pressure, and wind. Of all the **controls** that are responsible for causing variations in the weather elements, the amount of solar radiation received at any location is the most important.

179

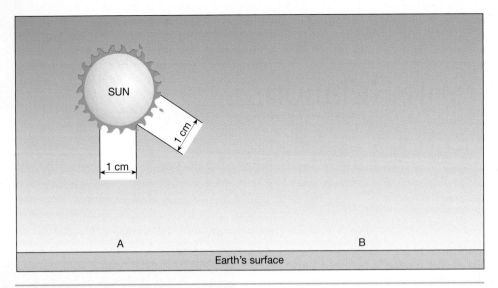

Figure 12.2 Vertical and oblique Sun beams.

Solar Radiation and the Seasons

The amount of solar energy (radiation) striking the outer edge of the atmosphere is not uniform over the face of Earth at any one time, nor is it constant throughout the year at any particular place. Rather, solar radiation at any location and time is determined by the Sun's **intensity** and **duration**. Intensity is the angle at which the rays of sunlight strike a surface, whereas duration refers to the length of daylight.

The standard unit of solar radiation is the **langley**, equal to one **calorie**[1] per square centimeter. The **solar constant**, or average intensity of solar radiation falling on a surface perpendicular to the solar beam at the outer edge of the atmosphere, is about 2 langleys per minute. As the radiation passes through the atmosphere, it undergoes absorption, reflection, and scattering. Therefore, at any one location, less radiation reaches Earth's surface than was originally intercepted at the upper atmosphere.

Solar Radiation and Latitude

The amount of radiation striking a square meter at the outer edge of the atmosphere, and eventually Earth's surface, varies with latitude because of a changing Sun angle (see Figure 12.1). To illustrate this fact, answer questions 1–11 using the appropriate figure.

1. On Figure 12.2, extend the 1-cm-wide beam of sunlight from the Sun vertically to point A on the

surface. Extend the second 1-cm-wide beam, beginning at the Sun, to the surface at point B.

Notice in Figure 12.2 that the Sun is directly overhead (vertical) at point A and the beam of sunlight strikes the surface at a 90° angle above the horizon.

Using Figure 12.2, answer questions 2–5.

2. Using a protractor, measure the angle between the surface and the beam of sunlight coming from the Sun to point B.

   _____ ° = angle of the Sun above the surface (horizon) at point B.

3. What are the lengths of the line segments on the surface covered by the Sun beam at point A and point B?

   Point A: _____ mm   point B: _____ mm

4. Of the two beams, beam (A, B) is more spread out at the surface and covers a larger area. Circle your answer.

5. More langleys per minute would be received by a square centimeter on the surface at point (A, B). Circle your answer.

Use Figure 12.3 to answer questions 6–11 concerning the total amount of solar radiation intercepted by each 30° segment of latitude on Earth.

6. With a metric ruler, measure the total width of incoming rays from point x to point y in Figure 12.3. The total width is _____ centimeters (_____ millimeters). Fill in your answers.

7. Assume the total width of the incoming rays from point x to point y equals 100 percent of the solar radiation that is intercepted by Earth. Each

---

[1] The most familiar energy unit used to measure heat is the calorie, which is the quantity of heat energy needed to raise the temperature of one gram of water one degree Celsius. Do not confuse it with the so-called large Calorie (note the capital C), the kind counted by weight watchers. A Calorie is the amount of heat energy needed to raise the temperature of a kilogram (1000 grams) of water one degree Celsius.

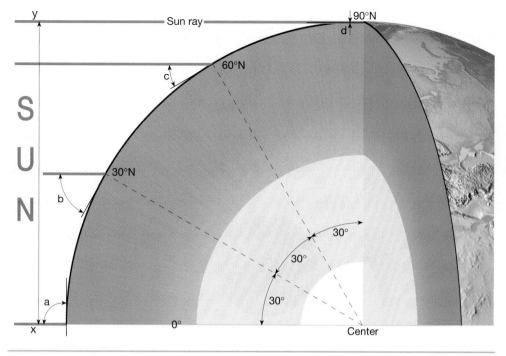

Figure 12.3 Distribution of solar radiation per 30° segment of latitude on Earth.

centimeter would equal _____ percent, and each millimeter would equal _____ percent. Fill in your answers.

8. What percentage of the total incoming radiation is concentrated in each of the following zones?

   0°–30° = _____ mm = _____ %

   30°–60° = _____ mm = _____ %

   60°–90° = _____ mm = _____ %

9. Use a protractor to measure the angle between the surface and Sun ray at each of the following locations. (Angle b is already done as an example.)

   Angle a: _____ °   angle c: _____ °

   Angle b: ___60___ °   angle d: _____ °

10. What is the general relation between the amount of radiation received in each 30° segment and the angle of the Sun's rays?

   _____

   _____

11. Explain in your own words what fact about Earth creates the unequal distribution of solar energy, even though each zone represents an equal 30° segment of latitude.

   _____

   _____

   _____

Yearly Variation in Solar Energy

The amount of solar radiation received at a particular place would remain constant throughout the year if it were not for these facts:

- Earth rotates on its axis and revolves around the Sun.

- The axis of Earth is inclined 23.5° from the perpendicular to the plane of its orbit.

- Throughout the year, the axis of Earth points to the same place in the sky, which causes the overhead (vertical or 90°) noon Sun to cross over the **equator** twice as it migrates from the **Tropic of Cancer** (23.5°N latitude) to the **Tropic of Capricorn** (23.5°S latitude) and back to the Tropic of Cancer.

As a consequence, the position of the vertical or overhead noon Sun shifts between the hemispheres, causing variations in the intensity of solar radiation and changes in the length of daylight and darkness. *The seasons are the result of this changing intensity and duration of solar energy and subsequent heating of the atmosphere.*

To help understand how the intensity and duration of solar radiation varies throughout the year, answer questions 12–31 after you have examined the location of the Tropic of Cancer, Tropic of Capricorn, **Arctic Circle**, and **Antarctic Circle** on a globe or world map.

12. List some of the countries each of the following special parallels of latitude passes through.

   Tropic of Cancer: _____

   Tropic of Capricorn: _____

   Arctic Circle: _____

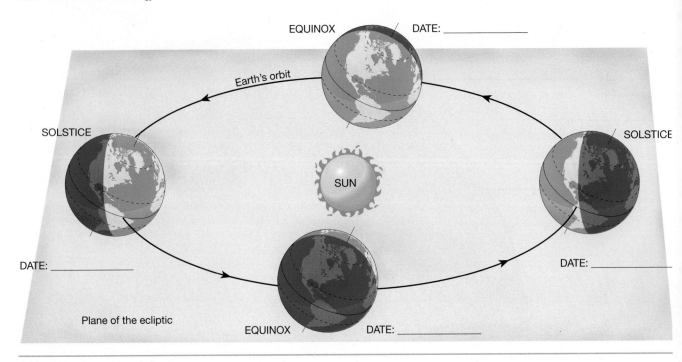

Figure 12.4 Earth-Sun relations.

13. Write the date represented by each position of Earth at the appropriate place in Figure 12.4. Then label the following on Earth at an equinox AND a solstice position.

North Pole and South Pole

Axis of Earth

Equator, Tropic of Cancer, Tropic of Capricorn

Arctic Circle and Antarctic Circle

Circle of illumination (day-night line)

Questions 14–19 refer to the June **solstice** position of Earth in Figure 12.4.

14. What term is used to describe the June 21–22 date in each hemisphere?

Northern Hemisphere: _____ solstice

Southern Hemisphere: _____ solstice

15. On June 21–22 the Sun's rays are perpendicular to Earth's surface at noon at the (Tropic of Cancer, equator, Tropic of Capricorn). Circle your answer.

16. What latitude is receiving the most intense solar energy on June 21–22?

Latitude: _____

17. Toward what direction, north or south, would you look to see the Sun at noon on June 21–22 if you lived at the following latitudes?

40°N latitude: _____

10°N latitude: _____

18. Position a rubber band, string, or pieces of tape on a globe corresponding to the *circle of illumination* on June 21–22. Then determine the approximate length of daylight at the following latitudes by examining the proportionate number of degrees of longitude a place located at each latitude spends in daylight as Earth rotates. (*Note:* Earth rotates a total of 360° of longitude per day. Therefore, each 15° of longitude is equivalent to one hour.)

70°N latitude: _____ hrs _____ min

40°S latitude: _____ hrs _____ min

40°N latitude: _____ hrs _____ min

90°S latitude: _____ hrs _____ min

0° latitude: _____ hrs _____ min

19. On June 21–22, latitudes north of the Arctic Circle are receiving (6, 12, 24) hours of daylight, while latitudes south of the Antarctic Circle are experiencing (6, 12, 24) hours of darkness. Circle your answers.

Questions 20–24 refer to the December solstice position of Earth in Figure 12.4.

20. What name is used to describe the December 21–22 date in each hemisphere?

Northern Hemisphere: _____ solstice

Southern Hemisphere: _____ solstice

21. On December 21–22 the Sun's rays are perpendicular to Earth's surface at noon on the (Tropic of

Table 12.1 **Length of daylight**

LATITUDE (DEGREES)	SUMMER SOLSTICE	WINTER SOLSTICE	EQUINOXES
0	12 h	12 h	12 h
10	12 h 35 min	11 h 25 min	12
20	13 12	10 48	12
30	13 56	10 04	12
40	14 52	9 08	12
50	16 18	7 42	12
60	18 27	5 33	12
66.5	24 h	0 00	12
70	24 h (for 2 mo)	0 00	12
80	24 h (for 4 mo)	0 00	12
90	24 h (for 6 mo)	0 00	12

Cancer, equator, Tropic of Capricorn). Circle your answer.

22. On December 21–22 the (Northern, Southern) Hemisphere is receiving the most intense solar energy. Circle your answer.

23. If you lived at the equator, on December 21–22 you would look (north, south) to see the Sun at noon.

24. Refer to Table 12.1, "Length of daylight." What is the length of daylight at each of the following latitudes on December 21–22?

90°N latitude: _____ hrs _____ min

40°S latitude: _____ hrs _____ min

40°N latitude: _____ hrs _____ min

90°S latitude: _____ hrs _____ min

0° latitude: _____ hrs _____ min

Questions 25–31 refer to the March and September **equinox** positions of Earth in Figure 12.4.

25. For those living in the Northern Hemisphere, what terms are used to describe the following dates?

March 21: _____ equinox

September 22: _____ equinox

26. For those living in the Southern Hemisphere, what terms are used to describe the following dates?

March 21: _____ equinox

September 22: _____ equinox

27. On March 21 and September 22 the Sun's rays are perpendicular to Earth's surface at noon at the (Tropic of Cancer, equator, Tropic of Capricorn). Circle your answer.

28. What latitude is receiving the most intense solar energy on March 21 and September 22?

Latitude: _____

29. If you lived at 20°S latitude, you would look (north, south) to see the Sun at noon on March 21 and September 22. Circle your answer.

30. What is the relation between the North and South Poles and the circle of illumination on March 21 and September 22?

_____

31. Write a brief statement describing the length of daylight everywhere on Earth on March 21 and September 22.

_____

As you have seen, the latitude where the noon Sun is directly overhead (vertical, or 90° above the horizon) is easily determined for the solstices and equinoxes.

Figure 12.5 is a graph, called an **analemma**, that can be used to determine the latitude where the overhead noon Sun is located for any date. To determine

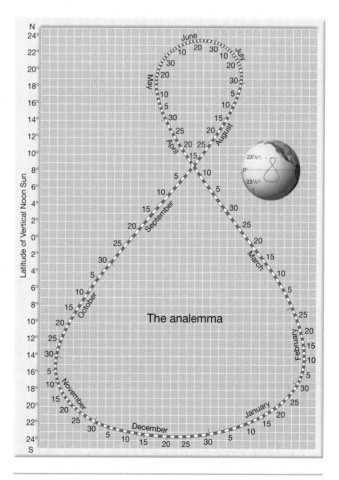

Figure 12.5 The analemma, a graph illustrating the latitude of the overhead (vertical) noon Sun throughout the year.

the latitude of the overhead noon Sun from the analemma, find the desired date on the graph and read the coinciding latitude along the left axis. Don't forget to indicate North or South when writing latitude.

32. Using a colored pencil, draw lines on Figure 12.5 that correspond to the equator, Tropic of Cancer, and Tropic of Capricorn. Label each of these special parallels of latitude on the figure.

33. Using the analemma, Figure 12.5, determine where the Sun is overhead at noon on the following dates.

    December 10: _____

    March 21: _____

    May 5: _____

    June 22: _____

    August 10: _____

    October 15: _____

34. The position of the overhead noon Sun is always located on or between which two parallels of latitude?

    _____ °N (named the Tropic of _____) and _____ °S (named the Tropic of _____)

35. The overhead noon Sun is located at the equator on September _____ and March _____. Together, these two days are called the _____. Fill in your answers.

36. Refer to Figure 12.5 and write a brief paragraph summarizing the yearly movement of the overhead noon Sun and how the intensity and duration of solar radiation varies over Earth's surface throughout the year.

    _____

    _____

    _____

    _____

Calculating Noon Sun Angle

Knowing where the noon Sun is overhead on any given date (the analemma), you can determine the angle above the horizon of the noon Sun at any other latitude on that same day. The relation between latitude and **noon Sun angle** is

> For each degree of latitude that the place is away from the latitude where the noon Sun is overhead, the angle of the noon Sun becomes one degree *lower* from being vertical (or 90°) above the horizon (Figure 12.6).

37. Complete Table 12.2 by calculating the noon Sun angle for each of the indicated latitudes on the dates

Table 12.2   Noon Sun angle calculations

LATITUDE OF OVERHEAD NOON SUN	MAR 21 (___)	APR 11 (___)	JUN 21 (___)	DEC 22 (___)
		Noon Sun Angle		
90°N	___	___	___	___
40°N	50°	___	___	26½°
0°	___	___	66½°	___
20°S	___	62°	___	___

given. Some of the calculations have already been done.

38. From Table 12.2, the highest average noon Sun angle occurs at (40°N, 0°, 20°S). Circle your answer.

39. Calculate the noon Sun angle for your latitude on today's date.

    Date: _____

    Latitude of overhead noon Sun: _____

    Your latitude: _____

    Your noon Sun angle: _____

    (*Note*: You may want to compare your calculated noon Sun angle with a measured noon Sun angle obtained by using the technique described in Exercise 17, *Astronomical Observations*.)

40. Calculate the maximum and minimum noon Sun angles for your latitude.

MAXIMUM NOON SUN ANGLE	**MINIMUM NOON SUN ANGLE**
Date: _____	Date: _____
Angle: _____°	Angle: _____°

41. Calculate the average noon Sun angle (maximum plus minimum, divided by 2) and the range of the noon Sun angle (maximum minus minimum) for your location.

    Average noon Sun angle = _____°

    Range of the noon Sun angle = _____°

42. Describe some situations in which knowing the noon Sun angle might be useful.

    _____

    _____

Using Noon Sun Angle

One very practical use of noon Sun angle is in navigation. Like a navigator, you can determine your latitude if the date and angle of the noon Sun at your location are known. As you answer questions 43 and 44, keep in mind the relation between latitude and noon Sun angle (Figure 12.6).

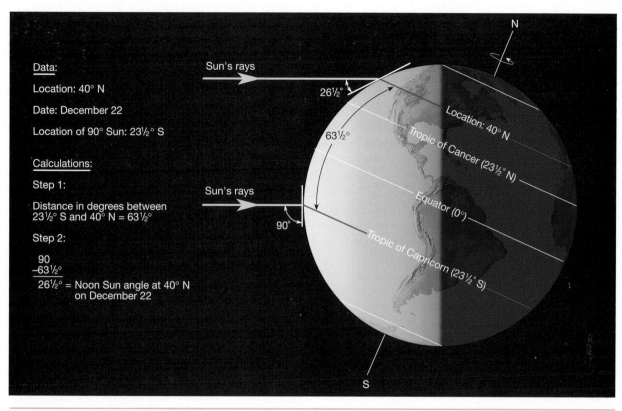

Figure 12.6 Calculating the noon Sun angle. Recall that on any given day, only one latitude receives vertical (90°) rays of the Sun. A place located 1° away (either north or south) receives an 89° angle at any location; a place 2° away, an 88° angle, and so forth. To calculate the noon Sun angle, simply find the number of degrees of latitude separating that location from the latitude that is receiving the vertical rays of the Sun. Then subtract that value from 90°. The example in this figure illustrates how to calculate the noon Sun angle for a city located at 40° north latitude on December 22 (winter solstice).

43. What is your latitude if, on March 21, you observe the noon Sun to the north at 18° above the horizon?

    Latitude: _____

44. What is your latitude if, on October 16, you observe the noon Sun to the south at 39° above the horizon?

    Latitude: _____

Solar Radiation at the Outer Edge of the Atmosphere

Table 12.3 on the following page shows the average daily radiation received at the outer edge of the atmosphere at select latitudes for different months.

To help visualize the pattern, plot the data from Table 12.3 on the graph in Figure 12.7. Using a different color for each latitude, draw lines through the monthly values to obtain yearly curves. Then answer questions 45–48.

45. Why do two periods of maximum solar radiation occur at the equator?

    _____

46. In June, why does the outer edge of the atmosphere at the equator receive less solar radiation than both the North Pole and 40°N latitude?

    _____

    _____

47. Why does the outer edge of the atmosphere at the North Pole receive no solar radiation in December?

    _____

48. What would be the approximate monthly values for solar radiation at the outer edge of the atmosphere at 40°S latitude? Explain how you arrived at the values.

    March: _____

    June: _____

    September: _____

    December: _____

    Explanation: _____

Table 12.3 **Solar radiation at the outer edge of the atmosphere (langleys/day) at various latitudes during select months**

LATITUDE	MARCH	JUNE	SEPTEMBER	DECEMBER
90°N	50	1050	50	0
40°N	700	950	720	325
0°	890	780	880	840

Earth–Sun Relations on the Internet

Apply the concepts from this exercise to an examination of solar and terrestrial radiation by completing the corresponding online activity on the *Applications & Investigations in Earth Science* website at http://prenhall.com/earthsciencelab

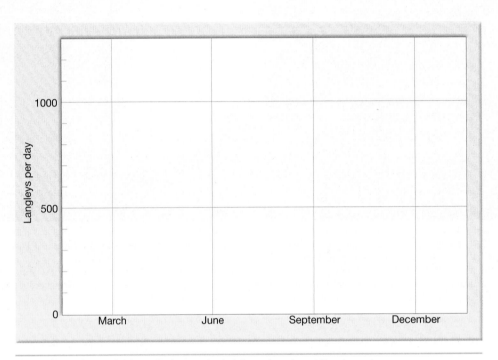

Figure 12.7 Graph of solar radiation received at the outer edge of the atmosphere.

Earth–Sun Relations

Date Due: _____

Name: _____

Date: _____

Class: _____

After you have finished Exercise 12, complete the following questions. You may have to refer to the exercise for assistance or to locate specific answers. Be prepared to submit this summary/report to your instructor at the designated time.

1. From Figure 12.3, what was the calculated percentage of solar radiation that is intercepted by each of the following 30° segments of latitude?

    0°–30° _____ %

    30°–60° _____ %

    60°–90° _____ %

2. How many hours of daylight occur at the following locations on the specified dates?

	MARCH 22	DECEMBER 22
40°N	_____ hrs	_____ hrs
0°	_____ hrs	_____ hrs
90°S	_____ hrs	_____ hrs

3. What is the noon Sun angle at these latitudes on April 11?

    40°N _____ °   0° _____ °

4. What is the relation between the angle of the noon Sun and the quantity of solar radiation received per square centimeter at the outer edge of the atmosphere?

    _____

    _____

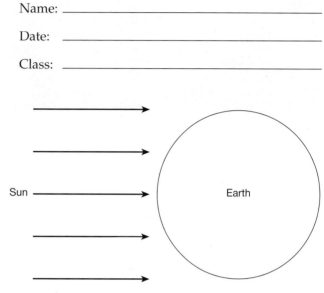

Figure 12.8   Earth's relation to the Sun on June 22.

5. Complete Figure 12.8 showing Earth's relation to the Sun on June 22. On the Earth, accurately draw and label the following:

    Axis
    Equator
    Tropic of Cancer
    Tropic of Capricorn
    Antarctic Circle
    Arctic Circle
    Circle of illumination
    Location of the overhead noon Sun

6. What causes the intensity and duration of solar radiation received at any place to vary throughout the year?

    _____

    _____

    _____

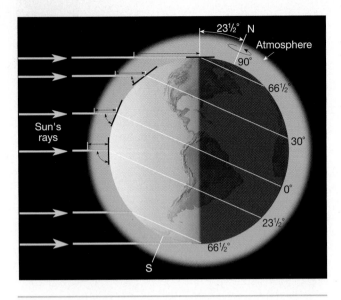

Figure 12.9 Earth-Sun relation diagram.

7. What is the date illustrated by the diagram in Figure 12.9? Calculate the noon Sun angle at 30° N latitude on this date and write a paragraph describing the distribution of solar radiation over Earth on this date.

_____

_____

_____

_____

8. What are the maximum and minimum noon Sun angles at your latitude?

Maximum noon Sun angle = _____ ° on _____ (date)

Minimum noon Sun angle = _____ ° on _____ (date)

9. What are the maximum and minimum durations of daylight at your latitude?

Maximum duration of daylight = _____ hrs

Minimum duration of daylight = _____ hrs

10. Write a brief statement describing how the intensity and duration of solar radiation change at your location throughout the year.

_____

_____

_____

11. The day is March 22. You view the noon Sun to the south at 35° above the horizon. What is your latitude?

Latitude: _____

Atmospheric Heating

The quantity of radiation from the Sun that strikes the outer edge of Earth's atmosphere at any one place is not constant but varies with the seasons. This exercise examines, step-by-step, what happens to **solar radiation** as it passes through the atmosphere, is absorbed at Earth's surface, and is reradiated by land and water back to the atmosphere (Figure 13.1). Investigating the journey of solar radiation and how it is influenced and modified by air, land, and water will provide a better understanding of one of the most basic weather elements, atmospheric temperature.

Objectives

After you have completed this exercise, you should be able to:

1. Explain how Earth's atmosphere is heated.

2. Describe the effect that the atmosphere has on absorbing, scattering, and reflecting incoming solar radiation.

3. List the gases in the atmosphere that are responsible for absorbing long-wave radiation.

4. Explain how the heating of a surface is related to its albedo.

5. Discuss the differences in the heating and cooling of land and water.

6. Summarize the global pattern of surface temperatures for January and July.

7. Describe how the temperature of the atmosphere changes with increasing altitude.

8. List the cause of a surface temperature inversion.

9. Determine the effect that wind speed has on the windchill equivalent temperature.

Materials

calculator
colored pencils

Figure 13.1 Solar radiation and atmospheric heating. (Photo by E. J. Tarbuck)

Materials Supplied by Your Instructor

light source	wood splints
black and silver	beaker of sand
containers	beaker of water
two thermometers	

Terms

solar radiation	albedo	isotherm
greenhouse	environmental	windchill
effect	lapse rate	equivalent
terrestrial	temperature	temperature
radiation	inversion	

Introduction

Temperature is an important element of weather and climate because it greatly influences air pressure, wind, and the amount of moisture in the air. The unequal heating that takes place over the surface of Earth is what sets the atmosphere in motion, and the movement of air is what brings changes in our weather.

The single greatest cause for temperature variations over the surface of Earth is differences in the reception of solar radiation. Secondary factors such as the differential heating of land and water, ocean currents, and altitude can modify local temperatures.

The amount of solar energy (radiation) striking Earth is not constant throughout the year at any particular place, nor is it uniform over the face of Earth at any one time. However, the total amount of radiation that the planet intercepts from the Sun equals the total radiation that it loses back to space. It is this balance between incoming and outgoing radiation that keeps Earth from becoming continuously hotter or colder.

Solar Radiation at the Outer Edge of the Atmosphere

The two factors that control the amount of solar radiation that a square meter receives at the outer edge of the atmosphere, and eventually Earth's surface, are the Sun's *intensity* and its *duration*. These variables were examined in detail in Exercise 12, "Earth–Sun Relations." Answer questions 1–3 after you have reviewed Exercise 12.

1. Briefly define solar intensity and duration.

   Intensity of solar radiation: _____

   _____

   Duration of solar radiation: _____

   _____

2. Complete Table 13.1 by calculating the angle that the noon Sun would strike the outer edge of the atmosphere at each of the indicated latitudes on the specified date. How many hours of daylight would each place experience on these dates? (*Hint:* You may find Tables 12.1 and 12.2 in Exercise 12 helpful.)

Table 13.1 Noon Sun angle and length of day

	March 21		June 21	
	NOON SUN ANGLE	LENGTH OF DAY	NOON SUN ANGLE	LENGTH OF DAY
40°N:	____°	____ hrs	____°	____ hrs
0°:	____°	____ hrs	____°	____ hrs
40°S:	____°	____ hrs	____°	____ hrs

3. Explain the reason why the intensity and duration of solar radiation received at the outer edge of the atmosphere is not constant at any particular latitude throughout the year.

_____

_____

Atmospheric Heating

Atmospheric heating is a function of (1) the ability of atmospheric gases to absorb radiation, (2) the amount of solar radiation that reaches Earth's surface, and (3) the nature of the surface material. Of the three, selective absorption of radiation by the atmosphere provides an insight into the mechanism of atmospheric heating. The quantity of radiation that reaches Earth's surface and the ability of the surface to absorb and reradiate the radiation determine the extent of atmospheric heating.

The atmosphere is rather selective and efficiently absorbs long-wave radiation that we detect as heat while allowing the transmission of most of the short wavelengths—a process called the **greenhouse effect**. The short-wave radiation that reaches Earth's surface and is absorbed ultimately returns to the atmosphere in the form of long-wave, **terrestrial radiation**. As the radiation travels up from the surface through the atmosphere, it is absorbed by atmospheric gases, heating the atmosphere from below. Since terrestrial radiation supplies most of the long-wave radiation to the atmosphere, it is the primary source of heat. The fact that temperature typically decreases with an increase in altitude in the lower atmosphere is clear evidence supporting this mechanism of atmospheric heating.

Solar Radiation Received at Earth's Surface

As solar radiation travels through the atmosphere, it may be reflected, scattered, or absorbed. The effect of the atmosphere on incoming solar radiation and the amount of radiation that ultimately reaches the surface is primarily dependent upon the angle at which the solar beam passes through the atmosphere and strikes Earth's surface.

Figure 13.2 illustrates the atmospheric effects on incoming solar radiation for an average noon Sun angle. Answer questions 4—7 by examining the figure and supplying the correct response.

4. _____ percent of the incoming solar radiation is reflected and scattered back to space.

5. _____ percent of the incoming solar radiation is absorbed by gases in the atmosphere and clouds.

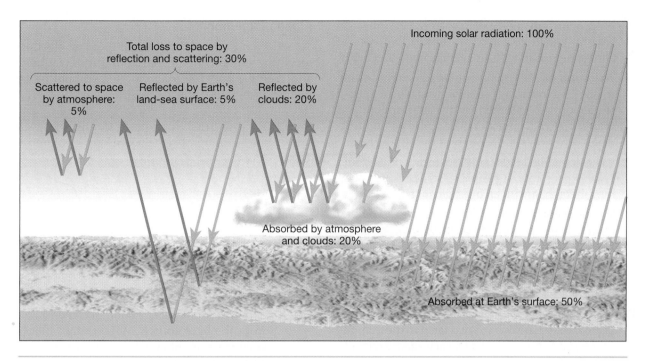

Figure 13.2 Solar radiation budget of the atmosphere and Earth.

6. _____ percent of the incoming solar radiation is absorbed at Earth's surface.

7. (Two and a half, Four) times as much incoming radiation is absorbed by Earth's surface than by the atmosphere and clouds. Circle your answer.

Figure 13.3 illustrates the effects of the atmosphere on various wavelengths of radiation. Use Figure 13.3 to answer questions 8–11 by circling the correct response.

8. The incoming solar radiation that passes through the atmosphere and is absorbed at Earth's surface is primarily in the form of (ultraviolet, visible, infrared) wavelengths.

9. When the surface releases the solar radiation it has absorbed, this terrestrial radiation is primarily (ultraviolet, visible, infrared) wavelengths.

10. (Ultraviolet, Visible, Infrared) wavelengths of radiation are absorbed efficiently by oxygen and ozone in the atmosphere.

11. Oxygen and ozone are (good, poor) absorbers of infrared radiation.

12. (Nitrogen, Carbon dioxide) and (water vapor, ozone) are the two principal gases that absorb most of the terrestrial radiation in the atmosphere.

Assume Figure 13.2 represents the atmospheric effects on incoming solar radiation for an average noon Sun angle of about 50°. Answer questions 13–16 concerning other noon Sun angles by circling the appropriate responses.

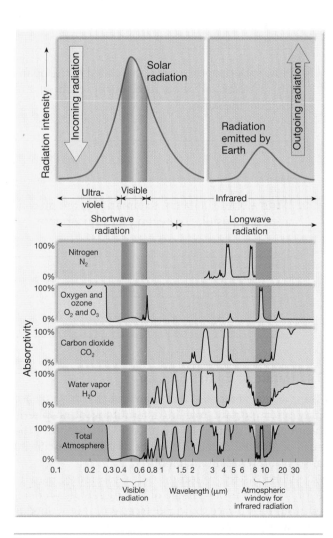

Figure 13.3 The absorptivity of selected gases of the atmosphere and the atmosphere as a whole.

13. If the noon Sun angle is 90°, solar radiation would have to penetrate a (greater, lesser) thickness of atmosphere than with an average noon Sun angle.

14. The result of a 90° noon Sun angle would be that (more, less) incoming radiation would be reflected, scattered, and absorbed by the atmosphere and (more, less) radiation would be absorbed and reradiated by Earth's surface to heat the atmosphere.

15. If the noon Sun angle is 20°, solar radiation would have to penetrate a (greater, lesser) thickness of atmosphere than with an average noon Sun angle.

16. The result of a 20° noon Sun angle would be that (more, less) incoming radiation would be reflected, scattered, and absorbed by the atmosphere and (more, less) radiation would be absorbed and reradiated by Earth's surface to heat the atmosphere.

17. How is the angle (intensity) at which the solar beam strikes Earth's surface related to the quantity of solar radiation received by each square meter?

_____

_____

18. How is the length of daylight related to the quantity of solar radiation received by each square meter at the surface?

_____

_____

19. Write a brief statement summarizing the mechanism responsible for heating the atmosphere.

_____

_____

_____

The Nature of Earth's Surface

The various materials that comprise Earth's surface play an important role in determining atmospheric heating. Two significant factors are the **albedo** of the surface and the different abilities of land and water to absorb and reradiate radiation.

Albedo is the reflectivity of a substance, usually expressed as the percentage of radiation that is reflected from the surface. Since surfaces with high albedos are not efficient absorbers of radiation, they cannot return much long-wave radiation to the atmosphere for heating.

Albedo Experiment

To better understand the effect of color on albedo, observe the equipment in the laboratory (Figure 13.4) and then conduct the following experiment by completing each of the indicated steps.

Step 1: Write a brief hypothesis stating the heating and cooling of light versus dark colored surfaces.

_____

_____

_____

Step 2: Place the black and silver containers (with lids and thermometers) about six inches away from the light source. Make certain that both containers are of equal distance from the light and are not touching one another.

Step 3: Record the starting temperature of both containers on the albedo experiment data table, Table 13.2.

Step 4: Turn on the light and record the temperature of both containers on the data table at about 30-second intervals for 5 minutes.

Step 5: Turn off the light and continue to record the temperatures at 30-second intervals for another 5 minutes.

Step 6: Plot the temperatures from the data table on the albedo experiment graph, Figure 13.5. Use a different color line to connect the points for each container.

Figure 13.4 Albedo experiment lab equipment.

Table 13.2 **Albedo experiment data table**

	STARTING TEMPERATURE	30 SEC	1 MIN	1.5 MIN	2 MIN	2.5 MIN	3 MIN	3.5 MIN	4 MIN	4.5 MIN	5 MIN	5.5 MIN	6 MIN	6.5 MIN	7 MIN	7.5 MIN	8 MIN	8.5 MIN	9 MIN	9.5 MIN	10 MIN
Black Container																					
Silver Container																					

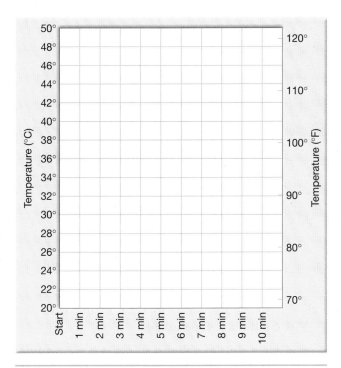

Figure 13.5 Albedo experiment graph.

20. Write a statement that summarizes and explains the results of your albedo experiment.

_____

_____

_____

_____

_____

21. What are some Earth surfaces that have high albedos and some that have low albedos?

High albedos: _____

Low albedos: _____

22. Given equal amounts of radiation reaching the surface, the air over a snow-covered surface will be (warmer, colder) than air above a dark-colored, barren field. Circle your answer. Then explain your choice fully in terms of what you have learned about albedo.

_____

_____

23. Why is it wise to wear light-colored clothes on a sunny summer day?

_____

24. If you lived in an area with long, cold winters, a (light-, dark-) colored roof would be the best choice for your house. Circle your answer. Explain the reasons for your choice.

_____

_____

Land and Water Heating Experiment

Land and water influence the air temperatures above them in different manners because they do not absorb and reradiate energy equally.

Investigate the differential heating of land and water by observing the equipment in the laboratory (Figure 13.6) and conducting the following experiment by completing each of the indicated steps.

Step 1: Fill one beaker three-quarters full with dry sand and a second beaker three-quarters full with water at room temperature.

Step 2: Using a wood splint, suspend a thermometer in each beaker so that the bulbs are _just below_ the surfaces of the sand and water.

Step 3: Hang a light from a stand so it is equally as close as possible to the top of the two beakers.

Step 4: Record the starting temperatures for both the dry sand and water on the land and water heating data table, Table 13.3.

Figure 13.6 Land and water heating experiment lab equipment.

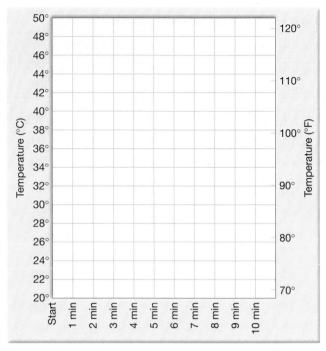

Figure 13.7 Land and water heating graph.

Step 5: Turn on the light and record the temperature on the data table at about one-minute intervals for 10 minutes.

Step 6: Turn off the light for several minutes. Dampen the sand with water and record the starting temperature of the damp sand on the data table. Turn on the light and record the temperature of the damp sand on the data table at about one-minute intervals for 10 minutes.

Step 7: Plot the temperatures for the water, dry sand, and damp sand from the data table on the land and water heating graph, Figure 13.7. Use a different color line to connect the points for each material.

25. Questions 25a and 25b refer to the land and water heating experiment.

   a. How do the abilities to change temperature differ for dry sand and water when they are exposed to equal quantities of radiation?

   _____

   _____

   b. How do the abilities to change temperature differ for dry sand and damp sand when they are exposed to equal quantities of radiation?

   _____

   _____

Table 13.3 **Land and water heating data table**

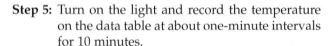

	STARTING TEMPERATURE	1 MIN	2 MIN	3 MIN	4 MIN	5 MIN	6 MIN	7 MIN	8 MIN	9 MIN	10 MIN
Water											
Dry sand											
Damp sand											

26. Suggest several reasons for the differential heating of land and water.

_____

_____

_____

_____

Figure 13.8 presents the annual temperature curves for two cities, A and B, that are located in North America at approximately 37°N latitude. On any date both cities receive the same intensity and duration of solar radiation. One city is in the center of the continent, while the other is on the west coast. Use Figure 13.8 to answer questions 27–34.

27. In Figure 13.8, city (A, B) has the highest mean monthly temperature. Circle your answer.

28. City (A, B) has the lowest mean monthly temperature.

29. The greatest _annual temperature range_ (difference between highest and lowest mean monthly temperatures) occurs at city (A, B).

30. City (A, B) reaches its maximum mean monthly temperature at an earlier date.

31. City (A, B) maintains a more uniform temperature throughout the year.

32. Of the two cities, city A is most likely located (along a coast, in the center of a continent).

33. The most likely location for city B is (coastal, mid-continent).

34. Describe the effect that the location, along the coast or in the center of a continent, has on the temperature of a city.

_____

_____

Atmospheric Temperatures

Air temperatures are not constant. They normally change (1) through time at any one location, (2) with latitude because of the changing sun angle and length of daylight, and (3) with increasing altitude in the lower atmosphere because the atmosphere is primarily heated from the bottom up.

Daily Temperatures

In general, the daily temperatures that occur at any particular place are the result of long-wave radiation being released at Earth's surface. However, secondary factors, such as cloud cover and cold air moving into the area, can also cause significant variations.

Questions 35–42 refer to the daily temperature graph, Figure 13.9.

35. The coolest temperature of the day occurs at _____. Fill in your answer.

36. The warmest temperature occurs at _____.

37. What is the _daily temperature range_ (difference between maximum and minimum temperatures for the day)?

Daily temperature range: _____ °F (_____ °C).

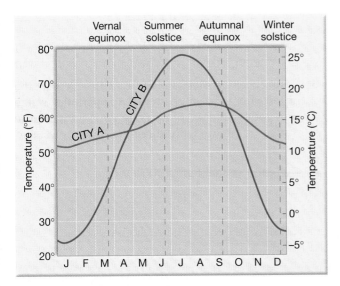

Figure 13.8 Mean monthly temperatures for two North American cities located at approximately 37°N latitude.

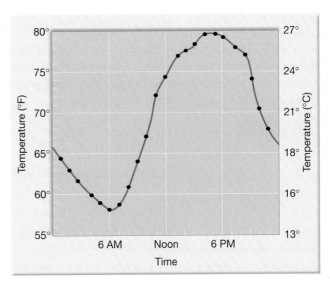

Figure 13.9 Typical daily temperature graph for a mid-latitude city during the summer.

38. What is the *daily temperature mean* (average of the maximum and minimum temperatures)?

    Daily temperature mean: _____ °F (_____ °C).

39. Refer to the mechanism for heating the atmosphere. Why does the warmest daily temperature occur in mid-to-late afternoon rather than at the time of the highest Sun angle?

    _____

    _____

40. Why does the coolest temperature of the day occur about sunrise?

    _____

    _____

41. How would cloud cover influence daily maximum and minimum temperatures?

    _____

    _____

42. On Figure 13.9 sketch and label a colored line that best represents a daily temperature graph for a typical cloudy day.

Global Pattern of Temperature

The primary reason for global variations in surface temperatures is the unequal distribution of radiation over the Earth. Among the most important secondary factors are differential heating of land and water, ocean currents, and differences in altitude.

Questions 43–55 refer to Figure 13.10, "World Distribution of Mean Surface Temperatures (°C) for January and July." The lines on the maps, called **isotherms**, connect places of equal surface temperature.

43. The general trend of the isotherms on the maps is (north-south, east-west). Circle your answer.

44. In general, how do surface temperatures vary from the equator toward the poles? Why does this variation occur?

    _____

    _____

    _____

45. The warmest and coldest temperatures occur over which countries or oceans?

    Warmest global temperature: _____

    _____

    Coldest global temperature: _____

    _____

46. The locations of the warmest and coldest temperatures are over (land, water).

47. Calculate the *annual temperature range* at each of the following locations:

    Coastal Norway at 60°N: _____ °C (_____ °F)

    Siberia at 60°N, 120°E: _____ °C (_____ °F)

    On the equator over the center of the Atlantic Ocean: _____ °C (_____ °F)

48. Explain the large annual range of temperature in Siberia.

    _____

    _____

49. Why is the annual temperature range smaller along the coast of Norway than at the same latitude in Siberia?

    _____

    _____

50. Why is temperature relatively uniform throughout the year in the tropics?

    _____

    _____

51. Using the two maps in Figure 13.10, calculate the approximate average annual temperature range for your location. How does your temperature range compare with those in the tropics and Siberia?

    Average annual temperature range:

    _____ °C (_____ °F)

    _____

52. Trace the path of the 5°C isotherm over North America in January. Explain why the isotherm

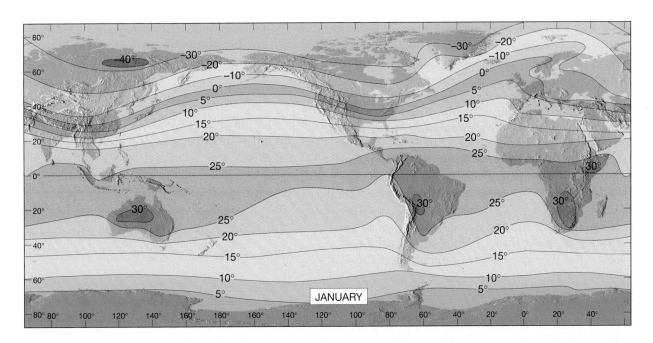

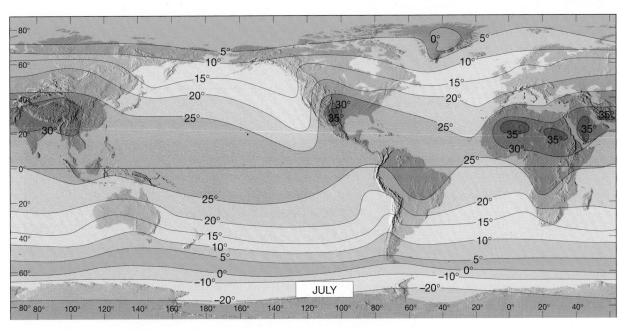

Figure 13.10 World distribution of mean surface temperatures (°C) for January and July.

deviates from a true east-west trend where it crosses from the Pacific Ocean onto the continent.

_____

_____

53. Trace the path of the 20°C isotherm over North America in July. Explain why the isotherm deviates from a true east-west trend where it crosses from the Pacific Ocean onto the continent.

_____

_____

54. Why do the isotherms in the Southern Hemisphere follow a true east-west trend more closely than those in the Northern Hemisphere?

_____

_____

_____

55. Why does the entire pattern of isotherms shift northward from the January map to the July map?

_____

_____

_____

Temperature Changes with Altitude

Since the primary source of heat for the lower atmosphere is Earth's surface, the normal situation found in the lower 12 kilometers of the atmosphere is a decrease in temperature with increasing altitude. This temperature decrease with altitude in the lower atmosphere is called the **environmental lapse rate**. However, at altitudes from about 12 to 45 kilometers, the atmospheric absorption of incoming solar radiation causes temperature to increase.

Use Figure 13.11 to answer questions 56–60.

56. Using the temperature curve as a guide, label the *troposphere, mesosphere, stratosphere,* and *thermosphere* on the atmospheric temperature curve, Figure 13.11.

57. On Figure 13.11, mark with a line and label the *tropopause, mesopause,* and *stratopause.*

58. What is the approximate temperature of the atmosphere at each of the following altitudes?

10 km: _____ °C (_____ °F)

50 km: _____ °C (_____ °F)

80 km: _____ °C (_____ °F)

59. Using Figure 13.11, calculate the average decrease in temperature with altitude of the troposphere in both °C/km and °F/mi.

_____

60. Explain the reason for each of the following:

Temperature decrease with altitude in the troposphere:

_____

_____

Temperature increase in the stratosphere:

_____

_____

Temperature increase in the thermosphere:

_____

_____

Figure 13.11 Atmospheric temperature curve.

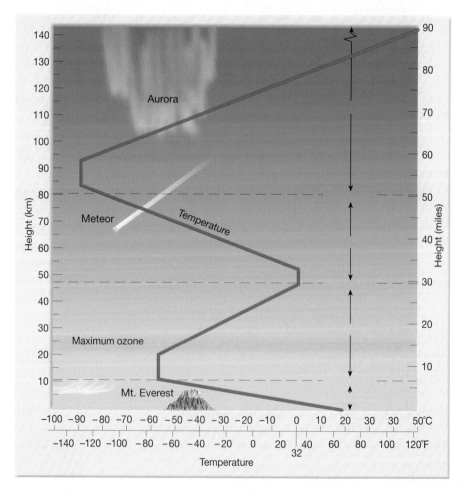

61. Of what importance is the gas *ozone* in the stratosphere? What will be the effect on radiation received at Earth's surface of a decrease of ozone in the stratosphere?

_____

_____

_____

Assume the average, or normal, environmental lapse rate (temperature decrease with altitude) in the troposphere is 3.5°F per 1,000 feet (6.5°C per kilometer).

62. If the surface temperature is 60°F (16°C), what would be the approximate temperature at 20,000 feet (6,000 meters)?

_____ °F (_____ °C)

63. If the surface temperature is 80°F (27°C), at approximately what altitude would a pilot expect to find each of the following atmospheric temperatures?

50°F: _____ feet (10°C: _____ meters)

0°C: _____ meters (32°F: _____ feet)

Periodically, the temperature near the surface of Earth increases with altitude. This situation, which is opposite from the normal condition, is called a **temperature inversion**.

64. Suggest a possible cause for a surface temperature inversion.

_____

_____

Windchill Equivalent Temperature

Windchill equivalent temperature is the term applied to the sensation of temperature that the human body feels, in contrast to the actual temperature of the air as recorded by a thermometer. Wind cools by evaporating perspiration and carrying heat away from the body. When temperatures are cool and the wind speed increases, the body reacts as if it were being subjected to increasingly lower temperatures—a phenomenon known as *windchill.*

65. Refer to the windchill equivalent temperature chart, Figure 13.12. What is the windchill equivalent temperature sensed by the human body in the following situations?

Air Temperature (°F)	Wind Speed (mph)	Windchill Equivalent Temperature (°F)
30°	10	_____
−5°	20	_____
−20°	30	_____

66. Write a brief summary of the effect of wind speed on how long a person can be exposed before frostbite develops.

_____

_____

_____

Atmospheric Heating on the Internet

Research current and historical atmospheric temperatures at your location by completing the corresponding online activity on the *Applications & Investigations in Earth Science* website at http://prenhall.com/earthsciencelab

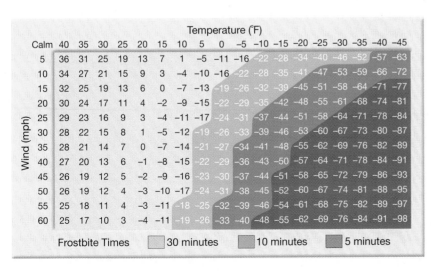

Figure 13.12 This windchill chart came into use in November 2001. Fahrenheit temperatures are used here because this is how the National Weather Service and the news media in the United States commonly report windchill information. The shaded areas on the chart indicate frostbite danger. Each shaded zone shows how long a person can be exposed before frostbite develops. (*After NOAA, National Weather Service*)

Temperature (°F)

Wind (mph)	Calm / 40	35	30	25	20	15	10	5	0	−5	−10	−15	−20	−25	−30	−35	−40	−45
5	36	31	25	19	13	7	1	−5	−11	−16	−22	−28	−34	−40	−46	−52	−57	−63
10	34	27	21	15	9	3	−4	−10	−16	−22	−28	−35	−41	−47	−53	−59	−66	−72
15	32	25	19	13	6	0	−7	−13	−19	−26	−32	−39	−45	−51	−58	−64	−71	−77
20	30	24	17	11	4	−2	−9	−15	−22	−29	−35	−42	−48	−55	−61	−68	−74	−81
25	29	23	16	9	3	−4	−11	−17	−24	−31	−37	−44	−51	−58	−64	−71	−78	−84
30	28	22	15	8	1	−5	−12	−19	−26	−33	−39	−46	−53	−60	−67	−73	−80	−87
35	28	21	14	7	0	−7	−14	−21	−27	−34	−41	−48	−55	−62	−69	−76	−82	−89
40	27	20	13	6	−1	−8	−15	−22	−29	−36	−43	−50	−57	−64	−71	−78	−84	−91
45	26	19	12	5	−2	−9	−16	−23	−30	−37	−44	−51	−58	−65	−72	−79	−86	−93
50	26	19	12	4	−3	−10	−17	−24	−31	−38	−45	−52	−60	−67	−74	−81	−88	−95
55	25	18	11	4	−3	−11	−18	−25	−32	−39	−46	−54	−61	−68	−75	−82	−89	−97
60	25	17	10	3	−4	−11	−19	−26	−33	−40	−48	−55	−62	−69	−76	−84	−91	−98

Frostbite Times ☐ 30 minutes ☐ 10 minutes ☐ 5 minutes

Notes and calculations.

Atmospheric Heating

Date Due: _____

Name: _____

Date: _____

Class: _____

After you have finished Exercise 13, complete the following questions. You may have to refer to the exercise for assistance or to locate specific answers. Be prepared to submit this summary/report to your instructor at the designated time.

1. Assume an average noon Sun angle. What percentage of the solar radiation will be absorbed by the atmosphere and what percentage will be absorbed by Earth's surface?

   Atmospheric absorption: _____ %

   Absorption by Earth's surface: _____ %

2. What will be the atmospheric effect of each of the following?

   Less ozone in the stratosphere:

   _____

   _____

   More carbon dioxide in the atmosphere:

   _____

   _____

   A surface with a high albedo:

   _____

   _____

3. Briefly explain how Earth's atmosphere is heated.

   _____

   _____

   _____

4. What are the primary heat-absorbing gases in the atmosphere? In general, what wavelength of radiation do they absorb?

   _____

   _____

5. What were the starting and 5-minute temperatures you obtained for the black and silver containers in the albedo experiment?

	STARTING TEMPERATURE	5-MINUTE TEMPERATURE
Black container:	_____	_____
Silver container:	_____	_____

6. Summarize the effect of color on the heating of an object.

   _____

   _____

7. What were the starting and ending temperatures you obtained for the water and dry sand in the land and water heating experiment?

	STARTING TEMPERATURE	ENDING TEMPERATURE
Water:	_____	_____
Dry sand:	_____	_____

8. Summarize the effects that equal amounts of radiation have on the heating of land and water.

_____

_____

9. Where are the highest and lowest average monthly temperatures located on Earth?

Highest average monthly temperature:

_____

_____

Lowest average monthly temperature:

_____

_____

10. Why does the Northern Hemisphere experience a greater annual range of temperature than the Southern Hemisphere?

_____

_____

11. Define each of the following:

Environmental lapse rate:

_____

_____

Windchill equivalent temperature:

_____

_____

Troposphere:

_____

_____

12. Referring to the average temperature graphs for Spokane and Seattle, Washington shown in Figure 13.13, discuss the reason(s) why the two graphs are dissimilar, even though both cities are at about the same latitude.

_____

_____

_____

_____

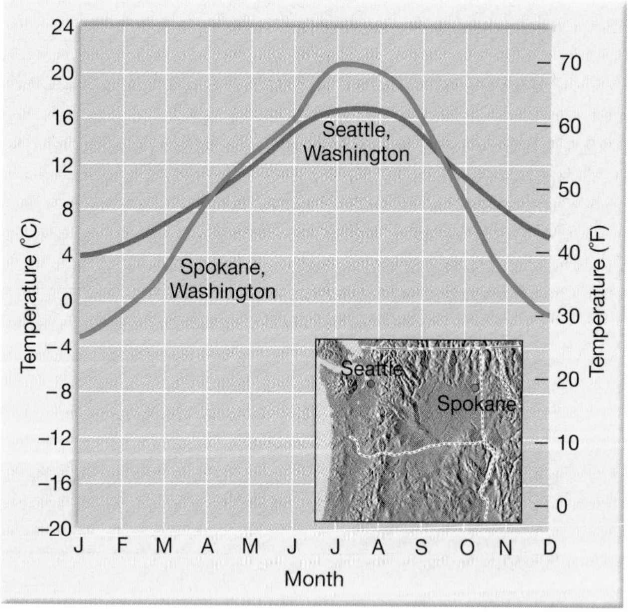

Figure 13.13   Temperature graphs for Spokane and Seattle, WA.

Atmospheric Moisture, Pressure, and Wind

By observing, recording, and analyzing weather conditions, meteorologists attempt to define the principles that control the complex interactions that occur in the atmosphere (Figure 14.1). One important element, temperature, has already been examined in Exercises 12 and 13. However, no analysis of the atmosphere is complete without an investigation of the remaining variables—humidity, precipitation, pressure, and wind.

This exercise examines the changes of state of water, how the water vapor content of the air is measured, and the sequence of events necessary to cause cloud formation. Global patterns of precipitation, pressure, and wind are also reviewed. Although the elements are presented separately, keep in mind that all are very much interrelated. A change in any one element often brings about changes in the others.

Objectives

After you have completed this exercise, you should be able to:

1. Explain the processes involved when water changes state.
2. Use a psychrometer or hygrometer and appropriate tables to determine the relative humidity and dew-point temperature of air.
3. Explain the adiabatic process and its effect on cooling and warming the air.
4. Calculate the temperature and relative humidity changes that take place in air as the result of adiabatic cooling.
5. Describe the relation between pressure and wind.
6. Describe the global patterns of surface pressure and wind.

Materials

calculator   ruler   colored pencils

Figure 14.1   Developing storm clouds. (Photo by E. J. Tarbuck)

Materials Supplied by Your Instructor

psychrometer or hygrometer	hot plate
beaker, ice, thermometer	thumbtacks
barometer	cardboard
atlas	tape

Terms

water vapor	relative humidity	psychrometer/
evaporation	saturated	hygrometer
precipitation	dew-point	condensation
latent heat	temperature	nuclei
dry adiabatic rate	equatorial low	Coriolis effect
wet adiabatic rate	subtropical high	trade winds
atmospheric	subpolar low	westerlies
pressure	anticyclone	polar easterlies
barometer	cyclone	monsoon
isobar	wind	

Atmospheric Moisture and Precipitation

Water vapor, an odorless, colorless gas produced by the **evaporation** of water, comprises only a small percentage of the lower atmosphere. However, it is an important atmospheric gas because it is the source of all **precipitation**, aids in the heating of the atmosphere by absorbing radiation, and is the source of **latent heat** (hidden or stored heat).

Changes of State

The temperatures and pressures that occur at and near Earth's surface allow water to change readily from one state of matter to another. The fact that water can exist as a gas, liquid, or solid within the atmosphere makes it one of the most unique substances on Earth. Use Figure 14.2 to answer questions 1–4.

1. To help visualize the processes and heat requirements for changing the state of matter of water, write the name of the process involved (choose from the list) and whether heat is absorbed or released by the process at the indicated locations by each arrow in Figure 14.2.

PROCESSES

Freezing	Evaporation	Deposition
Sublimation	Melting	Condensation

2. To melt ice, heat energy must be (absorbed, released) by the water molecules. Circle your answer.

3. The process of condensation requires that water molecules (absorb, release) heat energy. Circle your answer.

4. The energy requirement for the process of deposition is the (same as, less than) the total energy required to condense water vapor and then freeze the water. Circle your answer.

Latent Heat Experiment

This experiment will help you gain a better understanding of the role of heat in changing the state of matter. You are going to heat a beaker that contains a mixture of ice and water (Figure 14.3). You will record temperature changes *as the ice melts* and continue to record the temperature changes *after the ice melts*. Conduct the experiment by completing the following steps.

Step 1: Write a brief hypothesis as to how you expect the temperature of the ice-water mixture to change as heat energy is added.

_____

_____

_____

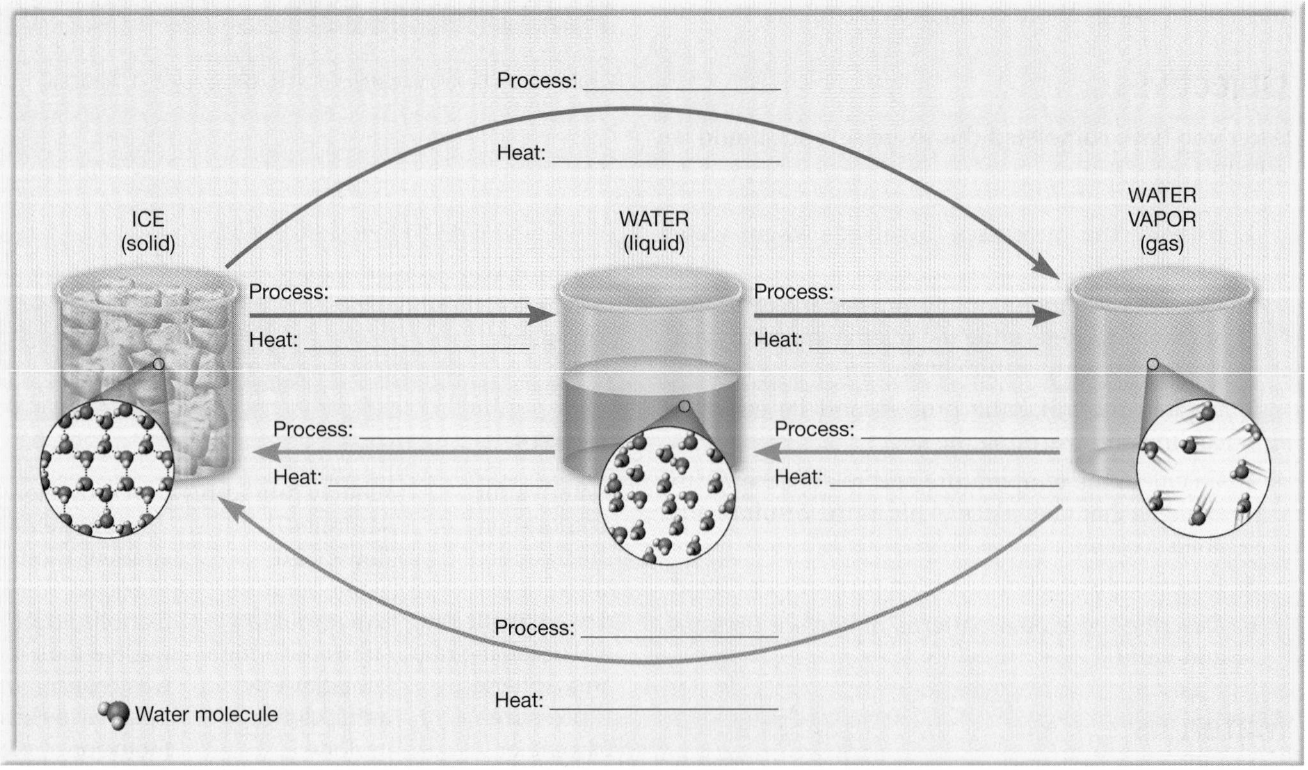

Figure 14.2 Changes of state of water.

Thermometer →

Figure 14.3 Latent heat experiment equipment.

Table 14.1 Latent Heat Data Table

TIME (MINUTES)	TEMPERATURE (° ____)
Starting	____
1	____
2	____
3	____
4	____
5	____
6	____
7	____
8	____
9	____
10	____
11	____
12	____
13	____
14	____
15	____

Questions 5 through 8 refer to your latent heat experiment.

5. How did the temperature of the mixture change prior to, and after, the ice had melted?

_____

_____

6. Calculate the average temperature change per minute of the ice-water mixture prior to the ice

Step 2: Turn on the hot plate and set the temperature setting to about three-fourths maximum (7 on a scale of 10).

CAUTION: The hot plate will become hot quickly. Do not touch the heating surface.

Step 3: Fill a 400-ml or larger beaker approximately half full with ice and add enough COLD water to cover the ice.

Step 4: Gently stir the ice-water mixture about 15 seconds with the thermometer and record the temperature in the "Starting" temperature space on the data table, Table 14.1.

Step 5: Place the beaker with the ice-water mixture and thermometer on the hot plate, and while STIRRING THE MIXTURE CONSTANTLY, record the temperature of the mixture at *one-minute intervals* on the data table. Watch the ice closely as it melts. *Note the exact time on the data table when all the ice has melted.*

Step 6: Continue stirring the mixture and recording its temperature for at least 3 or 4 minutes after all the ice has melted.

Step 7: Plot the temperatures from the data table on the graph, Figure 14.4.

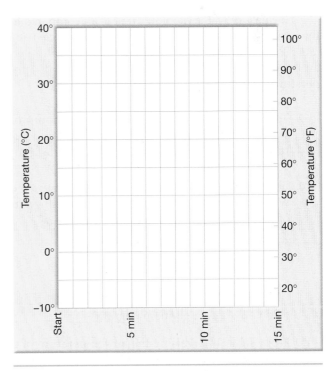

Figure 14.4 Latent heat experiment graph.

melting and the average rate after the ice had melted.

Average rate prior to melting: _____

Average rate after melting: _____

7. With your answers to questions 5 and 6 in mind, write a statement comparing your results to your hypothesis in **Step 1**.

_____

_____

8. With reference to the absorption or release of latent (hidden) heat, explain why the temperature changed at a different rate after the ice melted as compared to before all the ice had melted.

_____

_____

_____

Water-Vapor Capacity of Air

Any measure of water vapor in the air is referred to as *humidity*. The amount of water vapor required for saturation is directly related to temperature.

The mass of water vapor in a unit of air compared to the remaining mass of dry air is referred to as the *mixing ratio*. Table 14.2 presents the mixing ratios of saturated air (water vapor needed for saturation) at various temperatures. Use the table to answer questions 9–12.

9. To illustrate the relation between the amount of water vapor needed for saturation and temperature, prepare a graph by plotting the data from Table 14.2 on Figure 14.5.

10. From Table 14.2 and/or Figure 14.5, what is the water vapor content at saturation of a kilogram of air at each of the following temperatures?

    40°C: _____ grams/kilogram

    68°F: _____ grams/kilogram

    0°C: _____ grams/kilogram

   −20°C: _____ grams/kilogram

11. From Table 14.2, raising the air temperature of a kilogram of air 5°C, from 10°C to 15°C, (increases, decreases) the amount of water vapor needed

Table 14.2 Amount of water vapor needed to saturate a kilogram of air at various temperatures, the saturation mixing ratio.

TEMPERATURE		WATER VAPOR CONTENT AT SATURATION (G/KG)
(°C)	(°F)	
−40	−40	0.1
−30	−22	0.3
−20	−4	0.75
−10	14	2
0	32	3.5
5	41	5
10	50	7
15	59	10
20	68	14
25	77	20
30	86	26.5
35	95	35
40	104	47

for saturation by (3, 6) grams. However, raising the temperature from 35°C to 40°C (increases, decreases) the amount by (8, 12) grams. Circle your answers.

12. Using Table 14.2 and/or Figure 14.5, write a statement that relates the amount of water vapor needed for saturation to temperature.

_____

_____

_____

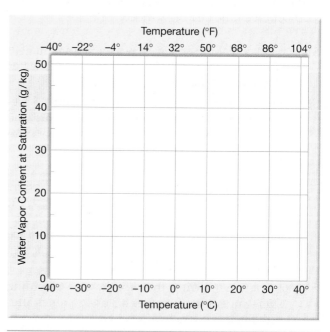

Figure 14.5 Graph of water vapor content at saturation of a kilogram of air versus temperature.

Measuring Humidity

Relative humidity is the most common measurement used to describe water vapor in the air. In general, it expresses how close the air is to reaching saturation at that temperature. Relative humidity is a *ratio* of the air's actual water vapor *content* (amount actually in the air) compared with the amount of water vapor required for saturation at that temperature (saturation mixing ratio), expressed as a percent. The general formula is

$$\text{Relative humidity (\%)} = \frac{\text{Water vapor content}}{\text{Saturation mixing ratio}} \times 100$$

For example, from Table 14.2, the saturation mixing ratio of a kilogram of air at 25°C would be 20 grams per kilogram. If the actual amount of water vapor in the air was 5 grams per kilogram (the water vapor content), the relative humidity of the air would be calculated as follows:

$$\text{Relative humidity (\%)} = \frac{5 \text{ g/kg}}{20 \text{ g/kg}} \times 100 = 25\%$$

13. Use Table 14.2 and the formula for relative humidity to determine the relative humidity for each of the following situations of identical temperature.

AIR TEMPERATURE	WATER VAPOR CONTENT	RELATIVE HUMIDITY
15°C	2 g/kg	____%
15°C	5 g/kg	____%
15°C	7 g/kg	____%

14. From question 13, if the temperature remains constant, adding water vapor will (raise, lower) the relative humidity, while removing water vapor will (raise, lower) the relative humidity. Circle your answers.

15. Use Table 14.2 and the formula for relative humidity to determine the relative humidity for each of the following situations of identical water vapor content.

AIR TEMPERATURE	WATER VAPOR CONTENT	RELATIVE HUMIDITY
25°C	5 g/kg	____%
15°C	5 g/kg	____%
5°C	5 g/kg	____%

16. From question 15, if the amount of water vapor in the air remains constant, cooling will (raise, lower) the relative humidity, while warming will (raise, lower) the relative humidity. Circle your answers.

17. In the winter, air is heated in homes. What effect does heating have on the relative humidity inside the home? What can be done to lessen this effect?

_____

_____

18. Explain why the air in a cool basement is humid (damp) in the summer.

_____

_____

19. Write brief statements describing each of the two ways that the relative humidity of air can be changed.

1. _____

2. _____

One of the misconceptions concerning relative humidity is that it alone gives an accurate indication of the actual quantity of water vapor in the air. For example, on a winter day if you hear on the radio that the relative humidity is 90 percent, can you conclude that the air contains more water vapor than on a summer day that records a 40 percent relative humidity? Completing question 20 will help you find the answer.

20. Use Table 14.2 to determine the water vapor content for each of the following situations. As you do the calculations, keep in mind the definition of relative humidity.

SUMMER	WINTER
Air temperature = 77°F	Air temperature = 41°F
Relative humidity = 40%	Relative humidity = 90%
Content = _____ g/kg	Content = _____ g/kg

21. Explain why relative humidity does *not* give an accurate indication of the actual amount of water vapor in the air.

_____

_____

_____

Dew-Point Temperature

Air is **saturated** when it contains all the water vapor that it can hold at a particular temperature. The temperature at which saturation occurs is called the **dew-point temperature**. Put another way, the dew point is the temperature at which the relative humidity of the air is 100 percent.

Previously, in question 15, you determined that a kilogram of air at 25°C, containing 5 grams of water vapor had a relative humidity of 25 percent and was not saturated. However, when the temperature was lowered to 5°C, the air had a relative humidity of 100 percent and was saturated. Therefore, 5°C is the dew-point temperature of the air in that example.

22. By referring to Table 14.2, what is the dew-point temperature of a kilogram of air that contains 7 grams of water vapor?

    Dew-point temperature = _____ °C

23. What is the relative humidity and dew-point temperature of a kilogram of 25°C air that contains 10 grams of water vapor?

    Relative humidity = _____ %

    Dew-point temperature = _____ °C

Using a Psychrometer or Hygrometer

The relative humidity and dew-point temperature of air can be determined by using a **psychrometer** (Figure 14.6) or **hygrometer** and appropriate charts. The psychrometer consists of two thermometers mounted side by side. One of the thermometers, the *dry-bulb* thermometer, measures the air temperature. The other thermometer, the *wet-bulb thermometer*, has a piece of wet cloth wrapped around its bulb. As the psychrometer is spun for approximately one minute, water on the wet-bulb thermometer evaporates and cooling results. In dry air, the rate of evaporation will be high, and a low wet-bulb temperature will be recorded. After using the instrument and recording both the dry- and wet-bulb temperatures, the relative humidity and dew-point temperature are determined using Table 14.3, "Relative humidity (percent)" and Table 14.4, "Dew-point temperature." With a hygrometer, relative humidity can be read directly, without the use of tables.

24. Use Table 14.3 to determine the relative humidity for each of the following psychrometer readings.

	READING 1	READING 2
Dry-bulb temperature:	20°C	32°C
Wet-bulb temperature:	18°C	25°C
Difference between dry- and wet-bulb temperatures:	_____	_____
Relative humidity:	_____%	_____%

Figure 14.6 Sling psychrometer. The sling psychrometer is one instrument that is used to determine relative humidity and dew-point temperature. (Photo by E. J. Tarbuck)

25. From question 24, what is the relation between the *difference* in the dry-bulb and wet-bulb temperatures and the relative humidity of the air?

    _____

    _____

26. Use Table 14.4 to determine the dew-point temperature for each of the following two psychrometer readings.

	READING 1	READING 2
Dry-bulb temperature:	8°C	30°C
Wet-bulb temperature:	6°C	24°C
Difference between dry- and wet-bulb temperatures:	_____	_____
Dew-point temperature:	_____°C	_____°C

Table 14.3 Relative humidity (percent)*

DRY-BULB TEMPERA-TURE (°C)	Depression of Wet-bulb Temperature (Dry-bulb Temperature − Wet-bulb Temperature = Depression of the Wet Bulb)																					
	1	2	3	4	5	6	7	8	9	10	11	12	13	14	15	16	17	18	19	20	21	22
−20	28																					
−18	40																					
−16	48	0																				
−14	55	11																				
−12	61	23																				
−10	66	33	0																			
−8	71	41	13																			
−6	73	48	20	0																		
−4	77	54	43	11																		
−2	79	58	37	20	1																	
0	81	63	45	28	11																	
2	83	67	51	36	20	6																
4	85	70	56	42	27	14																
6	86	72	59	46	35	22	10	0														
8	87	74	62	51	39	28	17	6														
10	88	76	65	54	43	33	24	13	4													
12	88	78	67	57	48	38	28	19	10	2												
14	89	79	69	60	50	41	33	25	16	8	1											
16	90	80	71	62	54	45	37	29	21	14	7	1										
18	91	81	72	64	56	48	40	33	26	19	12	6	0									
20	91	82	74	66	58	51	44	36	30	23	17	11	5	0								
22	92	83	75	68	60	53	46	40	33	27	21	15	10	4	0							
24	92	84	76	69	62	55	49	42	36	30	25	20	14	9	4	0						
26	92	85	77	70	64	57	51	45	39	34	28	23	18	13	9	5						
28	93	86	78	71	65	59	53	47	42	36	31	26	21	17	12	8	2					
30	93	86	79	72	66	61	55	49	44	39	34	29	25	20	16	12	8	4				
32	93	86	80	73	68	62	56	51	46	41	36	32	27	22	19	14	11	8	4			
34	93	86	81	74	69	63	58	52	48	43	38	34	30	26	22	18	14	11	8	5		
36	94	87	81	75	69	64	59	54	50	44	40	36	32	28	24	21	17	13	10	7	4	
38	94	87	82	76	70	66	60	55	51	46	42	38	34	30	26	23	20	16	13	10	7	
40	94	89	82	76	71	67	61	57	52	48	44	40	36	33	29	25	22	19	16	13	10	7

Dry-bulb (Air) Temperature

Relative Humidity Values

*To determine the relative humidity and dew point, find the air (dry-bulb) temperature on the vertical axis (far left) and the depression of the wet bulb on the horizontal axis (top). Where the two meet, the relative humidity or dew point is found. For example, use a dry-bulb temperature of 20°C and a wet-bulb temperature of 14°C. From Table 14.3, the relative humidity is 51 percent, and from Table 14.4, the dew point is 10°C.

If a psychrometer or hygrometer is available in the laboratory, your instructor will explain the procedure for using the instrument. FOLLOW THE SPECIFIC DIRECTIONS OF YOUR INSTRUCTOR to complete question 27.

27. Use the psychrometer (or hygrometer) to determine the relative humidity and dew-point temperature of the air in the room and outside the building. If you use a psychrometer, record your information in the following spaces.

	ROOM	OUTSIDE
Dry-bulb temperature:	_____	_____
Wet-bulb temperature:	_____	_____
Difference between dry- and wet-bulb temperatures:	_____	_____
Relative humidity:	_____%	_____%
Dew-point temperature:	_____	_____

Table 14.4 Dew-point temperature (°C)*

DRY-BULB TEMPERA-TURE (°C)	Depression of Wet-bulb Temperature (Dry-bulb Temperature − Wet-bulb Temperature = Depression of the Wet Bulb)																					
	1	**2**	**3**	**4**	**5**	**6**	**7**	**8**	**9**	**10**	**11**	**12**	**13**	**14**	**15**	**16**	**17**	**18**	**19**	**20**	**21**	**22**
−20	−33																					
−18	−28																					
−16	−24																					
−14	−21	−36																				
−12	−18	−28																				
−10	−14	−22																				
−8	−12	−18	−29																			
−6	−10	−14	−22																			
−4	−7	−22	−17	−29																		
−2	−5	−8	−13	−20																		
0	−3	−6	−9	−15	−24																	
2	−1	−3	−6	−11	−17																	
4	1	−1	−4	−7	−11	−19																
6	4	1	−1	−4	−7	−13	−21															
8	6	3	1	−2	−5	−9	−14															
10	8	6	4	1	−2	−5	−9	−14														
12	10	8	6	4	1	−2	−5	−9	−16													
14	12	11	9	6	4	1	−2	−5	−10	−17												
16	14	13	11	9	7	4	1	−1	−6	−10	−17											
18	16	15	13	11	9	7	4	2	−2	−5	−10	−19										
20	19	17	15	14	12	10	7	4	2	−2	−5	−10	−19									
22	21	19	17	16	14	12	10	8	5	3	−1	−5	−10	−19								
24	23	21	20	18	16	14	12	10	8	6	2	−1	−5	−10	−18							
26	25	23	22	20	18	17	15	13	11	9	6	3	0	−4	−9	−18						
28	27	25	24	22	21	19	17	16	14	11	9	7	4	1	−3	−9	−16					
30	29	27	26	24	23	21	19	18	16	14	12	10	8	5	1	−2	−8	−15				
32	31	29	28	27	25	24	22	21	19	17	15	13	11	8	5	2	−2	−7	−14			
34	33	31	30	29	27	26	24	23	21	20	18	16	14	12	9	6	3	−1	−5	−12	−29	
36	35	33	32	31	29	28	27	25	24	22	20	19	17	15	13	10	7	4	0	−4	−10	
38	37	35	34	33	32	30	29	28	26	25	23	21	19	17	15	13	11	8	5	1	−3	−9
40	39	37	36	35	34	32	31	30	28	27	25	24	22	20	18	16	14	12	9	6	2	−2

Dew-Point Temperature Values

*See footnote to Table 14.3

Condensation

If air is cooled below the dew-point temperature, water will condense (change from vapor to liquid) on available surfaces. In the atmosphere, the particles on which water condenses are called **condensation nuclei**. Condensation may result in the formation of dew or frost on the ground and clouds or fog in the atmosphere.

28. Examine the process of condensation by gradually adding ice to a beaker approximately one-third full of water. As you add the ice, stir the water-ice mixture gently with a thermometer. Note the temperature at the moment water begins to condense on the outside surface of the beaker. After you complete your observations, answer questions 28a and 28b.

   a. The temperature at which water began condensing on the outside surface of the beaker was _____.

b. How does the temperature at which water began to condense compare to the dew-point temperature of the air in the room you determined using the psychrometer (or hygrometer)?

_____

29. Refer to Table 14.2. How many grams of water vapor will condense on a surface if a kilogram of 50°F air with a relative humidity of 100 percent is cooled to 41°F?

   _____ grams of water will condense

30. Assume a kilogram of 25°C air that contains 10 grams of water vapor. Use Table 14.2. How many grams of water will condense if the air's temperature is lowered to each of the following temperatures?

   5°C: _____ grams of condensed water

   −10°C: _____ grams of condensed water

31. Considering your answers to the previous questions, what relation exists between the altitude of the base of a cloud, which consists of very small droplets of water, and the dew-point temperature of the air at that altitude?

Daily Temperature and Relative Humidity

Figure 14.7 shows the typical daily variations in air temperature, relative humidity, and dew-point temperature during two consecutive spring days at a middle latitude city. Use the figure to answer questions 32–36.

32. Relative humidity is at its maximum at (6 A.M., 3 P.M.) on day (1, 2). Circle your answers.

33. The lowest temperature over the two-day period occurs at (6 A.M., noon, 3 P.M.) on day (1, 2).

34. The lowest relative humidity occurs at (6 A.M., noon, 4 P.M.) on day (1, 2).

35. Write a general statement describing the relation between temperature and relative humidity throughout the time period shown in the figure.

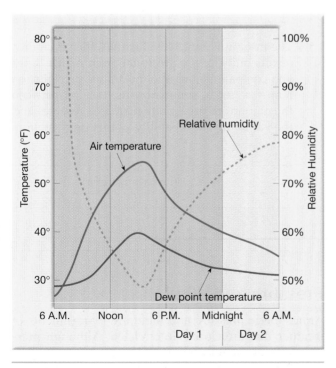

Figure 14.7 Typical variations in air temperatures, relative humidity, and dew-point temperature during two consecutive spring days at a middle latitude city.

36. Did a dew or frost occur on either of the two days represented in the figure? If so, list when and explain how you arrived at your answer.

Adiabatic Processes

As you have seen, the key to causing water vapor to condense, which is necessary before precipitation can occur, is to reach the dew-point temperature. In nature, when air rises and experiences a decrease in pressure, the air expands and cools. The reverse is also true. Air that is compressed will warm. Temperature changes brought about solely by expansion or compression are called *adiabatic temperature changes*. Air with a temperature above its dew point (unsaturated air) cools by expansion or warms by compression at a rate of 10°C per 1000 meters (1°C per 100 meters) of changing altitude—the dry **adiabatic rate**. After the dew-point temperature is reached, and as condensation occurs, latent heat that has been stored in the water vapor will be liberated. The heat being released by the condensing water slows down the rate of cooling of the air. Rising saturated air will continue to cool by expansion, but at a lesser rate of about 5°C per 1000 meters (0.5°C per 100 meters) of changing altitude—the **wet adiabatic rate**.

Figure 14.8 illustrates a kilogram of air at sea level with a temperature of 25°C and a relative humidity of 50 percent. The air is forced to rise over a 5,000-meter mountain and descend to a plateau 2,000 meters above sea level on the opposite (leeward) side. To help understand the adiabatic process, answer questions 37–49 by referring to Figure 14.8.

37. What is the saturation mixing ratio, content, and dew-point temperature of the air at sea level?

Saturation mixing ratio: _____ g/kg of air

Content: _____ g/kg of air

Dew-point temperature: _____°C

38. The air at sea level is (saturated, unsaturated). Circle your answer.

39. The air will initially (warm, cool) as it rises over the windward side of the mountain at the (wet, dry) adiabatic rate, which is (1, 0.5)°C per 100 meters. Circle the correct responses.

40. What will be the air's temperature at 500 meters?
_____ °C at 500 meters

41. Condensation (will, will not) take place at 500 meters. Circle your answer.

42. The rising air will reach its dew-point temperature at _____ meters and water vapor will begin to (condense, evaporate). Circle your answer.

Figure 14.8 Adiabatic processes associated with a mountain barrier.

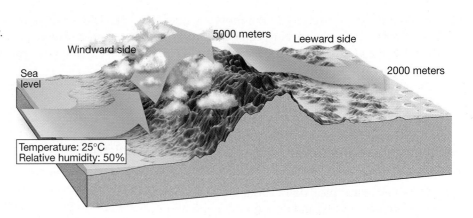

43. From the altitude where condensation begins to occur, to the summit of the mountain, the rising air will continue to expand and will (warm, cool) at the (wet, dry) adiabatic rate of about _____ °C per 100 meters.

44. The temperature of the rising air at the summit of the mountain will be _____ °C.

45. Assuming the air begins to descend on the leeward side of the mountain, it will be compressed and its temperature will (increase, decrease).

46. Assume the relative humidity of the air is below 100% during its entire descent to the plateau. The air will be (saturated, unsaturated) and will warm at the (wet, dry) adiabatic rate of about _____ °C per 100 meters.

47. As the air descends and warms on the leeward side of the mountain, its relative humidity will (increase, decrease).

48. The air's temperature when it reaches the plateau at 2,000 meters will be _____ °C.

49. Explain why mountains might cause dry conditions on their leeward sides.

_____

_____

Global Patterns of Precipitation

Use Figure 14.9 to answer questions 50–54.

50. List at least four areas of the world that receive the greatest average annual (over 160 cm) precipitation.

_____

_____

51. The polar regions of Earth have (high, low) average annual precipitation. Circle your answer.

52. What is the average annual precipitation at your location?

_____ centimeters per year, which is equivalent to _____ inches per year.

53. Describe the pattern of average annual precipitation in North America.

_____

_____

_____

54. The inset in Figure 14.9 illustrates the variability, or reliability, of the precipitation in Africa. After you compare the inset to the annual precipitation map for Africa, summarize the relation between the amount of precipitation an area receives and the variability, or reliability, of that precipitation.

_____

_____

_____

Pressure and Wind

Atmospheric pressure and wind are two elements of weather that are closely interrelated. Although pressure is the element least noticed by people in a weather report, it is pressure differences in the atmosphere that drive the winds that often bring changes in temperature and moisture.

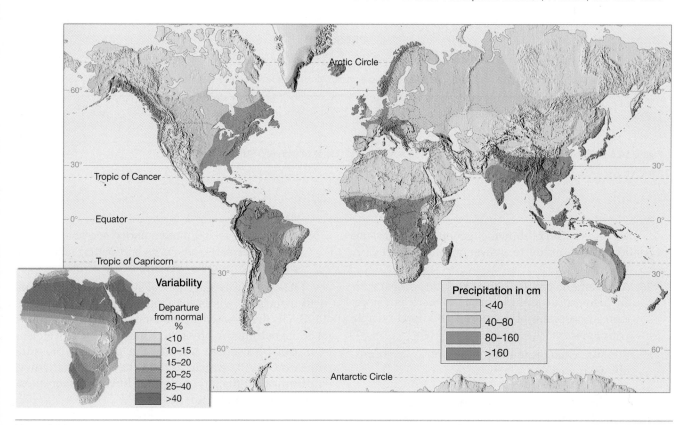

Figure 14.9 Average annual precipitation in centimeters.

Atmospheric Pressure

Atmospheric pressure is the force exerted by the weight of the atmosphere. It varies over the face of Earth, primarily because of temperature differences. Typically, warm air is less dense than cool air and, therefore, exerts less pressure. Cold, dense air is often associated with higher pressures. Pressure also changes with altitude.

The instrument used to determine atmospheric pressure is the **barometer** (Figure 14.10). Two units that can be used to measure air pressure are *inches of mercury* and *millibars*. Inches of mercury refers to the height to which a column of mercury will rise in a glass tube that has been inverted into a reservoir of mercury. The millibar is a unit that measures the actual force of the atmosphere pushing down on a surface. Standard pressure at sea level is 29.92 inches of mercury or 1,013.2 millibars (Figure 14.11). A pressure greater than 29.92 inches or 1,013.2 millibars is called *high pressure*. A pressure less than the standards is called *low pressure*.

55. If a barometer is located in the laboratory, record the current atmospheric pressure in both inches of mercury and millibars. If necessary, use Figure 14.11 to convert the units.

Inches of mercury: _____ inches of mercury

Millibars: _____ millibars (mb)

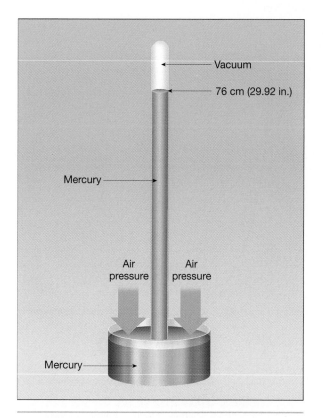

Figure 14.10 Simple mercury barometer. The weight of the column of mercury is balanced by the pressure exerted on the dish of mercury by the air above. If the pressure decreases, the column of mercury falls; if the pressure increases, the column rises.

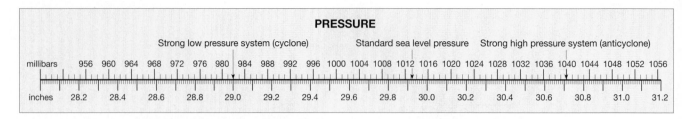

Figure 14.11 Scale for comparing pressure readings in millibars and inches of mercury. (After NOAA)

Use Figure 14.12, showing generalized pressure variations with altitude, to answer questions 56 and 57.

56. Atmospheric pressure (increases, decreases) with an increase in altitude because there is (more, less) atmosphere above to exert a force. Circle your answers.

57. Pressure changes with altitude (most, least) rapidly near Earth's surface.

Since surface elevations vary, barometric readings are adjusted to indicate what the pressure would be if the barometer was located at sea level. This provides a common standard for mapping pressure, regardless of elevation.

58. A city that is 200 meters above sea level would (add, subtract) units to its barometric reading in order to correct its pressure to sea level.

Observe the average global surface pressure maps with associated winds for January and July in Figure 14.13. The lines shown are **isobars**, which connect points of equal barometric pressure, adjusted to sea level. Isobars can be used to identify the principal pressure zones on Earth, which include the **equatorial lows**, **subtropical highs**, and **subpolar lows**. Answer questions 59–65 using Figure 14.13.

59. The units used on the maps to indicate pressure are (inches of mercury, millibars). Circle your answer.

60. By writing the word "HIGH" or "LOW," indicate on the maps the general pressure at each of the following latitudes: 60°N, 30°N, 0°, 30°S, 60°S.

61. Write on the maps the names (equatorial low, subtropical high, or subpolar low) of each of the pressure zones you identified in question 60.

62. During the summer months, January in the Southern Hemisphere and July in the Northern Hemisphere, (high, low) pressure is more common over land. Circle your answer.

63. (High, Low) pressure is most associated with the land in the winter months.

64. Considering what you know about the unequal heating of land and water and the influence of air temperature on pressure, why does the pressure over continents change with the seasons?

_____

_____

65. Why does the air over the oceans maintain a more uniform pressure throughout the year?

_____

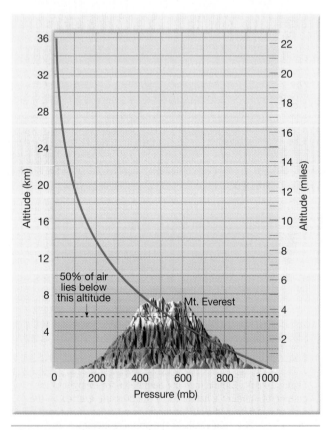

Figure 14.12 Pressure variations with altitude.

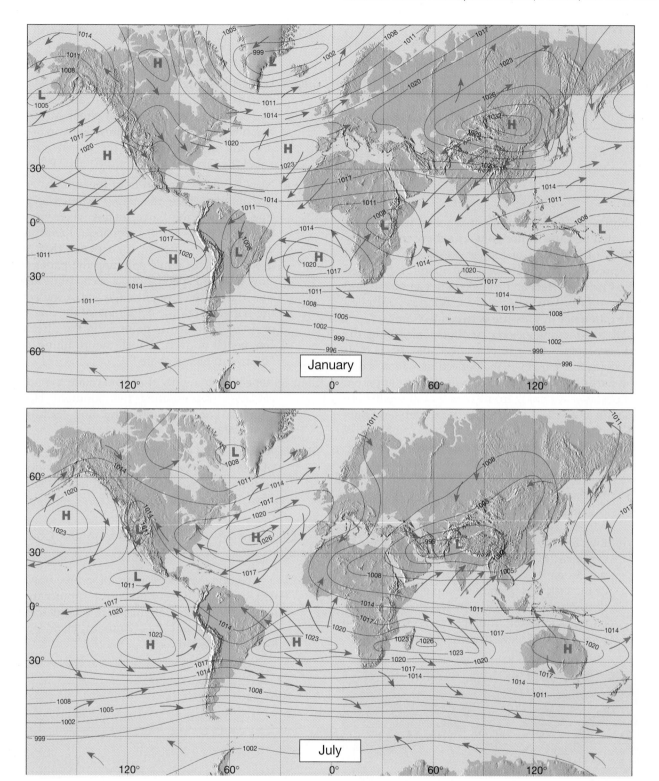

Figure 14.13 Average surface barometric pressure in millibars for January and July with associated winds.

The differential heating and cooling of land and water over most of Earth causes the zones of pressure to be broken into cells of pressure. Pressure cells are shown on a map as a system of closed, concentric isobars. **High pressure cells**, called **anticyclones**, are typically associated with descending (subsiding) air. **Low pressure cells**, called **cyclones**, have rising air in their centers. Locate

and examine some of the pressure cells on the maps in Figure 14.13. Then answer questions 66 and 67.

66. In an anticyclone, the (highest, lowest) pressure occurs at the center of the cell. In a cyclone, the (highest, lowest) pressure occurs at the center. Circle your answers.

67. With which pressure cell, anticyclone or cyclone, would the vertical movement of air be most favorable for cloud formation and precipitation? Explain your answer with reference to the adiabatic process.

_____

_____

Wind

The horizontal movement of air is called **wind**. Wind is initiated because of horizontal pressure differences. Air flows from areas of higher pressure to areas of lower pressure. The direction of the wind is influenced by the **Coriolis effect** and friction between the air and Earth's surface. The velocity of the wind is controlled by the difference in pressure between two areas.

All free-moving objects or fluids, including the wind, experience the Coriolis effect and have their paths deflected. However, the deflection, which is due to the Earth's rotation about an axis, is only apparent as an observer watching from space would see the object's path as a straight line. To better understand how the Coriolis effect influences the motion of objects as they move across Earth's surface, conduct the following experiment by completing the indicated steps.

Step 1: Working in groups of two or more, construct a rotating "table" that represents Earth's surface by first taping a thumbtack upside down on the table top (or inserting it through a piece of cardboard). Next, center a sheet of heavyweight paper or thin cardboard over the point of the tack and push the sharp end of the tack through the paper (Figure 14.14). Turning the paper about the thumbtack represents the rotating Earth viewed from above the pole.

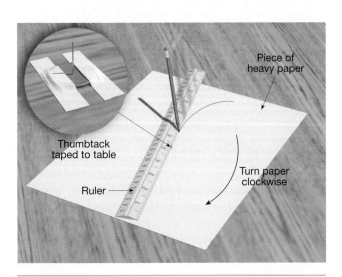

Figure 14.14 Coriolis experiment setup—Southern Hemisphere.

Step 2: Place a straight edge (a 12-inch ruler will do) across the paper, resting it on the sharp point of the thumbtack.

Step 3: Using the straight edge as a guide and beginning at the edge nearest you, draw a straight line across the entire piece of paper. Mark the beginning of the line you drew with an arrow pointing in the direction the pencil moved. Label the line "no rotation."

Step 4: Now have one person hold the straight edge and another turn the paper _counterclockwise_ (the direction the Earth rotates when viewed from above the North Pole) about the thumbtack. Using a different color pencil than you used in Step 3, while spinning the paper at a slow constant speed, draw a second line along the edge of the straight edge across the rotating paper. Mark the beginning of the line you drew with an arrow pointing in the direction the pencil moved. Label the resulting line "Northern Hemisphere."

Step 5: Repeat Step 4, only this time rotate the paper _clockwise_ (representing the Southern Hemisphere). Label the resulting line "Southern Hemisphere."

Step 6: Repeat Step 4 several times, varying the speed of rotation of the paper with each trial. Identify each new line by labeling it either "slow," "fast," or "very fast."

Questions 68 through 70 refer to the Coriolis experiment.

68. Describe the apparent path of a free-moving object over Earth's surface on a "nonrotating" Earth, the Northern Hemisphere (counterclockwise rotation), and the Southern Hemisphere (clockwise rotation).

_____

_____

_____

_____

69. In the Northern Hemisphere, the deflection of objects is to the (right, left), while in the Southern Hemisphere it is to the (right, left). Circle your answers.

70. Summarize your observations of the relation between speed of rotation and magnitude of the Coriolis effect you observed in **Step 6**.

_____

_____

_____

From your observations in **Step 6**, you would suspect that the deflection in the path of a free-moving object would be maximized near the equator where Earth's rotational velocity is greatest. However, it is NOT. Vector motion on the surface of a sphere is complex; however, the Coriolis effect is controlled primarily by rotation about a VERTICAL axis. In your experiment, all points on the paper were rotating about a vertical axis, the thumbtack. However, the flat paper does not represent the "real" spherical Earth. On Earth, the vertical axis of rotation is a line connecting the geographic poles that is only vertical to Earth's surface at each pole. On the Equator, the axis is *parallel* to the surface; therefore, there is no rotation about a vertical axis. As a consequence, the Coriolis effect is strongest at the poles and weakens equatorward, becoming nonexistent at the equator. (To help visualize this changing orientation, obtain a globe and hold a pencil parallel to Earth's axis at the North Pole. Notice the axis is vertical to the surface, just as in the experiment. Now, *while keeping the pencil parallel to Earth's axis of rotation*, slowly move it over the surface to the equator. Notice how the orientation of the pencil changes relative to the surface as it is moved. At the equator, the pencil is now parallel to the surface and rotation about the pencil is directed toward and away from Earth's center, not over the surface.)

71. Considering what you have learned about the Coriolis effect, write a brief statement describing the Coriolis effect on the atmosphere of the planet Venus, about the same size as Earth, but with a period of rotation of 244 Earth days. What about on Jupiter, a planet much larger than Earth, with a 10-hour day?

_____

_____

_____

_____

Examine the global wind pattern (shown with arrows) that is associated with the global pattern of pressure in Figure 14.13. The wind arrows can be used to identify the global wind belts, which include the **trade winds**, **westerlies**, and **polar easterlies**. Then answer questions 72–75.

72. Examine the pressure cells in Figure 14.13. Then, on Figure 14.15, complete the diagrams of the indicated pressure cells for each hemisphere. Label the isobars with appropriate pressures *and* use arrows to indicate the surface air movement in each pressure cell.

73. In the following spaces, indicate the movements of air in high and low pressure cells for each hemisphere. Write one of the two choices given in italics for each blank.

	NORTHERN HEMISPHERE		SOUTHERN HEMISPHERE	
	HIGH	LOW	HIGH	LOW
Surface air moves *into* or *out of:*	___	___	___	___
Surface air will *rise* or *subside* in the center:	___	___	___	___
Surface air motion is *clockwise* or *counterclockwise:*	___	___	___	___

Northern Hemisphere

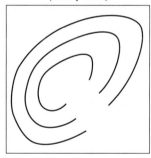
High
(Anticyclone)

Low
(Cyclone)

Southern Hemisphere

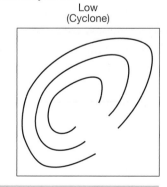
High
(Anticyclone)

Low
(Cyclone)

Figure 14.15 Northern and Southern Hemisphere pressure cells.

Figure 14.16 Global winds (generalized).

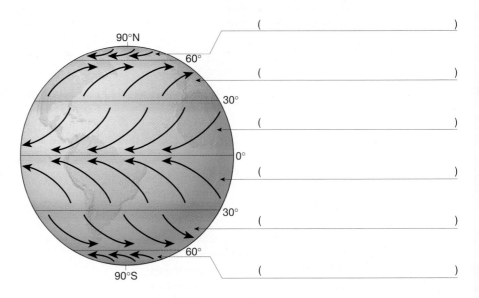

WIND BELT

90°N
60°
30°
0°
30°
60°
90°S

( )
( )
( )
( )
( )
( )

74. Write a brief statement that describes the difference in surface air movement between a Northern Hemisphere and Southern Hemisphere anticyclone.

75. Write the name of each global wind belt (trade winds, westerlies, or polar easterlies) at the appropriate location on Figure 14.16. Also indicate by name each global wind belt on the world maps in Figure 14.13.

In Figure 14.13, notice the seasonal changes in wind direction, called **monsoons**, over continents.

76. During the (summer, winter) season, the air is moving from the continent to the ocean. Circle your answer.

77. During the (summer, winter) season, the air is moving from the ocean to the continent.

78. In what way is the seasonal change in wind direction over continents related to pressure?

79. What effect will the seasonal reversal of wind have on moisture in the air and the potential for precipitation over the continents during the following seasons?

Summer season: _____

Winter season: _____

By examining Figure 14.13, you should notice that the systems of pressure belts and global winds change latitude with the seasons.

80. In what manner is the seasonal shift in pressure belts and global winds related to the movement of the overhead noon Sun throughout the year?

Atmospheric Moisture, Pressure, and Wind on the Internet

Using what you have learned in this exercise, investigate the current weather conditions at your location and in North America by completing the corresponding online activity on the *Applications & Investigations in Earth Science* website at http://prenhall.com/earthsciencelab

Atmospheric Moisture, Pressure, and Wind

Date Due: _____

Name: _____

Date: _____

Class: _____

After you have finished Exercise 14, complete the following questions. You may have to refer to the exercise for assistance or to locate specific answers. Be prepared to submit this summary/report to your instructor at the designated time.

1. Answer the following by circling the correct response.

   a. Liquid water changes to water vapor by the process called (condensation, evaporation, deposition).

   b. (Warm, Cold) air has the greatest saturation mixing ratio.

   c. Lowering the air temperature will (increase, decrease) the relative humidity.

   d. At the dew-point temperature, the relative humidity is (25%, 50%, 75%, 100%).

   e. When condensation occurs, heat is (absorbed, released) by water vapor.

   f. Rising air (warms, cools) by (expansion, compression).

   g. In the early morning hours when the daily air temperature is often coolest, relative humidity is generally at its (lowest, highest).

2. What is the dew-point temperature of a kilogram of air when a psychrometer measures an 8°C dry-bulb temperature and a 6°C wet-bulb reading?

   Dew-point temperature = _____ °C

3. Explain the principle that governs the operation of a psychrometer for determining relative humidity.

   _____

   _____

   _____

4. Describe the adiabatic process and how it is responsible for causing condensation in the atmosphere.

   _____

   _____

   _____

5. In the section of the exercise "Adiabatic Processes," question 42, what was the altitude where condensation occurred as the air was rising over the mountain?

   _____ meters

6. Place each of the following statements in proper sequence (1 for the first) leading to the development of clouds.

   _____: dew-point temperature reached

   _____: air begins to rise

   _____: condensation occurs

   _____: adiabatic cooling

7. Assume a parcel of air on the surface with a temperature of 29°C and a relative humidity of 50 percent. If the parcel rises, at what altitude should clouds form?

   _____

   _____

8. Describe the Coriolis effect and its influence in both the Northern and Southern Hemispheres.

   _____

   _____

   _____

PRESSURE ZONE

WIND BELT

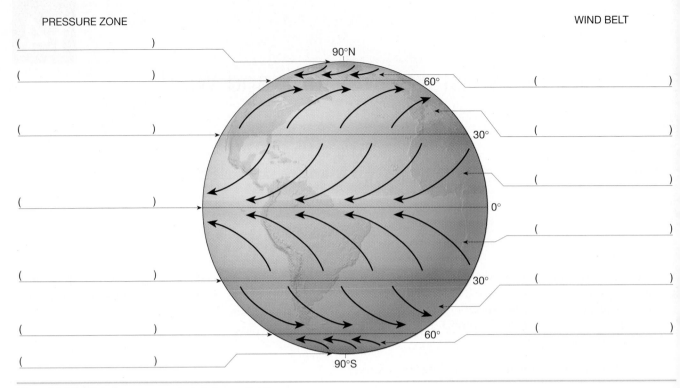

Figure 14.17   Global pressures and winds (generalized).

9. Refer to Figure 14.17. Use the indicated parallels of latitude as a guide to write the names of the global pressure zones and wind belts at their proper locations.

10. On Figure 14.18, illustrate, by labeling the isobars, a typical Northern Hemisphere cyclone and anticyclone. Use arrows to indicate the movement of surface air associated with each pressure cell.

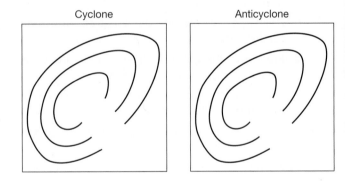

Figure 14.18   Pressures and winds in a typical Northern Hemisphere cyclone and anticyclone.

Air Masses, the Middle-Latitude Cyclone, and Weather Maps

For many people living in the middle latitudes, weather patterns are the result of the movements of large bodies of air and the associated interactions among the weather elements. Of particular importance are the boundaries between contrasting bodies of air, which are often associated with precipitation followed by a change in weather.

Exercise 15 investigates those atmospheric phenomena that most often influence our day-to-day weather—air masses, fronts, and traveling middle-latitude cyclones. Using the standard techniques for plotting weather station data, the exercise concludes with the preparation and analysis of a typical December surface weather map.

Objectives

After you have completed this exercise, you should be able to:

1. Discuss the characteristics, movements, and source regions of North American air masses.
2. Define and draw a profile of a typical warm front.
3. Define and draw a profile of a typical cold front.
4. Diagram and label all parts of an idealized, mature, middle-latitude cyclone.
5. Interpret the data presented on a surface weather map.
6. Prepare a simple surface weather map using standard techniques.
7. Use a surface weather map to forecast the weather for a city.

Materials

colored pencils

Materials Supplied by Your Instructor

United States map or atlas

Terms

air mass	occluded front	anticyclone
source region	adiabatic	middle-latitude
front	cooling	cyclone
warm front	polar front	wave cyclones
cold front	instability	

Air Masses

An **air mass** is a large body of air that has relatively uniform temperature and moisture characteristics. The area where an air mass acquires its traits is called a **source region**. For example, air with a source region over cool ocean water tends to become cool and moist, while air that stagnates over the American Southwest in summer becomes hot and dry.

Air masses are set into motion by passing high and low pressure cells. When the air mass moves out of its source region, its temperature and moisture conditions are carried with it.

1. Air masses are classified according to their source region: land vs. water and latitude of origin. Explain the meaning of each of the following air mass classification letters.

   c: _____   P: _____

   m: _____   T: _____

Figure 15.1 shows the source regions and directions of movement of the air masses that play an important role in the weather of North America. Use the figure to answer questions 2–8.

Figure 15.1 Source regions of North American air masses.

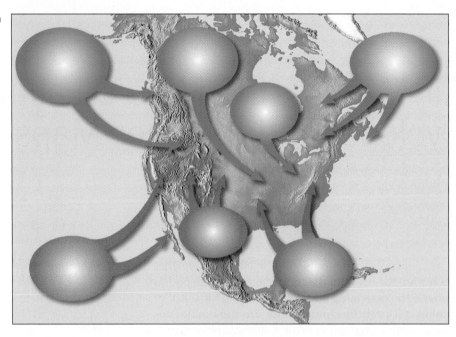

2. Label each of the following North American air masses on Figure 15.1. Then list the location of the source region of each in the following space.

SOURCE REGION

cP: _____

cT: _____

mP: _____

mT: _____

3. What would be the typical winter temperature and moisture characteristics of each of the following air masses?

TEMPERATURE   **MOISTURE**

cP: _____   _____

mP: _____   _____

mT: _____   _____

Notice the paths of air masses indicated by the arrows on Figure 15.1.

4. The general movement of air masses across North America is (east to west, west to east). Circle your answer.

5. How does the movement of air masses across North America correspond to the global flow of wind over the continent?

_____

6. Which air masses would have the greatest influence on the weather east of the Rocky Mountains?

_____

7. A (cP, mT) air mass would supply the greatest amount of moisture east of the Rocky Mountains. Circle your answer.

8. A (cP, mP) air mass has the greatest influence on the weather along the northwest Pacific coast. Circle your answer.

Fronts

A **front** is a surface of contact between air masses of different densities. One air mass is often warmer, less dense, and higher in moisture content than the other. There is little mixing of air across a front, and each air mass retains its basic characteristics. A **warm front**, shown on a weather map by the symbol ⬤⬤⬤⬤⬤ , occurs where warm air occupies an area formerly covered by cooler air. A **cold front**, indicated on a map with the symbol ▲▲▲▲▲ , forms when cold air actively advances into a region occupied by warmer air. An **occluded front**, shown on a weather map with the symbol ⬤▲⬤▲ , develops when a cold front overtakes a warm front and warm air is wedged above cold surface air.

Fronts typically act as barriers or walls over which air must rise. When it rises, air will expand and experience **adiabatic cooling**. As a consequence, clouds and precipitation often occur along fronts.

9. On Figure 15.1, draw a line where air masses are likely to collide and fronts develop. Where does this boundary occur?

_____

10. In the central United States, east of the Rocky Mountains, a (cP, mT) air mass will most likely be

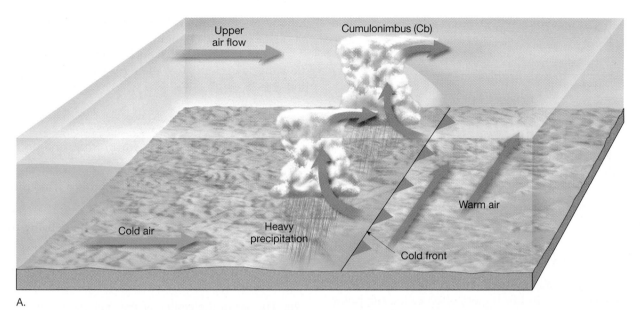

A.

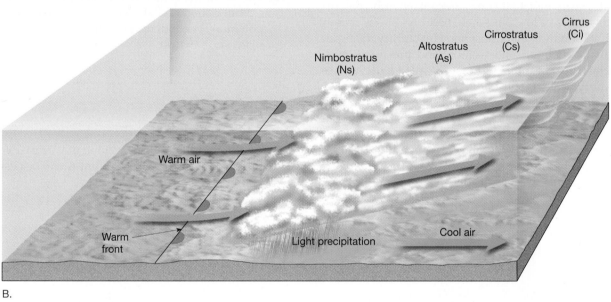

B.

Figure 15.2 **A.** Typical cold front profile. **B.** Typical warm-front profile.

found north of a front, and a (cP, mT) mass to the south. Circle your answers.

Figure 15.2 illustrates profiles through typical cold and warm fronts. Observe the profiles closely and then answer questions 11–17.

11. Along the (cold, warm) front, the cold air is the aggressive or "pushing" air. Circle your answer.

12. Along the (cold, warm) front, the warm air rises at the steepest angle.

13. Along which front are extensive areas of stratus clouds and periods of prolonged precipitation most probable? Explain why you expect longer

periods of precipitation to be associated with this type of front.

_____ front. Explanation: _____

_____

14. Assume that the fronts are moving from left to right in Figure 15.2. A drop in temperature is most likely to occur with the passing of a (cold, warm) front. Circle your answer.

The air following a cold front is frequently cold, dense, and subsiding.

15. (Clear, Cloudy) conditions are most likely to prevail after a cold front passes. Explain the reason

for your choice with reference to the adiabatic process.

_____

_____

16. Clouds of vertical development and perhaps thunderstorms are most likely to occur along a (cold, warm) front.

17. As a (cold, warm) front approaches, clouds become lower, thicker, and cover more of the sky.

Middle-Latitude Cyclone

Contrasting air masses frequently collide in the area of the _subpolar lows._ In this region, often called the **polar front**, warm, moist air comes in contact with cool, dry air in an area of low pressure. These conditions present an ideal situation for atmospheric **instability**, rising air, adiabatic cooling, condensation, and precipitation. In contrast, in areas of high pressure, called **anticyclones**, the air typically is subsiding.

In the Northern Hemisphere, the westerly winds to the south of the polar front and the easterly winds to the north cause a wave with counterclockwise (cyclonic) rotation to form along the frontal surface. As the low-pressure system called a **middle-latitude** (or **wave**) **cyclone** evolves, it follows a general eastward

path across the United States, bringing a sequence of passing fronts and changing weather.

Figure 15.3 illustrates an idealized, mature, middle-latitude cyclone. Use the figure to complete questions 18–28.

18. On Figure 15.3:

   a. Label the cold front, warm front, and occluded front.

   b. Draw arrows showing the surface wind directions at points A, C, E, F, and G.

   c. Label the sectors most likely experiencing precipitation with the word "precipitation."

19. The surface winds in the cyclone are (converging, diverging). Circle your answer.

20. The air in the center of the cyclone will be (subsiding, rising). What effect will this have on the potential for condensation and precipitation? Explain your answer.

   _____

   _____

21. As the middle-latitude cyclone moves eastward, the barometric pressure at point A will be (rising, falling). Circle your answer.

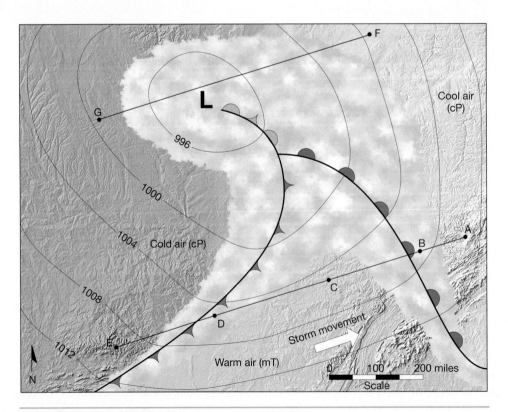

Figure 15.3 Mature, middle-latitude cyclone (idealized).

22. After the warm front passes, the wind at point B will be from the (south, north).

23. Describe the changes in wind direction and barometric pressure that will likely occur at point D after the cold front passes.

_____

_____

24. Considering the typical air mass types and their locations in a middle-latitude cyclone, the amount of water vapor in the air will most likely (increase, decrease) at A after the warm front passes.

25. The quantity of moisture in the air at point D will most likely (increase, decrease) after the cold front passes.

26. Use Figure 15.3 to describe the sequence of weather conditions—barometric changes, wind directions, humidity, precipitation, etc.—expected for a city as the cyclone moves and the city's relative position changes from location A to B, and then to C, D, and E.

_____

_____

_____

_____

27. Near the center of the low, a/an (warm, cold, occluded) front has formed where the cold front has overtaken the warm front. Circle your answer. Then answer questions 27a and 27b.

    a. What happens to the warm mT air in this type of front?

    _____

    b. With reference to the adiabatic process, why is there a good chance for precipitation with this type of front?

    _____

    _____

    After the entire wave cyclone passes, pressure will rise.

28. Describe the general weather often associated with a high pressure cell, called an anticyclone.

    _____

    _____

As mentioned previously, middle-latitude cyclones form in the belt of subpolar lows. After you have reviewed the subpolar lows in Exercise 14, answer the following question by circling the correct responses.

29. During the (summer, winter) season the belt of subpolar lows and the polar front are farthest south in North America, and the central United States will experience a (greater, lesser) frequency of passing middle-latitude cyclones.

Weather Station Analysis and Forecasting

In order to understand, analyze, and predict the weather, observers at hundreds of weather stations throughout the United States collect and record weather data several times a day. This information is forwarded to offices of the National Weather Service where it and satellite data are computer processed and mapped. Weather maps, containing data from throughout the country, are then distributed to any interested individual or agency.

Weather Station Data

To manage the great quantity of information necessary for accurate maps, meteorologists have developed a system for coding weather data. Figure 15.4 illustrates the system and many of the symbols that are used to record data for a weather station. (_Note:_ When plotting barometric pressure in millibars for a weather station, to conserve space, the initial number 9 or 10 is omitted and the last digit is tenths of a millibar. For example, on a map a barometric pressure of 216 for a station would be read as 1021.6 mb.)

Figure 15.5 is a coded weather station, shown as it would appear on a simplified surface weather map.

30. Using the specimen station model and explanations shown in Figure 15.4 as your guide, interpret the weather conditions reported at the station illustrated in Figure 15.5.

    Percent of sky cover: _____ %

    Wind direction: _____

    Wind speed: _____ mph

    Temperature: _____ °F

    Dew-point temperature: _____ °F

    Barometric pressure: _____ millibars

    Barometric change in past 3 hours: _____ mb

    Weather during the past 6 hours: _____

31. Encode and plot the weather conditions for the following weather station on the station symbol

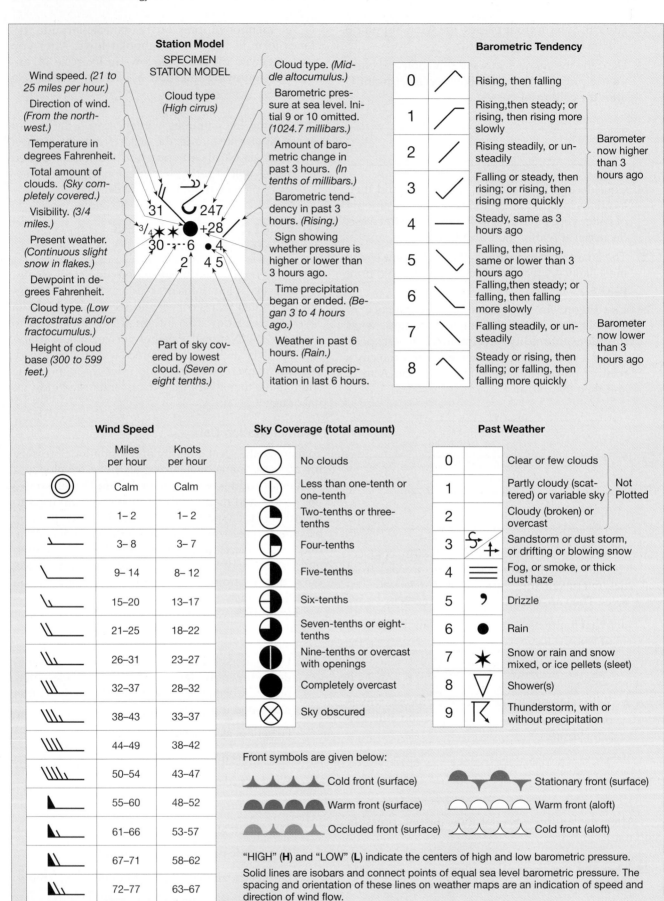

Station Model

SPECIMEN STATION MODEL

Cloud type (High cirrus)

Wind speed. (21 to 25 miles per hour.)

Direction of wind. (From the north-west.)

Temperature in degrees Fahrenheit.

Total amount of clouds. (Sky completely covered.)

Visibility. (3/4 miles.)

Present weather. (Continuous slight snow in flakes.)

Dewpoint in degrees Fahrenheit.

Cloud type. (Low fractostratus and/or fractocumulus.)

Height of cloud base (300 to 599 feet.)

Cloud type. (Middle altocumulus.)

Barometric pressure at sea level. Initial 9 or 10 omitted. (1024.7 millibars.)

Amount of barometric change in past 3 hours. (In tenths of millibars.)

Barometric tendency in past 3 hours. (Rising.)

Sign showing whether pressure is higher or lower than 3 hours ago.

Time precipitation began or ended. (Began 3 to 4 hours ago.)

Weather in past 6 hours. (Rain.)

Amount of precipitation in last 6 hours.

Part of sky covered by lowest cloud. (Seven or eight tenths.)

Barometric Tendency

0	/	Rising, then falling
1	/	Rising, then steady; or rising, then rising more slowly
2	/	Rising steadily, or unsteadily
3	✓	Falling or steady, then rising; or rising, then rising more quickly
4	—	Steady, same as 3 hours ago
5	\	Falling, then rising, same or lower than 3 hours ago
6	\	Falling, then steady; or falling, then falling more slowly
7	\	Falling steadily, or unsteadily
8	\	Steady or rising, then falling; or falling, then falling more quickly

Barometer now higher than 3 hours ago

Barometer now lower than 3 hours ago

Wind Speed

	Miles per hour	Knots per hour
◎	Calm	Calm
—	1–2	1–2
	3–8	3–7
	9–14	8–12
	15–20	13–17
	21–25	18–22
	26–31	23–27
	32–37	28–32
	38–43	33–37
	44–49	38–42
	50–54	43–47
	55–60	48–52
	61–66	53–57
	67–71	58–62
	72–77	63–67

Sky Coverage (total amount)

○	No clouds
◐	Less than one-tenth or one-tenth
◔	Two-tenths or three-tenths
◑	Four-tenths
◐	Five-tenths
◕	Six-tenths
◕	Seven-tenths or eight-tenths
◕	Nine-tenths or overcast with openings
●	Completely overcast
⊗	Sky obscured

Past Weather

0		Clear or few clouds
1		Partly cloudy (scattered) or variable sky
2		Cloudy (broken) or overcast
3		Sandstorm or dust storm, or drifting or blowing snow
4	≡	Fog, or smoke, or thick dust haze
5	,	Drizzle
6	●	Rain
7	✳	Snow or rain and snow mixed, or ice pellets (sleet)
8	▽	Shower(s)
9	⎡⟨	Thunderstorm, with or without precipitation

(0, 1, 2: Not Plotted)

Front symbols are given below:

Cold front (surface)

Warm front (surface)

Occluded front (surface)

Stationary front (surface)

Warm front (aloft)

Cold front (aloft)

"HIGH" (H) and "LOW" (L) indicate the centers of high and low barometric pressure.

Solid lines are isobars and connect points of equal sea level barometric pressure. The spacing and orientation of these lines on weather maps are an indication of speed and direction of wind flow.

Figure 15.4 Specimen weather station model and standard symbols. (Source: Daily Weather Maps, U.S. Department of Commerce)

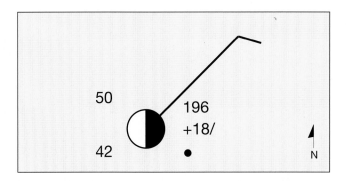

Figure 15.5 Coded weather station (abbreviated).

shown below. Use Figures 15.4 and 15.5 as your guides.

The sky is six-tenths covered by clouds. Air temperature is 82°F with a dew point of 50°F. Wind is from the south at 22 miles per hour. Barometric pressure is 1022.4 millibars and has fallen from 1023.0 millibars during the past three hours. There has been no precipitation during the past six hours.

Preparing a Weather Map and Forecast

Table 15.1 contains weather data for several cities in the central and eastern United States on a December day. Data for several of the cities have been plotted on the map, Figure 15.6. Use Table 15.1 and Figure 15.6 to complete questions 32 and 33.

32. Refer to stations where the data have already been plotted and the station model in Figure 15.4. Plot the data for the remaining stations on the map. Then complete the following steps.

Step 1: Beginning with 988 millibars, draw isobars as accurately as possible at four millibar intervals

(992 mb, 996 mb, 1000 mb, etc.). (*Note:* You will have to estimate pressures between cities to determine the location of isobars. Also, it may be a good idea to first sketch the isobars lightly in pencil.)

Step 2: By observing the data plotted on the map, determine the locations of the cold and warm fronts. Using the proper symbols, label the cold and warm fronts as accurately as possible on the map.

Step 3: Label the air mass types that are most likely to be located to the northwest and to the southeast of the cold front.

Step 4: Indicate areas of precipitation by lightly shading the map with a pencil.

33. Assume that the middle-latitude cyclone illustrated on the map was centered in Oklahoma the previous day and is moving northeastward. Indicate your forecast (temperature, wind direction, probability of precipitation, cloud cover, and barometric pressure) for the next 12–24 hours at the following locations.

Chattanooga, TN: _____

_____

_____

Little Rock, AR: _____

_____

_____

Washington, D.C.: _____

_____

_____

Raleigh, NC: _____

_____

_____

Weather Maps on the Internet

Apply the concepts from this exercise to an analysis of the current weather patterns in North America by completing the corresponding online activity on the *Applications & Investigations in Earth Science* website at http://prenhall.com/earthsciencelab

Table 15.1 December surface weather data for selected cities in the central and eastern United States

STATION	% CLOUD COVER	WIND DIRECTION	WIND SPEED (MPH)	TEMP.	DEW PT. TEMP.	PRESSURE (MB)	PRESSURE 3 HOUR + OR −	PRECIP.
Atlanta, GA	20	SE	18	61	56	1000.7	−2.0	
Birmingham, AL	80	SW	15	70	64	998.1	−1.4	
Charleston, SC	10	SE	12	63	58	1008.0	−2.0	
Charlotte, NC	70	SW	14	54	49	1003.4	−4.4	Drizzle
Chattanooga, TN	60	SW	12	66	60	995.3	−2.5	Drizzle
Chicago, IL	100	NE	13	34	21	1003.2	−2.2	
Columbus, OH	100	E	10	34	28	996.8	−5.8	Snow
Evansville, IN	100	NW	7	45	43	987.2	−2.7	Snow
Fort Worth, TX	0	NW	5	46	43	1002.6	+1.4	
Indianapolis, IN	100	NE	30	34	32	993.2	−5.6	Snow
Jackson, MS	40	SW	10	72	67	1001.3	+1.2	Thunderstorm
Kansas City, MO	30	N	18	30	27	1005.5	+1.7	
Little Rock, AR	0	NW	10	46	43	1001.5	+2.7	
Louisville, KY	100	E	12	34	34	993.0	−4.2	Snow
Memphis, TN	80	NW	12	50	45	996.7	+5.8	
Mobile, AL	60	SW	10	72	68	1004.3	−0.3	
Nashville, TN	100	SW	18	56	55	991.5	−0.1	Rain
New Orleans, LA	20	SW	11	75	70	1003.9	−0.1	
New York, NY	100	NE	23	36	18	1016.9	−2.1	
North Platte, NB	30	N	9	9	1	1017.4	+3.2	
Oklahoma City, OK	10	NW	13	41	37	1005.9	+1.5	
Richmond, VA	100	E	10	45	45	1010.9	−2.4	Rain
Roanoke, VA	100	SE	10	39	39	1007.5	−3.6	Snow
Savannah, GA	30	SE	7	61	55	1007.6	−2.0	
Shreveport, LA	0	NW	8	46	43	1002.6	+1.4	
St. Louis, MO	100	NW	10	32	32	999.7	+1.7	Showers

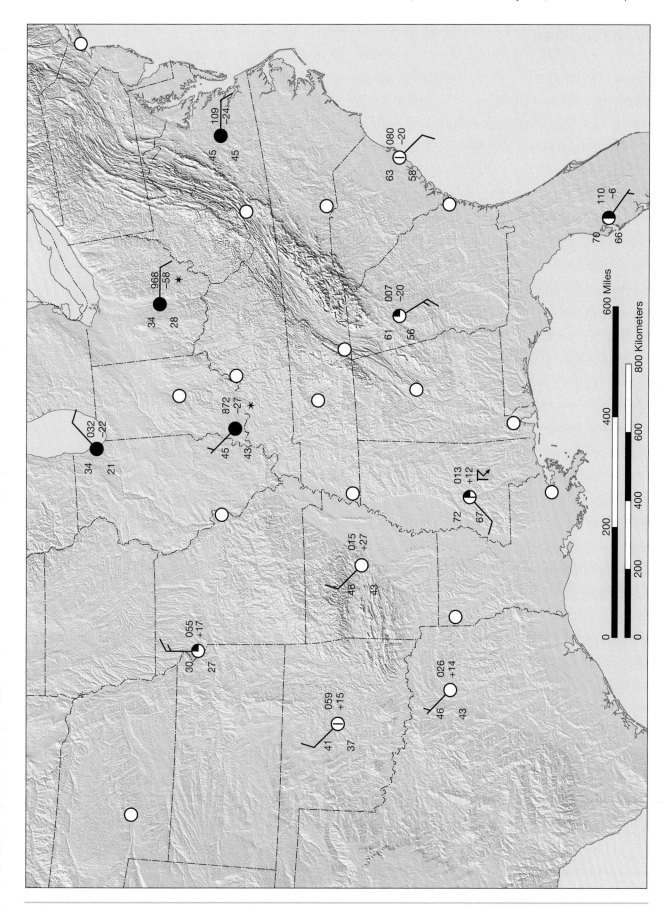

Figure 15.6 Map of December weather data for selected cities in the central and eastern United States.

Notes and calculations.

Air Masses, the Middle-Latitude Cyclone, and Weather Maps

Date Due: _____

Name: _____

Date: _____

Class: _____

After you have finished Exercise 15, complete the following questions. You may have to refer to the exercise for assistance or to locate specific answers. Be prepared to submit this summary/report to your instructor at the designated time.

1. List the source region(s) and winter temperature/moisture characteristics of each of the following North American air masses.

   cP: _____

   mT: _____

   mP: _____

2. In the following space, diagram a profile (side view) of an idealized cold front. Label the cold air, warm air, and sketch the probable cloud type at the appropriate location. Draw an arrow on the diagram showing the direction of movement of the front.

   Cold Front Profile

3. Indicate the type of front (cold, warm, or occluded) best described by each of the following statements.

   a. Steep wall of cold air: _____

   b. Warm air replaces cool air: _____

   c. Thunderstorms: _____

   d. Drop in temperature: _____

   e. After passing, wind comes from the south: _____

   f. Narrow belt of precipitation: _____

   g. Cold front overtakes warm front: _____

   h. Gradual rise of warm air over cool air: _____

4. Describe the sequence of weather events that a city would experience as an idealized, mature, middle-latitude cyclone that has not developed an occluded front passes over it. Assume that the center of the wave cyclone passes, west to east, 150 miles to the north of the city. Using a diagram may be helpful.

   _____

   _____

   _____

   _____

5. Refer to Figure 15.7. Draw a sketch of the December weather map that you prepared at the end of the exercise, question 32. Show isobars and wind direction arrows. Indicate and label the fronts.

6. Based on the December weather map that you constructed at the end of the exercise, from question 33, what was your forecast for Little Rock, AR?

_____

_____

7. Write a brief analysis of the weather map in Figure 15.8.

_____

_____

_____

_____

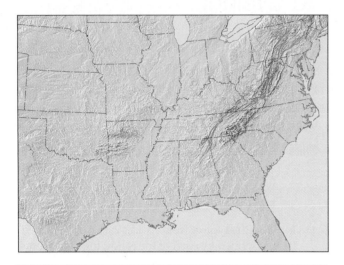

Figure 15.7 Sketch of the December weather map.

Figure 15.8 Surface weather map for a March day with a satellite image showing the cloud patterns on that day. (Courtesy of NOAA/Seattle)

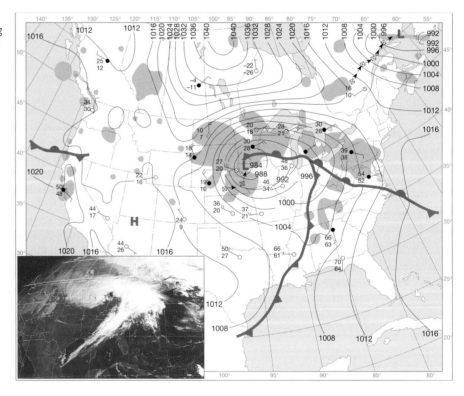

Global Climates

No investigation of the atmosphere is complete without examining the global distribution of the major atmospheric elements. To help understand this worldwide diversity, scientists have devised a variety of classification systems that simplify and describe the general weather conditions that occur at various places on Earth.

Exercise 16 investigates world climates using the system of climate classification devised by Wladimir Köppen (1846–1940). Climographs for several climatic types will be prepared and the global distribution of climates will be examined (Figure 16.1).

Objectives

After you have completed this exercise, you should be able to:

1. Understand the nature of classification systems.
2. Read and prepare a climograph.
3. List the criteria used to define each principal climatic type.
4. Describe the general location of each principal climate group.

Materials

ruler
calculator

Materials Supplied by Your Instructor

world map or atlas

Terms

climate   Köppen system
climatology   climograph

Climate

Climate may be defined as the synthesis, or summary, of weather conditions at a particular place over a long period of time. Climatic classification simplifies the complex distribution of the weather elements for analysis and explanation.

Climatology involves grouping those areas that have similar weather characteristics. Temperature and precipitation are the two elements most commonly used in climate classifications. However, other methods, using different criteria, have also been developed. It should be remembered that any classification is artificial, and its value depends upon the intended use.

Examining Global Temperatures

Temperature is one of the most essential elements in any climate classification. Completing questions 1–3 using the temperature data in Table 16.1 will help you gain a better understanding of the relationship between location and temperature.

1. Stations 1–3 in Table 16.1 are representative of three North American cities at approximately 40° North latitude. Choose which station represents each of the following locations, and give the reason for your selection.

   Interior of the continent? _____

   _____

   West coast of the continent? _____

   _____

   East coast of the continent? _____

   _____

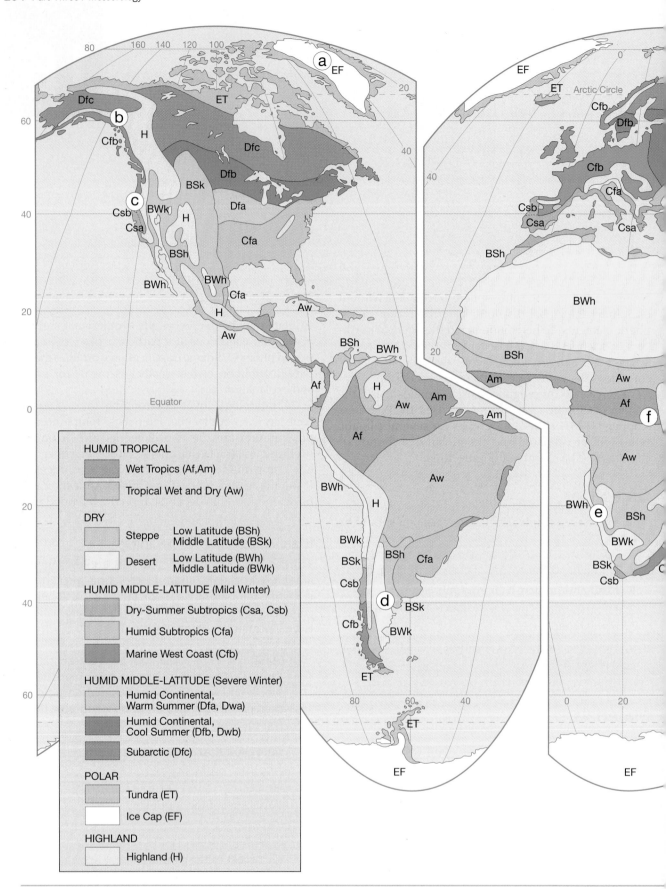

Figure 16.1 Climates of the world (Köppen). (Adapted from E. Willard Miller, *Physical Geography*, Columbus, Ohio, Macmillan/Merrill, 1985. Plate 2)

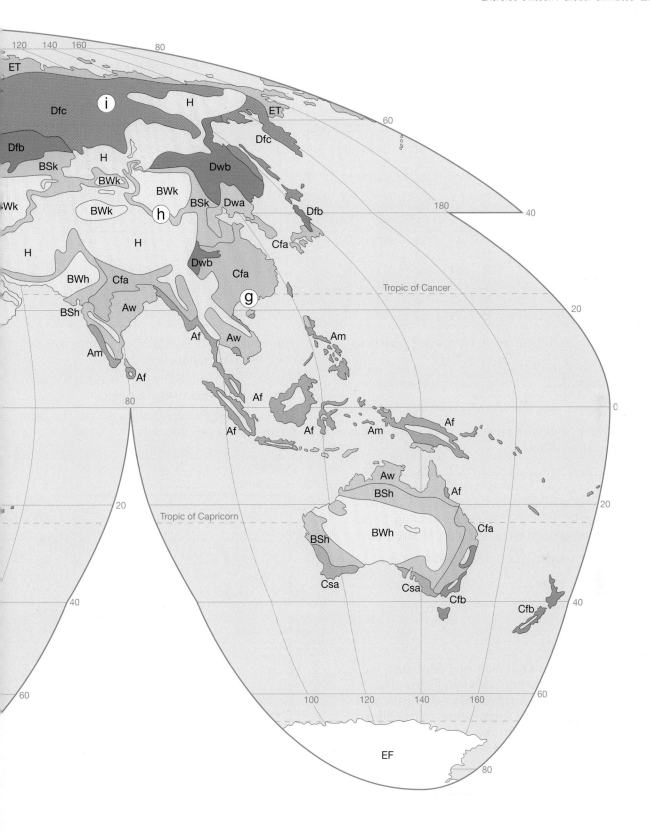

Table 16.1   Annual temperature data (°F)

STATION	J	F	M	A	M	J	J	A	S	O	N	D	ANNUAL MEAN
1	25	28	36	48	61	72	75	74	66	55	39	28	50
2	48	48	49	50	54	55	57	57	57	54	52	49	52
3	31	31	38	49	60	69	74	73	69	59	44	35	52
4	25	26	35	45	58	70	75	72	65	55	45	35	50
5	20	34	38	46	51	54	60	53	49	40	32	25	40
6	42	48	50	55	56	58	59	60	59	56	48	46	51
7	76	76	76	76	78	78	77	77	76	75	74	74	76
8	90	88	84	79	77	73	70	71	76	79	82	87	80
9	−2	11	25	36	48	54	70	60	50	36	18	10	34
10	59	58	59	59	60	58	60	59	60	58	59	60	59
11	70	69	69	66	61	57	56	60	66	70	71	70	66
12	−46	−35	−10	16	41	59	66	60	42	16	−21	−41	12
13	22	25	37	51	62	72	77	75	66	54	39	27	50
14	82	82	82	83	83	82	82	83	83	83	82	83	83
15	27	30	34	41	48	54	57	55	50	43	36	31	42
16	40	44	48	54	61	69	77	76	67	57	47	41	57

2. Selecting from Stations 4–10, answer questions 2a–c.

   a. Station _____ is in the Southern Hemisphere.

   b. Station _____ must be a high-altitude location.

   c. Which station(s) must be quite close to the equator? Why?

   _____

3. Match each of Stations 11–16 with its most likely location by selecting the corresponding lower-case letter (a–i) on the world map, Figure 16.1.

   Station 11: letter _____   Station 14: letter _____

   Station 12: letter _____   Station 15: letter _____

   Station 13: letter _____   Station 16: letter _____

Using a Climograph

Temperature and precipitation are presented on a **climograph** such as the one shown in Figure 16.2. Average monthly temperatures are connected with a single line and read from the temperature scale on the left axis. Average precipitation for each month is represented with a bar or line and read from the precipitation scale on the right axis.

   Refer to the climograph, Figure 16.2, to answer questions 4–9. Circle the correct response.

4. At the place represented by the climograph, the month of (May, June, January) receives the greatest amount of precipitation.

5. The lowest temperature occurs during the month of (July, December, January).

6. The approximate average annual temperature is (0, 10, 20)°C.

7. The total annual precipitation is approximately (240, 480, 720) millimeters.

8. The place represented by the climograph is in the (Northern, Southern) Hemisphere.

9. The (summer, winter) months receive the greatest amount of precipitation.

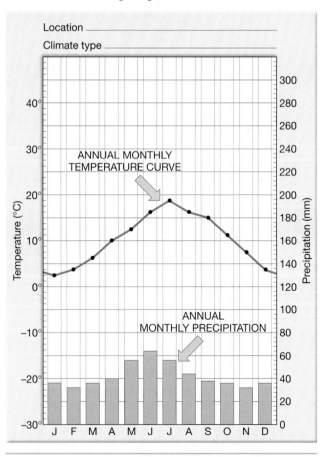

Figure 16.2   Typical climograph. Letters along the bottom margin represent the months. Average monthly temperatures (°C) are plotted with a single line using the scale on the left axis. Precipitation for each month (in mm) is plotted with a bar using the right axis.

The Nature of Classification

Classification by its nature is an artificial endeavor designed to simplify a large amount of data into manageable units. To accomplish this, decisions must be made as to which criteria and limits best serve the purpose of the classification. Completing questions 10–12 using the climographs in Figure 16.3 will give you some insight into the nature of conceiving a classification scheme.

10. Working in groups of 4 or 5, develop the best possible classification scheme for the stations represented by the climographs in Figure 16.3 by arranging them into groups with similar characteristics. When you have finished, describe your classification system, listing the criteria you established. (*Note*: Making a copy of Figure 16.3 and cutting out individual stations may make it easier to visualize different classification arrangements.)

   _____

   _____

   _____

   _____

   _____

   _____

11. Why is the classification scheme you presented in question 10 better than other possible systems you considered?

    _____

    _____

    _____

    _____

    _____

12. Compare the classification system your group devised with the systems developed by two other groups. Which is the best classification scheme? Why?

    _____

    _____

    _____

    _____

    _____

    _____

    _____

    _____

Köppen System of Climatic Classification

Table 16.2 presents the climatic classification system devised by Wladimir Köppen. Since its introduction, the **Köppen system**, with some modification, has become the best-known and most-used classification for presenting the general world pattern of climates.

The Köppen system of climatic classification employs five principal climate groups. Four of the groups are defined on the basis of temperature characteristics and the fifth has precipitation as its primary criterion. Further division of the groups into climatic types allows for a more detailed climatic description. Köppen believed that the distribution of natural vegetation was the best expression of the totality of climate. Therefore, the boundaries he chose were based largely on the limits of certain plant associations. Before you proceed, examine Table 16.2 closely.

13. On Table 16.3, list the names and general characteristics of each principal climate group next to its designated classification letter. Use Table 16.2 as a reference.

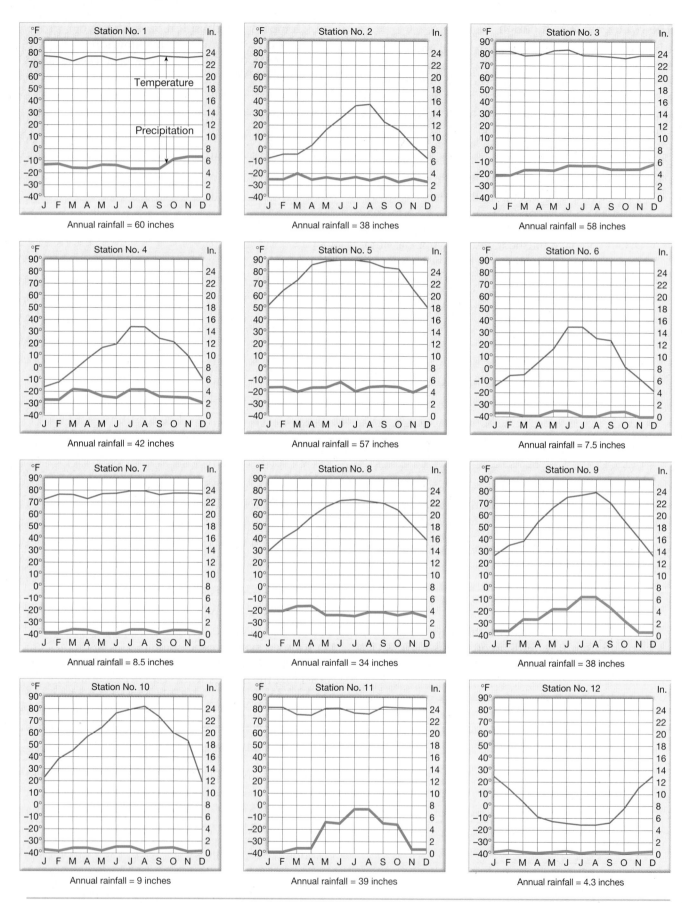

Figure 16.3 Generalized climographs.

Table 16.2 Köppen system of climatic classification

NAME	LETTER SYMBOL			CHARACTERISTICS
	1ST	**2ND**	**3RD**	
	A			Average temperature of the coldest month is 18°C or higher
		f		Every month has 6 cm of precipitation or more
Humid Tropical (type A) Climates		m		Short dry season; precipitation in driest month less than 6 cm but equal to or greater than $10 - R/25$ (R is the annual rainfall in cm)
		w		Well-defined winter dry season; precipitation in driest month less than $10 - R/25$
		s		Well-defined summer dry season (rare)
	B			Potential evaporation exceeds precipitation. The dry–humid boundary is defined by the following formulas: (NOTE: R is average annual precipitation in cm and T is average annual temperature in °C) $R < 2T+28$ when 70% or more of rain falls in warmer 6 months $R < 2T$ when 70% or more of rain falls in cooler 6 months $R < 2T+14$ when neither half year has 70% or more of rain
Dry (type B) Climates		S		Steppe ⎤ The BS–BW boundary is 1/2 the dry–humid boundary
		W		Desert ⎦
			h	Average annual temperature is 18°C or greater
			k	Average annual temperature is less than 18°C
	C			Average temperature of the coldest month is under 18°C and above −3°C
		w		At least ten times as much precipitation in a summer month as in the driest winter month
Humid Middle-latitude with Mild Winters (type C) Climates		s		At least three times as much precipitation in a winter month as in the driest summer month; precipitation in driest summer month less than 4 cm
		f		Criteria for w and s cannot be met
			a	Warmest month is over 22°C; at least 4 months over 10°C
			b	No month above 22°C; at least 4 months over 10°C
			c	One to 3 months above 10°C
	D			Average temperature of coldest month is −3°C or below; average temperature of warmest month is greater than 10°C
		s		Same as under C climates
Humid Middle-latitude with Severe Winters (type D) Climates		w		Same as under C climates
		f		Same as under C climates
			a	Same as under C climates
			b	Same as under C climates
			c	Same as under C climates
			d	Average temperature of the coldest month is −38°C or below
	E			Average temperature of the warmest month is below 10°C
Polar (type E) Climates		T		Average temperature of the warmest month is greater than 0°C and less than 10°C
		F		Average temperature of the warmest month is 0°C or below

Table 16.3 Characteristics of the principal climate groups

CLIMATE GROUP	NAME	TEMPERATURE AND/OR PRECIPITATION CHARACTERISTICS
A:	_____	_____
B:	_____	_____
C:	_____	_____
D:	_____	_____
E:	_____	_____

Table 16.4   Climatic data for representative stations

	J	F	M	A	M	J	J	A	S	O	N	D	YEAR
IQUITOS, PERU (AF); LAT. 3°39′S; 115 M													
Temp. (°C)	25.6	25.6	24.4	25.0	24.4	23.3	23.3	24.4	24.4	25.0	25.6	25.6	24.7
Precip. (mm)	259	249	310	165	254	188	168	117	221	183	213	292	2619
RIO DE JANEIRO, BRAZIL (AW); LAT. 22°50′S; 26 M													
Temp. (°C)	25.9	26.1	25.2	23.9	22.3	21.3	20.8	21.1	21.5	22.3	23.1	24.4	23.2
Precip. (mm)	137	137	143	116	73	43	43	43	53	74	97	127	1086
FAYA, CHAD (BWH); LAT. 18°00′N; 251 M													
Temp. (°C)	20.4	22.7	27.0	30.6	33.8	34.2	33.6	32.7	32.6	30.5	25.5	21.3	28.7
Precip. (mm)	0	0	0	0	0	2	1	11	2	0	0	0	16
SALT LAKE CITY, UTAH (BSK); LAT. 40°46′N; 1288 M													
Temp. (°C)	−2.1	0.9	4.7	9.9	14.7	19.4	24.7	23.6	18.3	11.5	3.4	−0.2	10.7
Precip. (mm)	34	30	40	45	36	25	15	22	13	29	33	31	353
WASHINGTON, D.C. (CFA); LAT. 38°50′N; 20 M													
Temp. (°C)	2.7	3.2	7.1	13.2	18.8	23.4	25.7	24.7	20.9	15.0	8.7	3.4	13.9
Precip. (mm)	77	63	82	80	105	82	105	124	97	78	72	71	1036
BREST, FRANCE (CFB); LAT. 48°24′N; 103 M													
Temp. (°C)	6.1	5.8	7.8	9.2	11.6	14.4	15.6	16.0	14.7	12.0	9.0	7.0	10.8
Precip. (mm)	133	96	83	69	68	56	62	80	87	104	138	150	1126
ROME, ITALY (CSA); LAT. 41°52′N; 3 M													
Temp. (°C)	8.0	9.0	10.9	13.7	17.5	21.6	24.4	24.2	21.5	17.2	12.7	9.5	15.9
Precip. (mm)	83	73	52	50	48	18	9	18	70	110	113	105	749
PEORIA, ILLINOIS (DFA); LAT. 40°45′N; 180 M													
Temp. (°C)	−4.4	−2.2	4.4	10.6	16.7	21.7	23.9	22.7	18.3	11.7	3.8	−2.2	10.4
Precip. (mm)	46	51	69	84	99	97	97	81	97	61	61	51	894
VERKHOYANSK, RUSSIA (DFD); LAT. 67°33′N; 137 M													
Temp. (°C)	−46.8	−43.1	−30.2	−13.5	2.7	12.9	15.7	11.4	2.7	−14.3	−35.7	−44.5	−15.2
Precip. (mm)	7	5	5	4	5	25	33	30	13	11	10	7	155
IVIGTUT, GREENLAND (ET); LAT. 61°12′N; 129 M													
Temp. (°C)	−7.2	−7.2	−4.4	−0.6	4.4	8.3	10.0	8.3	5.0	1.1	−3.3	−6.1	0.7
Precip. (mm)	84	66	86	62	89	81	79	94	150	145	117	79	1132
MCMURDO STATION, ANTARCTICA (EF); LAT. 77°53′S; 2 M													
Temp. (°C)	−4.4	−8.9	−15.5	−22.8	−23.9	−24.4	−26.1	−26.1	−24.4	−18.8	−10.0	−3.9	−17.4
Precip. (mm)	13	18	10	10	10	8	5	8	10	5	5	8	110

Table 16.4 contains climatic data for several stations that are representative of Köppen climatic types. Use the data in Tables 16.2 and 16.4 to answer the following questions.

Humid Tropical (type A) Climates

With the exception of the dry climates, no other climate covers as large an area on Earth as the humid tropical climates.

14. What temperature criterion is used for defining an A climate?

_____

15. Plot the monthly temperature and precipitation data for Iquitos, Peru, an A climate, given in Table 16.4 on the climate chart, Figure 16.4.

Use the Iquitos, Peru, climograph you prepared in question 15 to answer questions 16–19.

16. What is the *annual temperature range* (difference between highest and lowest monthly temperatures) for Iquitos?

_____ °C

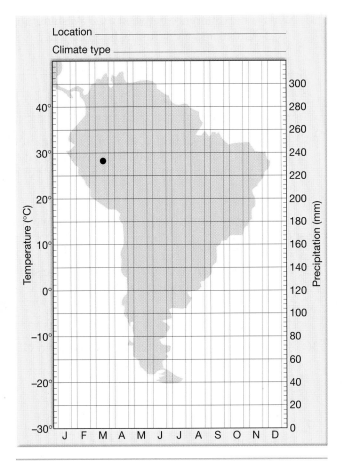

Location _____

Climate type _____

Figure 16.4 Climograph for Iquitos, Peru.

17. Describe the yearly variability of temperature for A climates.

_____

18. Notice that Iquitos receives an average of 2,619 mm of precipitation per year. How many inches of precipitation per year would this equal? (*Hint:* You may have to refer to the conversion tables located on the inside back cover of this manual.)

2,619 mm equals _____ inches

19. The precipitation at Iquitos is (concentrated in one season, distributed throughout the year). Circle your answer.

Use the world climate map, Figure 16.1, to answer questions 20–22.

20. In what latitude belt are A climates located?

_____

21. The most extensive areas of tropical rain forest (Af) climates are located (along coasts, in the interiors) of continents. Circle your answer.

22. Considering the locations of A climates, would weather fronts or columns of rapidly rising, hot,

surface air be most likely responsible for the precipitation? Explain your answer.

_____

_____

Dry (type B) Climates

Of all the climate groups, the dry climates cover the greatest portion of Earth's surface. To be classified as a dry climate does not necessarily imply little or no precipitation, but rather indicates that the yearly precipitation is not as great as the potential loss of moisture by evaporation.

23. What are three variables that the Köppen classification uses to establish the boundary between dry and humid climates?

1) _____

2) _____

3) _____

24. What name is applied to the following climatic types?

BW: _____

BS: _____

25. What is the primary cause of arid climates in the tropics?

_____

26. What factors contribute to the formation of arid climates in the middle latitudes?

_____

_____

To answer questions 27 and 28, refer to the world climate map, Figure 16.1.

27. At what latitudes, North and South, are the most extensive arid areas located?

_____

28. The Sahara desert in northern Africa is the largest area in the world with a BWh climate. What are some other regions that have the same climate?

_____

Humid Middle-Latitude Climates with Mild Winters (type C)

A large percentage of the world's population is located in areas with C climates. It is a climate characterized by weather contrasts brought about by changing seasons. On the average, the regions of C climates are dominated by contrasting air masses and associated middle-latitude cyclones.

29. What temperature criterion is used to define the boundary of C climates?

   _____

30. Using the data in Table 16.4, on Figure 16.5, prepare climographs for Washington, D.C., and Rome, Italy.

   Answer questions 31–34 using the climographs you have constructed for Washington, D.C., and Rome, Italy, Figure 16.5.

31. In what manner are the temperature curves for each of the two cities similar?

   _____

32. How does the annual distribution of precipitation vary between the two cities?

   _____

33. What is the difference between a Cf (Washington, D.C.) climate and a Cs (Rome, Italy) climate?

   _____

   _____

34. (Weather fronts, Columns of rapidly rising, hot surface air) are most likely responsible for the winter precipitation in Cf climates. Circle your answer.

   Use the world climate map, Figure 16.1, to answer questions 35–37.

35. What countries in Asia have areas of climate similar to that of Washington, D.C.?

   _____

36. What Southern Hemisphere countries have climates similar to that of Washington, D.C.?

   _____

37. Which U.S. state has a climate similar to that of Rome, Italy?

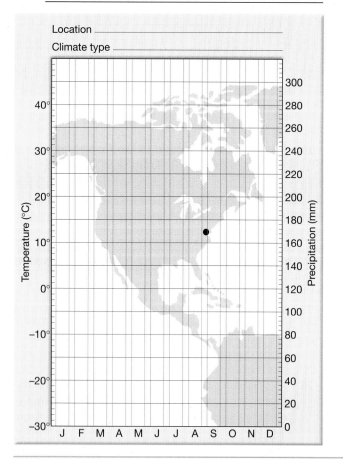
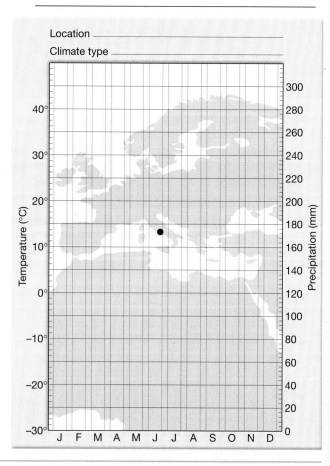

Figure 16.5 Climographs for Washington, D.C., and Rome, Italy.

Humid Middle-Latitude Climates with Severe Winters (type D)

The harsh winters and relatively short growing season restrict agricultural activity in much of the area of D climates. The northern portions of D climate regions are covered by coniferous forests, with lumbering being a significant economic activity.

38. What criteria are used for defining a D climate?

_____

_____

39. Use the data from Table 16.4 to plot a climograph for Peoria, Illinois, on Figure 16.6.

Use the climograph you have constructed for Peoria, Illinois, Figure 16.6, to answer questions 40–42.

40. What is the annual range of temperature in Peoria, Illinois?

_____ °C

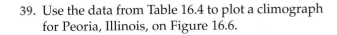

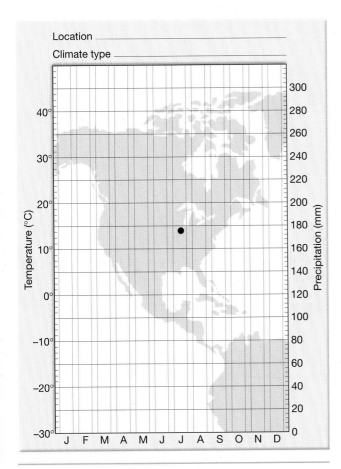

Figure 16.6 Climograph for Peoria, Illinois.

41. How does the annual range of temperature in Peoria, Illinois, compare to the temperature ranges in Iquitos, Peru, and Rome, Italy?

_____

_____

_____

_____

_____

42. During what season does Peoria receive its greatest precipitation and how does this compare with the seasonal distribution of precipitation for Rome, Italy?

_____

_____

_____

_____

Use the world climate map, Figure 16.1, to answer questions 43–45.

43. Which continent has the greatest continuous expanse of D climates?

_____

44. D climates are located only in the (Northern, Southern) Hemisphere. Circle your answer.

45. Suggest a reason why D climates are located in only the hemisphere you selected in question 44.

_____

_____

_____

_____

Table 16.5 Climatic data for Quito, Ecuador

		QUITO, ECUADOR LAT. 0°, LONG. 79°W; 2770 M											
	J	F	M	A	M	J	J	A	S	O	N	D	YEAR
Temp. (°C)	13	13	13	13	13	13	13	14	14	13	13	13	13
Precip. (mm)	99	110	142	175	137	43	20	30	69	113	95	93	1126

Polar (type E) Climates

The polar climates, found at high latitudes and scattered high altitudes in mountains, are regions of cold temperatures and sparse population. Low evaporation rates allow these areas to be classified as humid, even though the annual precipitation is modest.

46. What criterion is used for defining E climates?

_____

_____

47. Contrast the characteristics and locations of the two basic polar climates.

ET climates: _____

_____

_____

EF climates: _____

_____

_____

Climate and Altitude

Although often not included with the principal climate groups, *high-altitude*, or *highland* (type H) climates exist in all climatic regions. They are the result of the changes in radiation, temperature, humidity, precipitation, and atmospheric pressure that take place with elevation and orientation of mountain slopes.

Examine the climatic data for Quito, Ecuador, in Table 16.5. Quito is located in a region where A climates are expected. However, its altitude of 2,770 meters (9,086 feet) changes its Köppen classification.

48. Use Table 16.5 and Table 16.2 to determine the climatic classification of Quito, Ecuador.

Climate of Quito, Ecuador: _____

Criteria used for the selection of the climate type of Quito, Ecuador: _____

_____

_____

49. Locate Quito, Ecuador, on a map or globe. Considering its location, what effect has altitude had on the climatic classification?

_____

_____

50. Why would you expect the vegetation in the area of Quito to be different from that found at Guayaquil, Ecuador, a city located on the coast?

_____

_____

51. From Figure 16.1, where is the greatest continuous expanse of high-altitude (highland) climate located?

_____

_____

Climate on the Internet

Continue your analysis of the topics presented in this exercise by completing the corresponding online activity on the *Applications & Investigations in Earth Science* website at http://prenhall.com/earthsciencelab

Global Climates

Date Due: _____

Name: _____

Date: _____

Class: _____

After you have finished Exercise 16, complete the following questions. You may have to refer to the exercise for assistance or to locate specific answers. Be prepared to submit this summary/report to your instructor at the designated time.

1. Explain how a climograph is constructed.

_____

_____

_____

2. In your own words, provide some key words to describe the characteristics of each of the following Köppen climate groups.

A climates: _____

_____

B climates: _____

_____

C climates: _____

_____

D climates: _____

_____

E climates: _____

_____

H climates: _____

_____

3. List the general location of each of the following Köppen climate groups.

A climates: _____

_____

B climates: _____

_____

C climates: _____

_____

D climates: _____

_____

E climates: _____

_____

H climates: _____

_____

4. Indicate by name the Köppen climate group best described by each of the following statements.

Vast areas of northern coniferous forests: _____

_____

Smallest annual range of temperature: _____

_____

The highest annual precipitation: _____

_____

Mean temperature of the warmest month is below 10°C: _____

_____

The result of high elevation and mountain slope orientation: _____

_____

Potential evaporation exceeds precipitation: _____

_____

Very little change in the monthly precipitation and temperature throughout the year: _____

_____

Caused by the subsidence of air beneath high pressure cells: _____

_____

5. Using the data in Table 16.6, classify each station according to the most appropriate Köppen climate group. Where in North America is each likely to be located?

Station 1: _____

_____

Station 2: _____

_____

Station 3: _____

_____

Table 16.6   Climatic data for question 5

	J	F	M	A	M	J	J	A	S	O	N	D	YEAR
Station 1													
Temp. (°C)	1.7	4.4	7.9	13.2	18.4	23.8	25.8	24.8	21.4	14.7	6.7	2.8	13.8
Precip. (mm)	10	10	13	13	20	15	30	32	23	18	10	13	207
Station 2													
Temp. (°C)	−10.4	−8.3	−4.6	3.4	9.4	12.8	16.6	14.9	10.8	5.5	−2.3	−6.4	3.5
Precip. (mm)	18	25	25	30	51	89	64	71	33	20	18	15	459
Station 3													
Temp. (°C)	10.2	10.8	13.7	17.9	22.2	25.7	26.7	26.5	24.2	19.0	13.3	10.0	18.4
Precip. (mm)	66	84	99	74	91	127	196	168	147	71	53	71	1247

Source: Paul Souders/Danita Delimont Photography

Astronomy

Astronomical Observations

Scientific inquiry often starts with the systematic collection of data from which hypotheses and general theories are developed. The study of astronomy, an observational science, begins by carefully observing and recording the changing positions of the Sun, Moon, planets, and stars. To become proficient in astronomy requires developing the skills necessary to become a keen observer, using both the unaided eye and the telescope.

In this exercise you will observe several celestial objects. Observing and recording the changing positions of the Sun, Moon, and stars will aid in the interpretation and understanding of their movements in future exercises.

Objectives

After you have completed this exercise, you will have:

1. Records of the changing position of the Sun as it rises or sets on the horizon.
2. Measurements of the angle of the Sun above the horizon at noon on several days.
3. A record of the phases of the Moon over a period of several weeks.
4. Data on the times that the Moon rises and sets.
5. Records of the position and motion of stars.
6. An understanding of the parts of a telescope.

Materials

meterstick (or yardstick)	star chart	small weight
ruler	protractor	
calculator	string	

Materials Supplied by Your Instructor

telescope(s)   (optional)

Terms

revolution	altitude
astrolabe	rotation

Sun Observations

Many people are unaware that the Sun rises and sets at different locations on the horizon each day. As Earth **revolves** about the Sun, the orientation of its axis to the Sun continually changes. The result is that the location of the rising and setting Sun, as well as the **altitude** (angle above the horizon) of the Sun at noon, changes throughout the year.

Sunset (or Sunrise) Observations

Use the procedure presented in question 1 to make several observations and recordings of the Sun's location on the horizon at sunset or sunrise.

1. Following Steps 1–4, record at least four separate observations of the setting or rising Sun. Gather the data over a period of several weeks; *wait four or five days between each observation.* The directions are for sunset, although some minor adjustments will also allow their use for sunrise.

Step 1. Several minutes prior to sunset, estimate where the Sun will set on the western horizon. Draw the prominent features (buildings, trees, etc.) to the north and south of the Sun's approximate setting position on a sunset data sheet in Figure 17.1. (*Note:* As you observe the Sun setting in the west, south will be to your left and north to your right.)

CAUTION: Never look directly at the Sun; eye damage may result.

Step 2. As the Sun sets, draw its position on the data sheet relative to the fixed features on the horizon.

SUNSET (Sunrise) DATA SHEETS

HORIZON

Date of observation _____   Time of observation _____

HORIZON

Date of observation _____   Time of observation _____

HORIZON

Date of observation _____   Time of observation _____

HORIZON

Date of observation _____   Time of observation _____

Figure 17.1   Sunset (sunrise) data sheets.

Step 3. Note the date and time of your observation on the data sheet.

Step 4. Return to the same location several days later. Repeat your observation and record the results on a new data sheet.

2. After you complete your observations, describe the changing location of the Sun at sunset, or sunrise, that you have observed over the past several weeks.

_____

_____

_____

_____

Measuring the Noon Sun Angle

Observe the method for measuring the altitude (angle) of the noon Sun above the horizon illustrated in Figure 17.2. Then, following the steps listed in question 3, determine the altitude of the Sun at noon on several days.

3. To determine the altitude (angle) of the Sun at noon:

Step 1. Place a yardstick (a meterstick or ruler will do) perfectly vertical to the ground or a table top.

Step 2. When the Sun is at its highest position in the sky (noon, standard time, or 1 P.M. daylight savings time, will be close enough), accurately measure the length of the shadow.

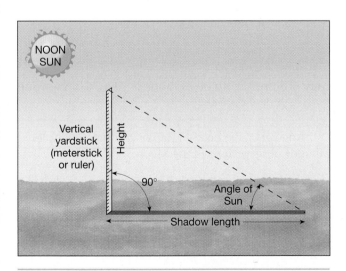

Figure 17.2 Illustration of the method for measuring the angle of the Sun above the horizon at noon. With each observation, be certain that the yardstick is perpendicular (at a 90° angle) to the ground or table top.

CAUTION: Never look directly at the Sun; eye damage may result.

Step 3. Divide the height of the stick by the length of the shadow.

Step 4. Consult Table 17.1 to determine the Sun angle. To find the Sun angle, locate the number on the table that comes closest to your answer from Step 3, and read the angle listed next to it.

Step 5. Repeat the measurement at exactly the same time on several different days over a period of four or five weeks. Record the dates and results of the measurements in the following spaces.

Date: _____   Noon Sun angle: _____°

Date: _____   Noon Sun angle: _____°

Date: _____   Noon Sun angle: _____°

Date: _____   Noon Sun angle: _____°

Answer questions 4–6 after you have completed your measurements of the altitude of the noon Sun.

4. The altitude of the noon Sun has (increased, decreased) over the period of the measurements. Circle your answer.

5. How many degrees has the noon Sun angle changed over the period of your observations?

_____°

6. What is the approximate average change of the noon Sun angle per day?

_____° per day

7. Based on your answer in question 6, how many degrees will the noon Sun angle change over a six-month period?

_____° over a six-month period

Moon Observations

Most people have noticed that the shape of the illuminated portion of the Moon as observed from Earth changes regularly. However, few take the time to systematically record and explain these changes. Following the procedure presented in question 8, begin your study of the Moon by observing its phases and recording your observations.

8. Record at least four observations of the Moon by completing each of the following steps regularly at a two- or three-day interval.

Step 1. On a Moon observation data sheet provided in Figure 17.3, indicate the approximate east-west

Table 17.1 Data table for determining the noon Sun angle. Select the nearest number to the quotient determined by dividing the height of the stick by the length of the shadow. Read the corresponding Sun angle.

IF $\dfrac{\text{HEIGHT OF STICK}}{\text{LENGTH OF SHADOW}}$	THEN SUN ANGLE IS	IF $\dfrac{\text{HEIGHT OF STICK}}{\text{LENGTH OF SHADOW}}$	THEN SUN ANGLE IS
0.2679	15°	1.235	51°
0.2867	16°	1.280	52°
0.3057	17°	1.327	53°
0.3249	18°	1.376	54°
0.3443	19°	1.428	55°
0.3640	20°	1.483	56°
0.3839	21°	1.540	57°
0.4040	22°	1.600	58°
0.4245	23°	1.664	59°
0.4452	24°	1.732	60°
0.4663	25°	1.804	61°
0.4877	26°	1.881	62°
0.5095	27°	1.963	63°
0.5317	28°	2.050	64°
0.5543	29°	2.145	65°
0.5774	30°	2.246	66°
0.6009	31°	2.356	67°
0.6249	32°	2.475	68°
0.6494	33°	2.605	69°
0.6745	34°	2.748	70°
0.7002	35°	2.904	71°
0.7265	36°	3.078	72°
0.7536	37°	3.271	73°
0.7813	38°	3.487	74°
0.8098	39°	3.732	75°
0.8391	40°	4.011	76°
0.8693	41°	4.332	77°
0.9004	42°	4.705	78°
0.9325	43°	5.145	79°
0.9657	44°	5.671	80°
1.0000	45°	6.314	81°
1.0360	46°	7.115	82°
1.0720	47°	8.144	83°
1.1110	48°	9.514	84°
1.1500	49°	11.430	85°
1.1920	50°		

position of the Moon in the sky by drawing a circle at the appropriate location. (*Note:* As you look to the south to observe the Moon, east will be to your left and west to your right.)

Step 2. By shading the circle, indicate the shape of the illuminated portion of the Moon you observe.

Step 3. Note the date and time of your observation on the data sheet.

Step 4. Keep in mind that the approximate time between moonrise on the eastern horizon and moonset on the western horizon is twelve hours. Estimate when the Moon may have risen and when it may set. Write your estimates on the data sheet.

Step 5. Repeat your observation in several days, using a new data sheet.

Answer questions 9–12 after you have completed all your observations of the Moon.

9. What happened to the size and shape of the illuminated portion of the Moon over the period of your observations?

_____

_____

MOON OBSERVATION DATA SHEETS

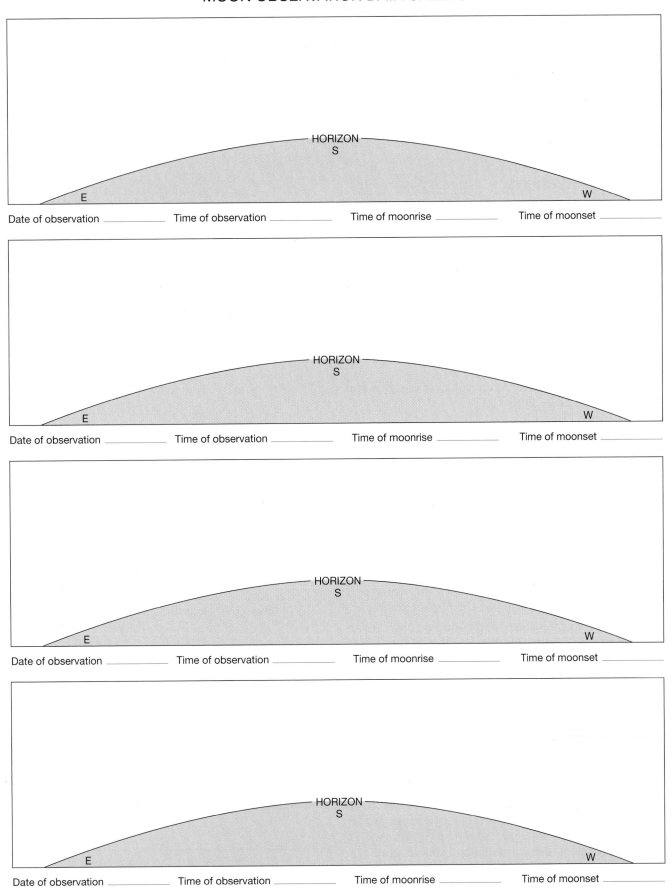

Figure 17.3 Moon observation data sheets.

10. The Moon moved farther (eastward, westward) in the sky with each successive observation. Circle your answer.

11. The times of moonrise and moonset became (earlier, later) with each successive observation. Circle your answer.

12. Based upon your observations and your answers to questions 10 and 11, the Moon revolves around Earth from (east to west, west to east).

Star Observations

Throughout history, people have been recording the positions and nightly movement of stars that result from Earth's **rotation**, as well as the seasonal changes in the constellations as Earth revolves about the Sun. Early astronomers offered many explanations for the changes before the true nature of the motions was understood in the 17th century.

To best observe the stars, select a suitable dark area on a clear, moonless night. Then complete questions 13 through 22.

13. Make a list of the different colors of the stars you can observe in the sky.

_____

Select one star that is overhead, or nearly so, and observe its movement over a period of one hour.

14. With your arm extended, approximately how many widths of your fist has the position of the star changed?

_____ fist widths

15. The star appears to move (eastward, westward) over a period of one hour. Circle your answer.

16. How is the movement of the star you observed in question 15 related to the direction of rotation of Earth?

_____

_____

Use a suitable star chart to locate several constellations and the North Star (Polaris).

17. Refer to Figure 17.4. Sketch the pattern of stars for two constellations you were able to locate. List the name of each constellation by its diagram.

18. Using Figure 17.5 as a guide, construct a simple **astrolabe** and measure the angle of the North

Constellation Star Pattern

Constellation name: _____

Constellation Star Pattern

Constellation name: _____

Figure 17.4 Constellation sketches.

Star (Polaris) above the horizon as accurately as possible.

_____° above the (north, south) horizon

Over a period of several hours, observe the motion of the stars in the vicinity of Polaris.

19. Write a brief summary of the motion of the stars in the vicinity of Polaris.

_____

_____

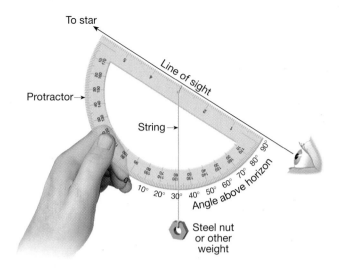

To star

Line of sight

Protractor→

String →

Angle above horizon

Steel nut
or other
weight

Figure 17.5 Simple astrolabe, an instrument used to measure the angle of an object above the horizon. The angle is read where the string crosses the outer edge of the protractor, 32° in the example illustrated. Notice that the angle above the horizon is the difference between 90° and the angle imprinted on the protractor.

If you can, come back to the same location at the exact same time, several weeks later.

20. The same star you observed overhead several weeks earlier (is still overhead, has moved to the east, has moved to the west). Circle your answer.

21. How is the change in position of the star you observed overhead several weeks earlier related to the revolution of Earth?

_____

_____

_____

22. Using the astrolabe you constructed in question 18, repeat your measurement of the angle of the North Star (Polaris) above the horizon. List your new measurement and compare it to the measurement you obtained several weeks earlier. Explain your result(s).

_____

_____

_____

Astronomical Observations on the Internet

Continue your analyses of the topics presented in this exercise by completing the corresponding online activity on the *Applications & Investigations in Earth Science* website at http://prenhall.com/earthsciencelab

Notes and calculations.

Astronomical Observations

Date Due: _____

Name: _____

Date: _____

Class: _____

After you have finished Exercise 17, complete the following questions. You may have to refer to the exercise for assistance or to locate specific answers. Be prepared to submit this summary/report to your instructor at the designated time.

1. On Figure 17.6, prepare a single sketch illustrating your observed positions of the setting (or rising) Sun on the horizon during the past several weeks. Show the reference features you used on the horizon. Label each position of the Sun with the date of the observation. Write a brief summary of your observations below the diagram.

2. From question 3, step 5, in the exercise, list the noon Sun angle that you calculated for the first and last day of your measurements.

    Noon Sun angle on the first day: _____°

    Date of observation: _____

    Noon Sun angle on the last day: _____°

    Date of observation: _____

3. Draw two sketches of the Moon—the first illustrating the Moon as you saw it on your first lunar observation, the second as you saw it on your last observation. Label the date and time of each observation.

FIRST MOON OBSERVATION	LAST MOON OBSERVATION
Date: JAN 12	Date: _____
Time: 9:23	Time: _____

4. Did the Moon rise earlier or later each night that you observed it?

_____

_____

Horizon

Summary: _____

_____

Figure 17.6 Sunset (sunrise) observations.

5. List the different colors of stars that you observed.

_____

6. Approximately how many widths of your fist, with your arm extended, will a star appear to move in one hour? Toward which direction do the stars appear to move throughout the night and what is the reason for the motion?

_____

_____

_____

7. What was your measured angle of the North Star (Polaris) above the horizon at your location? Did the angle change over a several-week period? Explain why.

_____

_____

_____

8. Refer to Figure 17.7. Sketch the pattern of stars for any constellation you have been able to locate in the sky. What is the name of the constellation?

Constellation Star Pattern

Constellation name: _____

Figure 17.7 Constellation sketch.

Patterns in the Solar System

Although composed of many diverse objects, the solar system (Figure 18.1) exhibits various degrees of order and several regular patterns. To simplify the investigation of planetary sizes, masses, etc., the planets can be arranged into two distinct groups, with the members of each displaying similar attributes. This exercise examines the physical properties and motions of the planets with the goal of summarizing these characteristics in a few general, easily remembered statements.

Objectives

After you have completed this exercise, you should be able to:

1. Describe the appearance of the solar system when it is viewed along the plane of the ecliptic.

2. Summarize the distances and spacing of the planets in the solar system.

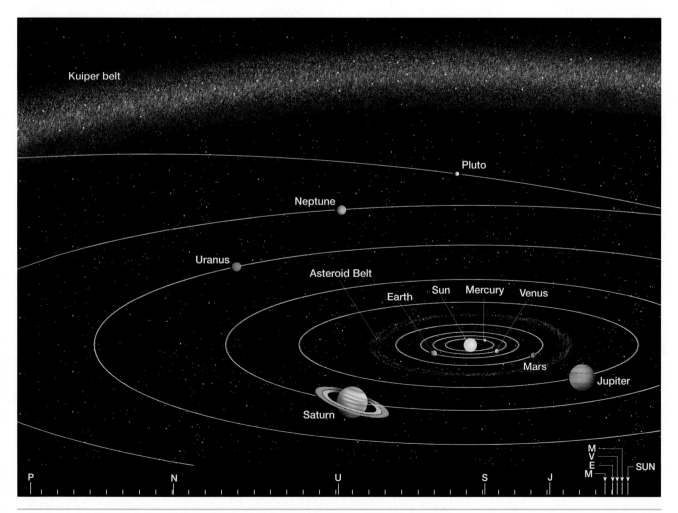

Figure 18.1   The solar system showing the orbits of the planets to scale. A different scale has been used for the sizes of the Sun and planets. Therefore, the diagram is not a true scale model representation of the solar system.

3. Summarize and compare the physical characteristics of the terrestrial and Jovian planets.
4. Describe the motions of the planets in the solar system.

Materials

ruler
colored pencils
calculator

Materials Supplied by Your Instructor

4-meter length of adding machine paper
meterstick
light source (150 watt bulb)
black containers with covers and thermometers

Terms

nebula
terrestrial planets
Jovian planets
plane of the ecliptic
mass
density
weight
rotation
revolution
Kepler's laws
astronomical unit

Introduction

The order that exists within the solar system is directly related to the laws of physics that governed its formation. Astronomers have determined that the Sun and planets originated approximately 4.6 billion years ago from an enormous cloud of dust and gas. As this **nebula** contracted, it began to rotate and flatten. Eventually the temperature and pressure in the center of the cloud was great enough to initiate nuclear fusion and form the Sun.

Near the center of the nebula, the planets Mercury, Venus, Earth, and Mars evolved under nearly the same conditions and consequently exhibit similar physical properties. Because these planets are rocky objects with solid surfaces, they are collectively called the **terrestrial** (Earth-like) **planets**.

The outer planets, Jupiter, Saturn, Uranus, and Neptune, being farther from the Sun than the terrestrial planets, formed under much colder conditions and are gaseous objects with central cores of ices and rock. Since the four planets are very similar, they are often grouped together and called the **Jovian** (Jupiterlike) **planets**. The planet Pluto is not included in either group since its great distance and small size make determining its complete characteristics impossible at the present time.

Table 18.1 illustrates many of the individual characteristics of the planets in the solar system.

1. Examine the data in Table 18.1, then
   a. Draw lines on the upper and lower parts of Table 18.1 that separate the terrestrial planets from the Jovian planets. Label the lines "Belt of Asteroids."

Table 18.1 **Planetary data.**

Planet	Symbol	Mean Distance from Sun: AU	Mean Distance from Sun: Millions of Miles	Mean Distance from Sun: Millions of Kilometers	Period of Revolution	Inclination of Orbit	Orbital Velocity: mi/s	Orbital Velocity: km/s
Mercury	☿	0.387	36	58	88^{d}	7°00′	29.5	47.5
Venus	♀	0.723	67	108	224.7^{d}	3°24′	21.8	35.0
Earth	⊕	1.000	93	150	365.25^{d}	0°00′	18.5	29.8
Mars	♂	1.524	142	228	687^{d}	1°51′	14.9	24.1
Jupiter	♃	5.203	483	778	11.86^{yr}	1°18′	8.1	13.1
Saturn	♄	9.539	886	1427	29.46^{yr}	2°29′	6.0	9.6
Uranus	⛢	19.180	1783	2870	84^{yr}	0°46′	4.2	6.8
Neptune	♆	30.060	2794	4497	165^{yr}	1°46′	3.3	5.3
Pluto	♇	39.440	3666	5900	248^{yr}	17°12′	2.9	4.7

Planet	Period of Rotation	Diameter: Miles	Diameter: Kilometers	Relative Mass (Earth = 1)	Average Density (g/cm^3)	Polar Flattening (%)	Mean Temperature (°C)	Number of Known Satellites
Mercury	59^{d}	3015	4854	0.056	5.4	0.0	167	0
Venus	244^{d}	7526	12,112	0.82	5.2	0.0	464	0
Earth	23^{h}56^{m}04^{s}	7920	12,751	1.00	5.5	0.3	15	1
Mars	24^{h}37^{m}23^{s}	4216	6788	0.108	3.9	0.5	−65	2
Jupiter	9^{h}50^{m}	88,700	143,000	317.87	1.3	6.7	−110	63
Saturn	10^{h}14^{m}	75,000	121,000	95.14	0.7	10.4	−140	37
Uranus	17^{h}14^{m}	29,000	47,000	14.56	1.2	2.3	−195	27
Neptune	16^{h}03^{m}	28,900	46,529	17.21	1.7	1.8	−200	13
Pluto	6.4^{d}	~1500	~2445	0.002	1.8	0.0	−225	1

b. On both parts of the table write the word "terrestrial" next to Mercury, Venus, Earth, and Mars and the word "Jovian" next to Jupiter, Saturn, Uranus, and Neptune.

The "Shape" of the Solar System

When the solar system is viewed from the side, the orbits of the planets all lie in nearly the same plane, called the **plane of the ecliptic** (Figure 18.1). The column labeled "Inclination of Orbit" in Table 18.1 lists how many degrees the orbit of each planet is inclined from the plane. Answer questions 2–4 by referring to Table 18.1.

2. The orbits of which two planets have the greatest inclination to the plane of the ecliptic?

   _____

3. With the exception of the two planets indicated in question 2, the orbits of the remaining planets all lie within (4, 6, 10) degrees of the plane. Circle your answer.

4. Considering the nebular origin of the solar system, suggest a reason why the orbits of the planets are nearly all in the same plane.

   _____

   _____

Distance and Spacing of the Planets

An examination of any scale-model solar system reveals that the distances from the Sun and the spacing between the planets appear to follow a regular pattern. Although many ancient astronomers were concerned with planetary distances and spacing, it was not until the mid-1700s that astronomers found a simple mathematical relation that described the arrangement of the planets known at the time.

A Scale Model of Planetary Distances

Perhaps the best way to examine distance and spacing of the planets in the solar system is to use a scale model.

5. Prepare a distance scale model of the solar system according to the following steps.

Step 1. Obtain a 4-meter length of adding machine paper and a meterstick from your instructor.

Step 2. Draw an "X" about 10 centimeters from one end of the adding machine paper and label it "Sun."

Step 3. Using the mean distances of the planets from the Sun in miles presented in Table 18.1 and the following scale, draw a small circle for each planet at its proper scale mile distance from the Sun. Use a different colored pencil for the terrestrial and Jovian planets and write the name of the planet next to its position.

SCALE

1 millimeter =	1 million miles
1 centimeter =	10 million miles
1 meter =	1,000 million miles

Step 4. Write the word "asteroids" 258 million scale miles from the Sun.

Answer questions 6–9 using the distance scale model you constructed in question 5.

6. What feature of the solar system separates the terrestrial planets from the Jovian planets?

   _____

7. Observe the scale model diagram and summarize the spacing for each of the two groups of planets.

   Spacing of the terrestrial planets: _____

   _____

   Spacing of the Jovian planets: _____

   _____

8. Write a brief statement that describes the spacing of the planets in the solar system.

   _____

   _____

9. Which planet(s) vary the most from the general pattern of spacing?

   _____

Comparing the Terrestrial and Jovian Planets

The physical characteristics such as diameter, density, and mass of the terrestrial planets are very similar and can be summarized in a few statements. Likewise, the characteristics exhibited by the Jovian planets as a group can also be generalized.

To gain an understanding of the similarities of the planets within each of the two groups and the contrasts between the two groups, complete the following sections using the planetary data presented in Table 18.1. Because of the lack of sufficient data, Pluto will not be used in determining group characteristics or for general comparisons between the groups.

Size of the Planets

The similarities in the diameters of the planets within each of the two groups and the contrast between the groups are perhaps the most obvious patterns in the solar system. The diameter of each planet is given in both miles and kilometers in Table 18.1.

10. To visually compare the relative sizes of the planets and Sun, complete the following steps using the unmarked side of your 4-meter length of adding machine paper.

Step 1. Determine the radius of each planet in kilometers by dividing its diameter (in kilometers) by 2. List your answers in the "Radius" column of Table 18.2.

Step 2. Use a scale of 1 cm = 2000 km. Determine the scale model radius of each planet and list your answer in the "Scale Model Radius" column of Table 18.2.

Step 3. Draw an "X" about 10 cm from one end of the adding machine paper and label it "Starting point."

Step 4. Using the scale model radius in Table 18.2, begin at the starting point and mark the radius of each planet with a line on the paper. Use a different colored pencil for the terrestrial and Jovian planets. Label each line with the planet's name.

Step 5. The diameter of the Sun is approximately 1,350,000 kilometers. Using the same scale as you used for the planets (1 cm = 2000 km), determine the scale model radius of the Sun. Mark the Sun's radius on the adding machine paper using a different colored pencil from the two planet groups. Label the line "Sun."

Table 18.2 Planetary radii with scale model equivalents.

Planet	Radius (in kilometers)	Scale Model Radius
Mercury	_____	_____ cm
Venus	_____	_____ cm
Earth	_____	_____ cm
Mars	_____	_____ cm
Jupiter	_____	_____ cm
Saturn	_____	_____ cm
Uranus	_____	_____ cm
Neptune	_____	_____ cm

Answer questions 11–17 using both Table 18.1 and the scale model radius diagram you constructed in question 10.

11. Which is the largest of the terrestrial planets and what is its diameter?

_____ , _____ miles

12. Which is the smallest Jovian planet and what is its diameter?

_____ , _____ miles

13. Complete the following statement.

The smallest Jovian planet, _____ , is _____ times larger than the largest terrestrial planet.

14. Summarize the sizes of the planets within each group.

The diameters of the terrestrial planets: _____

_____

The diameters of the Jovian planets: _____

_____

15. Write a general statement that compares the sizes of the terrestrial planets to those of the Jovian planets.

_____

_____

16. Complete the following statement.

The Sun is _____ times larger than Earth and _____ times larger than Jupiter.

17. Refer to Table 18.1. The diameter of Pluto is most like the diameters of the (terrestrial, Jovian) planets. Circle your answer.

Mass and Density of the Planets

Mass is a measure of the quantity of matter an object contains. In Table 18.1 the masses of the planets are given in relation to the mass of Earth. For example, the mass of Mercury is given as 0.056, which means that it consists of only a small fraction of the quantity of matter that Earth contains. On the other hand, the Jovian planets all contain several times more matter than Earth.

Density is the mass per unit volume of a substance. In Table 18.1 the average densities of the planets are expressed in grams per cubic centimeter (g/cm^3). As a reference, the density of water is approximately one gram per cubic centimeter.

Using the relative masses of the planets given in Table 18.1, answer questions 18–22.

18. Complete the following statements:

    a. The planet _____ is the most massive planet in the solar system. It is _____ times more massive than Earth.

    b. The least massive planet (excluding Pluto) is _____ , which contains only _____ as much mass as Earth.

    The gravitational attraction of a planet is directly related to its mass.

19. Which planet exerts the greatest pull of gravity? Explain your answer.

    _____

    _____

    Your **weight** is a function of the gravitational attraction of an object on your mass.

20. On which planet (excluding Pluto) would you weigh the least? Explain your answer.

    _____

    _____

21. Which of the two groups of planets would have the greatest ability to hold large quantities of gas as part of their compositions? Explain your answer.

    _____

    _____

22. Write a general statement comparing the masses of the terrestrial planets to the masses of the Jovian planets.

    _____

    _____

Diameter vs. Density

To visually compare the diameters and densities of the planets, use the data in Table 18.1 to complete the diameter vs. density graph, Figure 18.2, according to the procedure in question 23.

23. Plot a point on the diameter vs. density graph, Figure 18.2, for each planet (excluding Pluto) where its diameter intersects its density. Label each point with the planet's name. Use a different colored pencil for the terrestrial and Jovian planets.

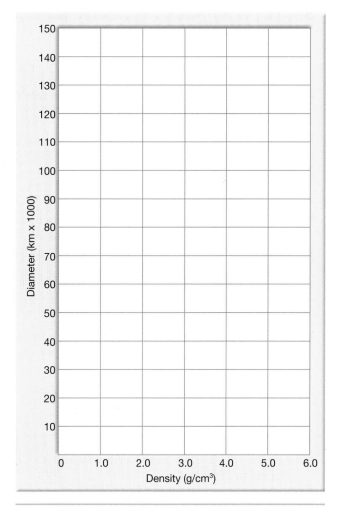

Figure 18.2   Diameter vs. density graph.

Answer questions 24–34 using Table 18.1 and the diameter vs. density graph you constructed in question 23.

24. What general relation exists between a planet's size and its density?

    _____

25. Consider the fact that the densities of the two rocks that form the majority of Earth's surface, the igneous rocks granite and basalt, are each about 3.0 g/cm^3. Therefore, the average density of the terrestrial planets is (greater, less than) the density of Earth's surface. Circle your answer.

26. The term (rocky, gaseous) best describes the terrestrial planets. Circle your answer.

27. The average density of Earth is about 5.5 g/cm^3. Considering that the densities of the surface rocks are much less than the average, what does this suggest about the density of Earth's interior?

    _____

28. Which of the planets has a density less than water and therefore would "float"?

_____

29. Write a brief statement comparing the densities of the Jovian planets to the density of water.

_____

_____

_____

30. The Jovian planets can be best described as (rocky, ice and gas) worlds. Circle your answer.

31. Explain why Jupiter can be such a massive object and yet have such a low density.

_____

_____

32. Write a general statement comparing the densities of the terrestrial planets to the Jovian planets.

_____

_____

_____

33. Why are the densities of the terrestrial and Jovian planets so different?

_____

_____

_____

34. Examine the estimated mass and density of Pluto presented in Table 18.1 and then complete the following statement by circling the correct responses.

   The mass of Pluto is most like the masses of the (terrestrial, Jovian) planets, while its density is similar to the (terrestrial, Jovian) planets. This suggests that Pluto is a (small, large) planet made of (rocky, ice and frozen gas) material.

Number of Moons of the Planets

The column labeled "Number of Known Satellites" in Table 18.1 indicates the number of known moons orbiting each planet.

35. Write a brief statement comparing the number of known moons of the terrestrial planets to the number orbiting the Jovian planets.

_____

_____

36. What is the general relation between the number of moons a planet has compared to its mass? Suggest a reason for the relation.

_____

_____

_____

Rotation and Revolution of the Planets

Rotation is the turning of a planet about its axis that is responsible for day and night. When the solar system is viewed from above the Northern Hemisphere of Earth, the planets, with the exception of Venus, rotate in a counterclockwise direction. Venus exhibits a very slow clockwise rotation. The time that it takes for a planet to complete one 360° rotation on its axis is called the _period of rotation_. The units used to measure a planet's period of rotation are Earth hours and days.

   Revolution is the motion of a planet around the Sun. The time that it takes a planet to complete one revolution about the Sun is the length of its year, called the _period of revolution_. The units used to measure a planet's period of revolution are Earth days and years. Without exception, the direction of revolution of the planets is counterclockwise around the Sun when the solar system is viewed from above the Northern Hemisphere of Earth.

   Use the planetary data in Table 18.1 to answer questions 37–46.

37. If you could live on Venus or Jupiter, approximately how long would you have to wait between sunrises?

   On Venus a sunrise would occur every _____ days.

   On Jupiter a sunrise would occur every _____ hours.

38. Write a statement comparing the periods of rotation of the terrestrial planets to those of the Jovian planets.

_____

_____

   The giant planet Jupiter rotates once on its axis approximately every 10 hours. If an object were on the equator of the planet and rotating with it, it would travel approximately 280,000 miles (the equatorial circumference or distance around the equator) in about 10 hours.

39. Calculate the equatorial rotational velocity of Jupiter using the following formula.

$$\text{Velocity} = \frac{\text{Distance}}{\text{Time}} = \underline{\hspace{2cm}} \frac{\text{mi}}{\text{hr}}$$

$$= \underline{\hspace{3cm}} \text{mi/hr}$$

40. The equatorial circumference of Earth is about 24,000 miles. What is the approximate equatorial rotational velocity of Earth?

    _____ miles/hour

41. How many times faster is Jupiter's equatorial rotational velocity than Earth's?

    _____ times faster

42. Compare the planets' periods of rotation to their periods of revolution and then complete the following statement by circling the correct responses.

    The terrestrial planets all have (long, short) days and (long, short) years, while the Jovian planets all have (long, short) days and (short, long) years.

43. In one Earth year, how many revolutions will the planet Mercury complete and what fraction of a revolution will Neptune accomplish?

    Mercury: _____ revolutions in one Earth year

    Neptune: _____ of a revolution in one Earth year

44. On Venus, how many days (sunrises) would there be in each of its years?

    _____ day(s) per year

45. How many days (rotations) will Mercury complete in one of its years?

    Mercury: _____ Mercury days in one Mercury year

46. Explain the relation between a planet's period of rotation and period of revolution that would cause one side of a planet to face the Sun throughout its year.

    _____

    _____

In the early 1600s Johannes Kepler set forth three laws of planetary motion. According to Kepler's third law, the period of revolution of a planet, measured in Earth years, is related to its distance from the Sun in astronomical units (one **astronomical unit (AU)** is defined as the average distance from the Sun to Earth—93 million miles or 150 million kilometers). The law states that a planet's orbital period squared is equal to its mean solar distance cubed ($p^2 = d^3$).

47. Applying Kepler's third law, what would be the period of revolution of a hypothetical planet that is 4 AUs from the Sun? Show your calculation in the following space.

    _____

    _____

    _____

Terrestrial Planet Temperatures

The temperature of an object is related to the intensity of the heat source, its distance from the source, and the nature of the material it is composed of. To better understand how these variables influence the temperatures of the terrestrial planets, observe the equipment in the laboratory (Figure 18.3) and then complete the following steps. Answer questions 48–51 after you complete your investigation.

Step 1. Working in groups of four or more, obtain four *identical* light (heat) sources and four *identical* black containers with covers and thermometers.

Step 2. Conduct four experiments simultaneously, one by each member of the group. Do one experiment with the covered can and thermometer 15 cm from the light source, another with the can 30 cm from the light source, the third 45 cm, and the fourth 60 cm.

Step 3. Note the starting temperature for each container on Table 18.3; the temperatures should all be the same.

Step 4. For each of the four setups, turn on the light and record the temperature of the container exactly 10 minutes later. Record the temperatures in Table 18.3.

Step 5. Using the temperature scale on the left axis of the graph, plot the temperatures from

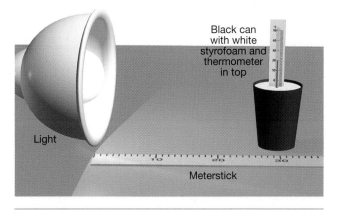

Figure 18.3 Terrestrial planet temperatures lab-equipment setup.

Table 18.3 Temperature data.

Distance from Light Source (cm)	Starting Temperature (°C)	10-minute Temperature
15		
30		
45		
60		

Table 18.3 on the graph in Figure 18.4. Connect the points and label the graph "temperature change with distance."

Step 6. In Table 18.1, notice the mean temperatures for the planets. Plot the mean temperatures of the terrestrial planets at their proper locations on the graph, Figure 18.4. Assume a scale of 40 cm equals 1 AU and use the temperature scale on the right axis of the graph. Label each point with the planet's name. Connect the points and label the graph "mean terrestrial planet temperatures."

48. The "temperature change with distance" graph represents how you would expect that, everything else being equal, the temperature of a planet would be related to its distance from the Sun.

In the following space, write a brief description of the "temperature change with distance" graph.

_____

_____

_____

49. The "mean terrestrial planet temperatures" graph represents the real mean temperatures of the planets. Compare this graph to the theoretical "temperature change with distance graph." How are the graphs similar? How are they different?

_____

_____

_____

_____

_____

50. Write a brief statement suggesting the reason(s) for the difference(s) between the two graphs you noted in question 49.

_____

_____

_____

_____

51. Complete your investigation by writing a statement describing the mean temperatures of the terrestrial planets and the variables that determine those temperatures.

_____

_____

_____

_____

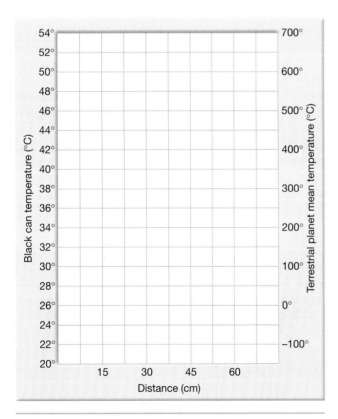

Figure 18.4 Terrestrial planet temperatures graph.

The Solar System on the Internet

Continue your analyses of the topics presented in this exercise by completing the corresponding online activity on the *Applications & Investigations in Earth Science* website at http://prenhall.com/earthsciencelab

Patterns in the Solar System

Date Due: _____

Name: _____

Date: _____

Class: _____

After you have finished Exercise 18, complete the following questions. You may have to refer to the exercise for assistance or to locate specific answers. Be prepared to submit this summary/report to your instructor at the designated time.

1. On Figure 18.5, prepare a sketch illustrating the planets Mercury, Venus, Earth, and Mars at their approximate distance from the Sun. View the solar system from above the Northern Hemisphere of Earth. Draw arrows around each planet to illustrate its direction of rotation. Also, draw an arrow in the orbit of each planet that shows the direction of revolution.

2. Briefly describe the spacing of the planets in the solar system.

_____

_____

_____

3. Define the following terms:

Terrestrial planets: _____

_____

Jovian planets: _____

_____

Plane of the ecliptic: _____

_____

Rotation: _____

_____

Mass: _____

_____

Astronomical unit: _____

_____

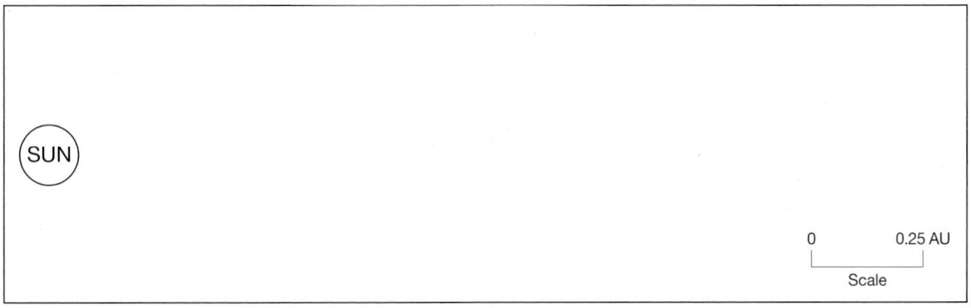

Figure 18.5 Spacing and motion of the terrestrial planets.

4. Referring to the nebular origin of the solar system, describe and explain the direction of revolution of the planets.

_____

_____

_____

5. Write a brief statement for each of the following characteristics that compares the terrestrial to the Jovian planets.

Diameter: _____

_____

_____

Density: _____

_____

_____

Period of rotation: _____

_____

_____

Number of moons: _____

_____

_____

Mass: _____

_____

_____

6. If you knew the distance of a planet from the Sun, explain how you would calculate its period of revolution.

_____

_____

_____

7. How does a planet's distance from the Sun affect the solar radiation the planet receives? Why?

_____

_____

_____

8. How are the mean temperatures of the terrestrial planets related to the solar radiation they intercept? What is the explanation for any discrepancy?

_____

_____

_____

Planet Positions

Generations of professional and amateur astronomers have noticed that certain celestial objects change position in the sky relative to the background of stars. This ability to recognize these "wanderers" is one of the first skills developed by any astronomer. Locating a planet begins with a knowledge of the motions of celestial objects and the proficient use of astronomical charts. This exercise examines a "working" scale model of the solar system. Using diagrams, you will investigate the movement of the planets in their orbits.

Objectives

After you have completed this exercise, you should be able to:

1. Explain the observed motion of a planet when it is viewed from Earth.

2. Give the position of a planet by listing the constellation in which it is located.

3. Prepare a scale model of the solar system for a specified date showing the positions of the five planets that can be viewed from Earth without a telescope.

4. Explain the conditions that determine whether or not a planet can be seen on a specified date.

5. Use a diagram of the solar system to estimate the time that a planet will rise and set.

Materials

ruler
colored pencils
calculator

Terms

period of revolution
constellations of the zodiac
retrograde motion

A Working Scale Model of the Solar System

Most scale models of the solar system, such as the one you may have constructed in Exercise 18, simply illustrate the planets in a line or do not incorporate the element of planetary motion. To accurately portray the solar system involves showing the planets at their correct scale model distances from the Sun as well as placing them in their proper position around the Sun.

Figures 19.1, 19.2, and 19.3 present the December 31, 1957, position of the five planets that can be seen from Earth without a telescope. The solar system is viewed from above the Northern Hemisphere of Earth. The small circle in the center of each figure represents the Sun. Surrounding each figure is a large circle, marked in degrees, which is used to reference the locations of the planets.

The zero point on each planet's orbit specifies its position on December 31, 1957. For Mercury, Venus, and Mars (Figures 19.1 and 19.2), the numbers marked on each orbit indicate the planet's position on successive days during its revolution. Since Jupiter and Saturn (Figure 19.3) have relatively long **periods of revolution**, their orbits have been divided into years. To simplify its location and allow for direct positioning, Earth's orbit has been divided into months and select days of the month.

The planets, Moon, and Sun lie in nearly the same plane. Therefore, when observed from Earth, they move along the same region of the sky, called the *zodiac* ("zone of animals"). Indicated on the outer reference circle of each figure are the twelve **constellations of the zodiac**, which form the background of stars.

Revolution of the Planets

The period of revolution of a planet is directly related to its distance from the Sun. The direction of revolution around the Sun is the same for all planets.

Examine each of the orbits of the planets shown in Figures 19.1–19.3. Then answer questions 1–4.

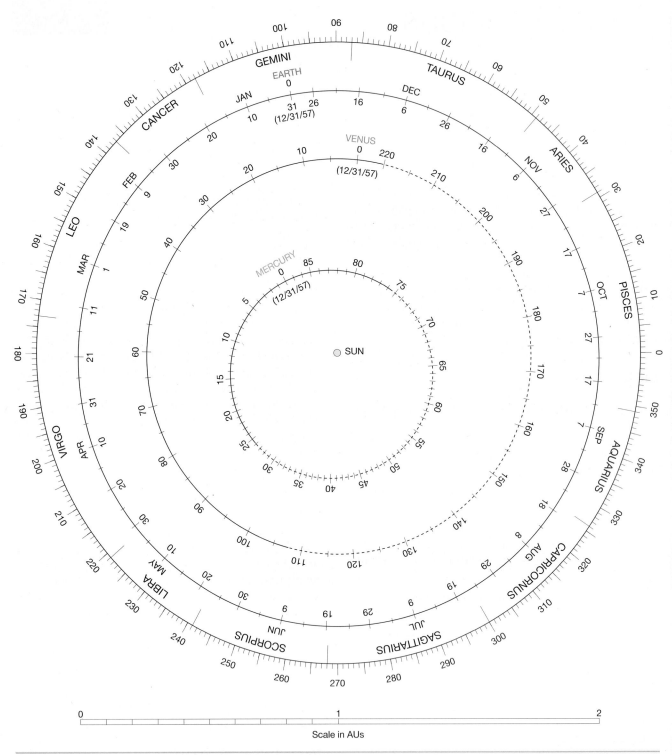

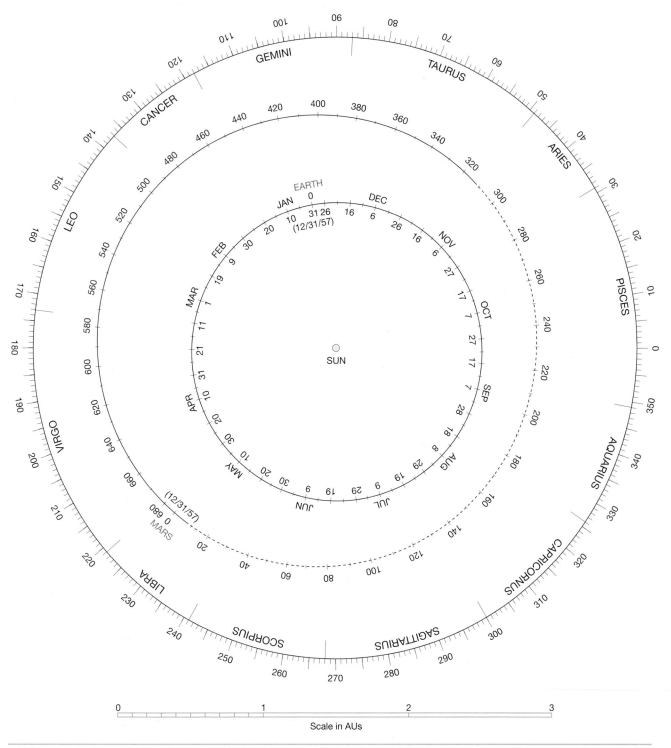

Figure 19.2 Positions of Earth and Mars on December 31, 1957. (Adapted from *Field Guide to the Stars and Planets* by Donald Menzel. Copyright 1964 by Donald Menzel. By permission of Houghton Mifflin Company)

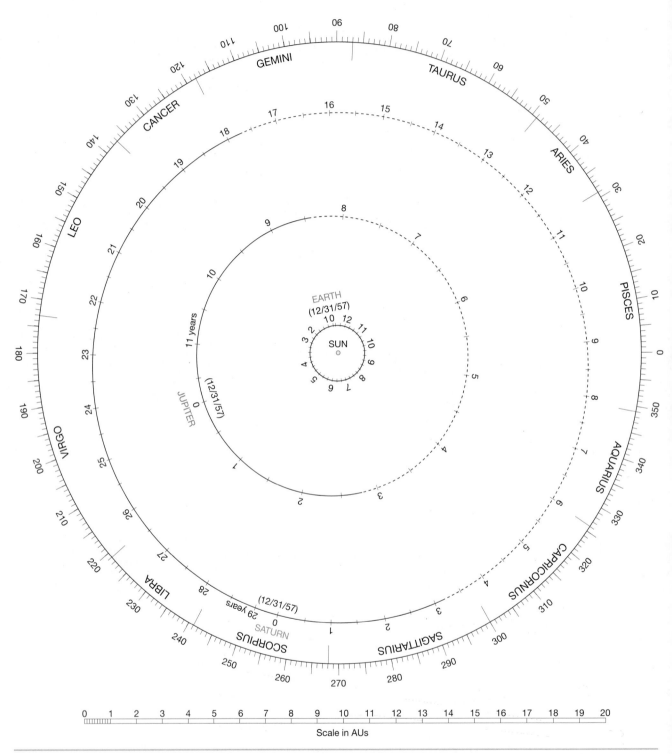

Figure 19.3 Positions of Earth, Jupiter, and Saturn on December 31, 1957. (Adapted from *Field Guide to the Stars and Planets* by Donald Menzel. Copyright 1964 by Donald Menzel. By permission of Houghton Mifflin Company)

1. On Figures 19.1–19.3, draw an arrow on each planet's orbit showing the direction of revolution of the planet.

2. As shown on Figures 19.1–19.3, the direction of revolution of the planets in the solar system, when viewed from above the Northern Hemisphere of Earth, is (clockwise, counterclockwise). Circle your answer.

3. Using Figures 19.1, 19.2, and 19.3, estimate the periods of revolution of Mercury, Mars, and Saturn.

   Mercury: _____ days

   Mars: _____ days

   Saturn: _____ years

4. Refer to Table 18.1, "Planetary Data," in Exercise 18, and record the *exact* period of revolution for the planets listed below. Also, write each planet's exact period of revolution by the zero point (December 31, 1957) on its orbit in Figures 19.1–19.3.

   PERIOD OF REVOLUTION

   Mercury: _____ days

   Venus: _____ . _____ days

   Mars: _____ days

   Jupiter: _____ . _____ years

   Saturn: _____ . _____ years

Determining the Location of a Planet Using the Reference Circle

The common reference circle surrounding Figures 19.1–19.3 is used to locate a **planet's** position around the Sun. The following steps are used to determine a planet's position on the reference circle.

Step 1. Draw a straight line that connects the center of the Sun (the small circle at the center of each figure) with the planet in its orbital position.

Step 2. Extend the line until it intersects the reference circle and note the degree.

5. By using the procedure for locating a planet's position on the reference circle, accurately determine the degree location of the following planets for December 31, 1957 (the zero location of each planet on the charts). As an example, Mercury has already been done.

PLANET	DEGREE LOCATION (ON DECEMBER 31, 1957)
Mercury (Figure 19.1)	(125°)
Venus (Figure 19.1)	_____

Earth (Figure 19.1)   _____

Mars (Figure 19.2)   _____

Jupiter (Figure 19.3)   _____

Saturn (Figure 19.3)   _____

Determining in Which Constellation a Planet Is Located

A planet is often referred to as being "in" a constellation. The reference is to the particular constellation that is located directly behind the planet when that planet is viewed from Earth on that same date. The following steps are used to determine in which constellation a planet is located.

Step 1. Position Earth at the appropriate date in its orbit.

Step 2. Draw a line beginning at the position of Earth through the position of the planet on the same date.

Step 3. Extend the line until it intersects the surrounding reference circle and note the constellation.

6. By following the procedure for determining in which constellation a planet is located, determine the constellation in which each of the planets was observed on December 31, 1957. As examples, Mercury and Jupiter have already been done.

PLANET	CONSTELLATION (ON DECEMBER 31, 1957)
Mercury (Figure 19.1)	(Scorpius)
Venus (Figure 19.1)	_____
Mars (Figure 19.2)	_____
Jupiter (Figure 19.3)	(Virgo)
Saturn (Figure 19.3)	_____

Relative Movement of the Planets

The planets that are farthest from the Sun take longer to complete one revolution than those that are nearest to the Sun (Kepler's third law). Therefore, given the same number of days, planets that are near the Sun will move farther around their orbits than planets that are at a great distance.

To aid in understanding the effects that different periods of revolution have on the location of planets in the sky, complete questions 7–11.

7. On Figures 19.1 and 19.3, advance Mercury, Earth, and Jupiter ninety days (0.25 year) beyond the December 31, 1957, position in their orbits. Place a dot in the orbit of each planet at the new

position. Use the new position to determine the constellations in which Mercury and Jupiter are located.

**CONSTELLATION
(90 DAYS AFTER DECEMBER 31, 1957)**

Mercury: _____

Jupiter: _____

8. Compare the constellation locations of Mercury and Jupiter ninety days after December 31, 1957, to their constellation locations on December 31, 1957 (question 6). What effect has ninety days of motion had on where each planet is seen from Earth?

_____

_____

9. How many revolutions will Mercury complete in one Earth year (365.25 days)? How will this motion influence how we see Mercury throughout the year?

_____

_____

10. What fraction of a revolution will Jupiter complete in one Earth year? How will this motion influence how we see Jupiter throughout the year?

_____

_____

11. Relative to the background of stars, throughout the year the positions of Mercury and Venus change (slightly, considerably), while the positions of Jupiter and Saturn change (slightly, considerably). Circle the correct responses.

Retrograde Motion

The fact that Earth periodically overtakes the more distant planets in their orbits makes the motion of the outer planets *appear*, on occasion, to move backward. The apparent backward, or westward, movement of a planet when viewed from Earth is called **retrograde motion**. (*Note:* A similar observation is made when one vehicle passes another going the same direction on a highway.)

Figure 19.4 illustrates six equally successive positions of Earth and Mars in their orbits over a period of approximately six months. Completing questions 12–16 will help explain retrograde motion.

12. To illustrate retrograde motion, complete the following steps using Figure 19.4.

Step 1. On Figure 19.4, accurately draw a line connecting Earth to Mars at each of the six positions in the orbits. Extend each line from Mars to the background of stars shown on the figure. Mark the place where each line intersects the background of stars and label it with the number of the position. (*Note:* Positions 1 and 2 have already been done and can serve as guides.)

Step 2. Using a continuous line, connect the six numbered positions on the background of stars in order, from 1 to 6.

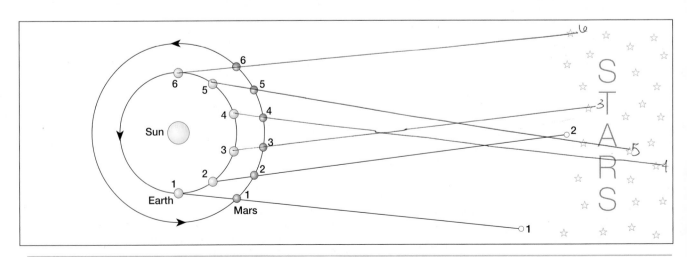

Figure 19.4   Diagram of the Sun, Earth, and Mars for illustrating retrograde motion. (Diagram not drawn to scale)

Use the diagram you have constructed in Figure 19.4 to answer questions 13–16.

13. Describe the apparent motion of Mars, as viewed from Earth, relative to the background stars.

_____

_____

14. At any time did Mars actually move backward in its orbit?

_____

15. Explain the reason for the observed motion of Mars, relative to the background of stars.

_____

_____

16. Assuming, as Ptolemy erroneously did in A.D. 141, that Earth does *not* move in an orbit, how must the observed motion of Mars you described in question 13 be explained?

_____

_____

Adjusting the Planets' Positions on the Orbital Charts

As you have seen, given the same number of days or years, each planet will move a varying amount in its orbit. Therefore, to use the orbital charts (Figures 19.1–19.3) for any date other than December 31, 1957, requires that the charts be adjusted.

The following steps are used to update the locations of the planets on Figures 19.1, 19.2, and 19.3 for any date.

Step 1. Accurately determine the total amount of time (in days and in years) that has elapsed between December 31, 1957 (the date that the current figures represent), and the date you are seeking. (*Note:* One year equals 365.25 days.)

Step 2. Advance each planet the number of days (for Mercury, Venus, and Mars) or years, including fractions (for Jupiter and Saturn), that have elapsed between December 31, 1957, and the new date. Mark the new position with a dot on the planet's orbit. Label the position with the new date.

Step 3. For Earth, locate the date on the orbit that corresponds to the date you are seeking and mark the new position. (The time that has elapsed since the figures were drawn can be disregarded because in any year Earth will be located at the specified date in its orbit.)

Step 4. Determine the position of the planet on the 360° reference circle by drawing a line from the Sun through the new position. To determine the constellation in which a planet is located, draw a line from the new position of Earth through the new position of the planet.

Using the four steps for adjusting the planets' positions on the charts for any date, complete question 17.

17. Determine the positions of the planets for today's date (or the date specified by your instructor). After calculating the number of days and years that have elapsed since the charts were drawn and entering your results below, complete Table 19.1 by listing the new location of each planet in degrees on the reference circle. Also, determine in which constellation the planet is located when viewed from Earth on the specified date and record your answer in Table 19.1.

Date of new positions of the planets: _____

Number of elapsed years since December 31, 1957 = _____ . _____ years

Number of elapsed days since December 31, 1957 = _____ days

Table 19.1 Planet positions on today's date (or the date specified by your instructor)

	LOCATION IN DEGREES	IN THE CONSTELLATION
Mercury:	_____	_____
Venus:	_____	_____
Earth:	_____	(not applicable)
Mars:	_____	_____
Jupiter:	_____	_____
Saturn:	_____	_____

Date of planet positions specified in the table: _____

Viewing a Planet from Earth

Earth rotates on its axis in a counterclockwise direction when viewed from above the Northern Hemisphere, the reference point of Figures 19.1–19.3. To see a planet, you must be in darkness and on the same side of Earth as the planet is located. Also, a planet that is opposite the Sun from Earth cannot be seen.

18. Label the positions of noon, midnight, sunrise, and sunset on Earth at its location for today's date (or the date specified by your instructor) in Figures 19.1, 19.2, and 19.3.

19. Using the locations of the planets for today's date (or the date specified by your instructor) in Figures 19.1–19.3, write a brief description of each planet's visibility on that same date. Indicate in which constellation the planet is located. Also, by examining the times marked on Earth, include, when appropriate, at what time the planet can first be seen (the time it rises) and at what time it sets.

Visibility from Earth of:

Mercury: _____

_____

_____

Venus: _____

_____

_____

Mars: _____

_____

_____

Jupiter: _____

_____

_____

Saturn: _____

_____

_____

Planet Positions on the Internet

Continue your analyses of the topics presented in this exercise by completing the corresponding online activity on the *Applications & Investigations in Earth Science* website at http://prenhall.com/earthsciencelab

Planet Positions

Date Due: _____

Name: _____

Date: _____

Class: _____

After you have finished Exercise 19 complete the following questions. You may have to refer to the exercise for assistance or to locate specific answers. Be prepared to submit this summary/report to your instructor at the designated time.

1. In Figure 19.5 prepare a diagram of the solar system showing the relative positions of the planets Mercury through Saturn for today's date (or the date specified by your instructor). View the solar system from above the Northern Hemisphere of Earth. Show the planets' locations as accurately as possible and label each.

2. Which, if any, planet(s) is/are visible from Earth on the diagram you prepared in Figure 19.5? At what times do each of the visible planets rise and set?

_____

_____

_____

3. What is the reason that a planet will periodically exhibit retrograde motion?

_____

_____

4. Describe what is meant when a planet is said to be "in" a particular constellation.

_____

_____

5. When viewed from Earth, why does the position of Saturn change only slightly from year to year?

_____

_____

6. Describe the relative positions of Earth and Mars when Mars is visible from Earth.

_____

_____

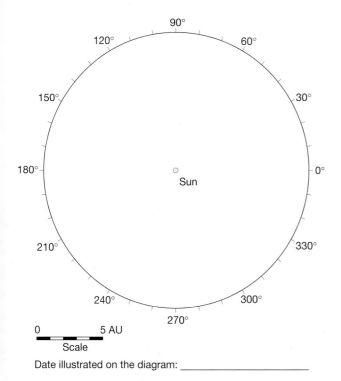

Date illustrated on the diagram: _____

Figure 19.5 Planet locations.

7. Figure 19.6 illustrates the Sun, Earth, and Venus viewed from above the Northern Hemisphere of Earth. Venus moves to the opposite side of the Sun in its orbit approximately every 112 days. The two positions of Venus represent the planet in its orbit on either side of the Sun.

Draw an arrow around Earth showing its direction of rotation. Label the location of noon, sunset, midnight, and sunrise on Earth. By circling your answers on the figure, indicate in which position Venus will be visible from Earth prior to sunrise and in which position it will be visible from Earth after sunset.

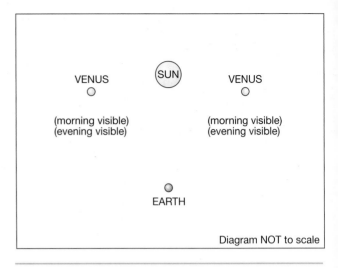

Figure 19.6 Visibility of Venus.

The Moon and Sun

The Moon and Sun are two of the nearest celestial objects to Earth, and each has inspired generations of observers to wonder about the nature of the cosmos (Figure 20.1). Many ancient civilizations revered these objects and devised methods for keeping track of their changing positions in the sky. Today, lunar and solar studies are providing astronomers with information that helps to answer questions about the evolution of the solar system and the source of energy of stars. This exercise investigates the motions and features of the Moon, as well as the structure and dynamic surface of our "nearest star," the Sun.

Objectives

After you have completed this exercise, you should be able to:

1. Recognize and name each of the phases of the Moon.

2. Diagram the Earth, Moon, and Sun in their proper relation for each of the phases of the Moon.

3. Explain the difference between the synodic and sidereal cycles of the Moon.

4. Diagram the Earth, Moon, and Sun in their proper relation during a solar and lunar eclipse.

5. Discuss the difference between lunar terrae and maria and be able to recognize each on a lunar map or photograph.

6. Determine the relative ages of lunar features.

7. Recognize and name the different types of lunar craters.

8. Describe several of the features found on the solar surface.

Materials

metric ruler   hand lens   calculator

Figure 20.1   Telescopic view of the lunar surface. (Image © UC Regents/Lick Observatory)

Materials Supplied by Your Instructor

stereoscope	lunar globe (optional)
sandbox	sand
meterstick	small balls (various masses)

Terms

synodic month	maria
sidereal month	new Moon
solar eclipse	crater
lunar eclipse	prominence
terrae	sunspot

The Moon

At an average distance of 384,401 kilometers (238,329 miles), Earth's nearest celestial neighbor and only natural satellite is the Moon. The monthly counterclockwise revolution around Earth and associated Earth-Moon revolution about the Sun change the Moon's position relative to the Sun, producing the phases viewed by Earthbound observers. Furthermore, as the Moon moves in its orbit, its slow counterclockwise rotation results in the same side continuously facing Earthward.

The Moon has no atmosphere, and therefore no landforms produced by wind or water erosion. Instead, the vast majority of lunar surface features are the result of ancient volcanic eruptions and impact cratering by *meteoroids*. Many features that formed on the Moon's surface over 3 billion years ago are still discernible and have been only slightly modified by the continuous bombardment of tiny *micrometeorites*.

Phases of the Moon

The changing phases of the Moon have been recorded throughout history and were among the first astronomical phenomena to be understood. The lunar phases observed from Earth are the result of the motion of the Moon and sunlight that is reflected from its surface. The half of the Moon facing directly toward the Sun is illuminated at all times. However, to an Earth-bound observer the percentage of the bright side that is visible depends on the location of the Moon with respect to the Sun and Earth. When the Moon lies between the Sun and Earth, none of the bright side can be seen by an Earthbound observer—a phase called *new-Moon* ("no-moon"). Conversely, when the Moon lies on the side of Earth opposite the Sun, all of its bright side is visible, producing the *full-Moon* phase. At any position between these extremes, only a fraction of the Moon's illuminated half is visible.

Figure 20.2 is a view of the Earth-Moon-Sun system from above the Northern Hemisphere of Earth. Eight positions of the Moon during its monthly journey around Earth are illustrated. Notice on the figure that the illuminated half of the Moon is always directly toward the Sun. However, the portion of the illuminated half visible from Earth changes during the lunar cycle. To help you understand the lunar phases, complete questions 1–13 using Figure 20.2.

1. For each of the eight numbered positions of the Moon, draw a line through the Moon which separates the half of the Moon that is *visible from Earth* (the half directly toward Earth) from the half which cannot be seen.

2. As the Moon journeys from position 1 to position 5, the proportion of its *illuminated side visible from Earth* is (increasing, decreasing). Circle your answer.

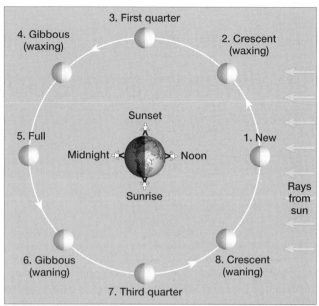

Diagram NOT drawn to scale

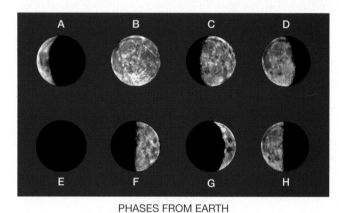

PHASES FROM EARTH

Figure 20.2 The lunar cycle as viewed from above the Northern Hemisphere of Earth, including phases of the Moon as observed from Earth (A–H).

3. As the Moon journeys from position 5 to position 8, the proportion of its *illuminated side visible from Earth* is (increasing, decreasing). Circle your answer.

4. Complete the following by matching each of the eight drawings of the lunar phases as seen from Earth (A–H) with the Moon's appropriate position (1–8) on the figure.

POSITION NUMBER	LETTER OF PHASE SEEN FROM EARTH	NAME OF PHASE
1.	_____	_____
2.	_____	_____
3.	_____	_____
4.	_____	_____
5.	_____	_____
6.	_____	_____
7.	_____	_____
8.	_____	_____

Observe the time of day (noon, sunset, etc.) represented by the four positions of the Earth-bound observer on Figure 20.2. Note that an observer can see the Moon approximately 90° on either side of his or her location—that is, about $\frac{1}{4}$ of a circle in both directions. Using the eight positions of the Moon in its orbit and the times represented on Earth, answer questions 5–13.

5. The new Moon is highest in the sky to an Earth-bound observer at (noon, sunset, midnight, sunrise). Circle your answer.

6. The full Moon appears highest in the sky to an Earth-bound observer at (noon, sunset, midnight, sunrise). Circle your answer.

7. Throughout the lunar cycle, the Moon moves further (eastward, westward) in the sky each day. Therefore, to an Earth-bound observer, the time of day when the Moon is highest in the sky becomes progressively (earlier, later). Circle the correct responses.

8. Can a full Moon be observed from Earth by an observer positioned at noon? Explain the reason for your answer.

_____

_____

9. A full Moon first becomes visible to an Earth-bound observer positioned at (noon, sunset, midnight, sunrise), and she or he must look (eastward, westward) to see the rising Moon. Circle the correct responses.

10. Can the first- and third-quarter lunar phases be observed during the daylight hours? Explain the reason for your answer.

_____

_____

11. At approximately what times will the first-quarter Moon rise and set?

    Rise: _____   Set: _____

12. At approximately what times will the third-quarter Moon rise and set?

    Rise: _____   Set: _____

13. Assume a crescent-phase Moon is observed in the early evening in the western sky. During the next few days the Moon will be rising (earlier, later) and the visible, illuminated portion of the Moon will become progressively (larger, smaller). Circle your answers.

Synodic and Sidereal Months

The time interval required for the Moon to complete a full cycle of phases is 29.5 days, a period of time called the **synodic month**. This complete cycle of the phases of the Moon (i.e. new-Moon to the next new-Moon) is the basis of the word "month" (or "moonth"). Although the cycle of phases requires 29.5 days, the true period of the Moon's 360° revolution around Earth takes only 27.3 days and is known as the **sidereal month**. The difference of approximately two days results from the fact that as the Moon revolves around Earth, the Earth-Moon system also is moving around the Sun.

Figure 20.3 illustrates an exaggerated month of motion of the Earth-Moon system around the Sun. Refer to the figure to answer questions 14–22.

14. On Month I of Figure 20.3, indicate the dark half of the Moon on each of the eight lunar positions by shading the appropriate area with a pencil.

15. Select from the eight lunar positions in Month I, and indicate which represents the following lunar phases.

PHASE	LUNAR POSITION (MONTH I)
New-Moon	_____
Third quarter	_____
Full-Moon	_____
First quarter	_____

16. On Month I, label the position of the new-Moon phase with the words "new Moon."

17. In Month II of Figure 20.3, lunar position number (1, 3, 5, 7) represents the new-Moon phase. Circle your answer and label the position of the new-Moon phase on Month II with the words "new Moon."

Begin with the position of the new-Moon phase in Month I, and imagine revolving the Moon 360° around Earth while at the same time moving it to Month II.

18. After a 360° revolution beginning at the new-Moon phase in Month I, the Moon is located at position (6, 7, 8) in Month II. Circle your answer.

19. A complete 360° revolution of the Moon around Earth is called a _____ month, which takes _____ days.

20. The position of the Moon you determined in question 18 occurs (before, after) the Moon completes a full cycle of phases from Month I to Month II. Circle your answer.

21. In Month II, when the Moon moves the additional distance in its orbit, from position 6 to 7, and again is at the new-Moon phase, it will have completed a _____ month, which takes _____ days.

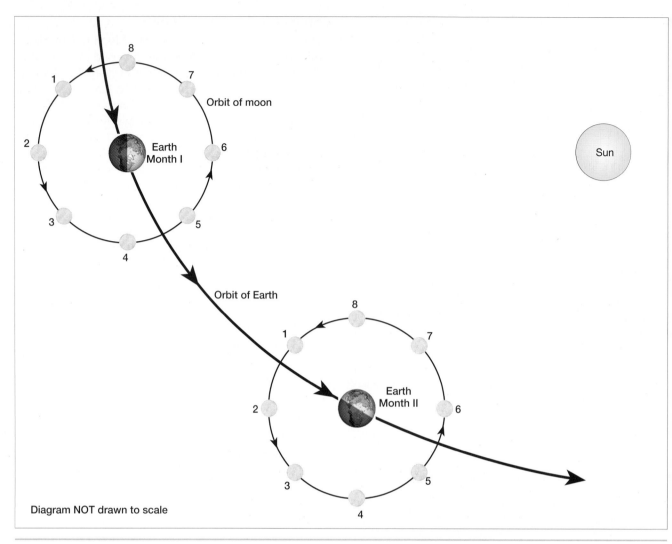

Diagram NOT drawn to scale

Figure 20.3 Monthly motion of the Earth-Moon system around the Sun viewed from above the Northern Hemisphere.

22. In your own words, explain the difference between a sidereal and synodic month.

_____

_____

Eclipses

Eclipses occur when the Sun, Moon, and Earth are in a direct line. An eclipse can be either a **solar eclipse** (when the Moon moves in a line directly between Earth and the Sun) or a **lunar eclipse** (when the Moon moves within Earth's shadow).

23. Refer to Figure 20.4. Describe a solar eclipse and prepare a "side view" diagram showing the relation of the Sun, Earth, and Moon during the eclipse.

Description: _____

_____

Figure 20.4 Solar eclipse diagram.

24. A solar eclipse occurs during the (new-Moon, first-quarter, full-Moon) phase of the Moon. Circle your answer.

25. Refer to Figure 20.5. Describe a lunar eclipse and prepare a "side view" diagram showing the relation of the Sun, Earth, and Moon during the eclipse.

Description: _____

_____

Figure 20.5 Lunar eclipse diagram.

26. A lunar eclipse occurs during the (new-Moon, third quarter, full-Moon) phase of the Moon. Circle your answer.

27. Suggest a reason(s) why Earth does not experience a solar and lunar eclipse during each synodic month.

_____

_____

_____

The Lunar Surface

Since the Moon lacks an atmosphere, its landscape has been shaped primarily by meteoroid impacts and ancient volcanic processes. In general, the Moon's surface can be classified as one of two types. As illustrated in Figure 20.6, **terrae** (plural for *terra*, the Latin word for land) are lunar highlands, which are the bright areas of the Moon seen from Earth. The dark areas of the Moon, called **maria** (plural for *mare*, the Latin word for sea), are flat lowland regions. Together, the arrangement of terrae and maria on the lunar surface result in the well-known "face on the Moon."

The most obvious features on the lunar surface are **craters**. Many different types of craters exist (see Figure 20.6), with most of them produced when rapidly moving debris impacted the lunar surface.

Figure 20.7 is a near-side, enhanced photograph of the Moon with many of the major features labeled. Lunar latitude and longitude are indicated along the edges of the photo. A graphic scale for determining distances is also included. Use Figures 20.6 and 20.7 to answer questions 28–43.

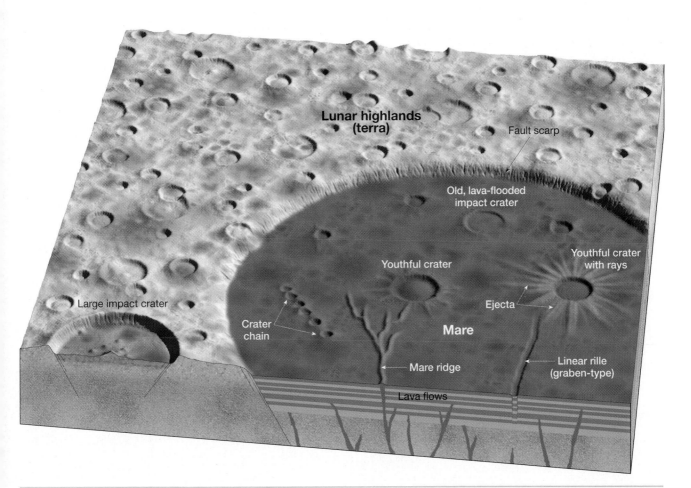

Figure 20.6 Block diagram illustrating major lunar features.

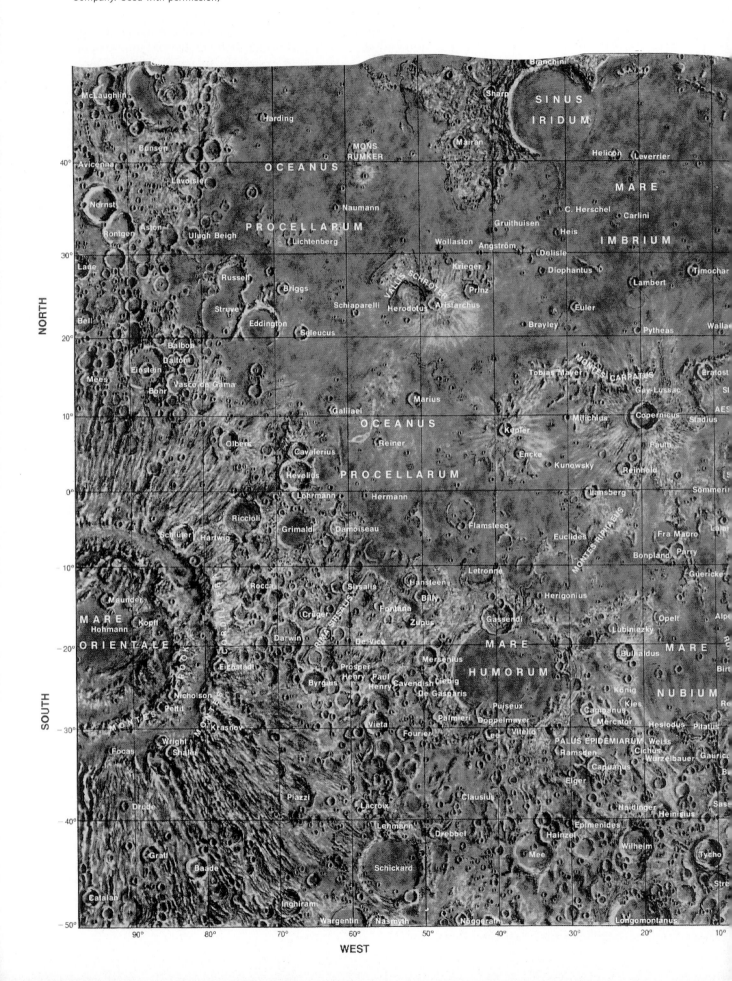

±50°
±40°
±30°
±20°
±10°
0°

0   100   200   300   400   500   600   700   800 km

| 10° | 20° | 30° | 40° | 50° | 60° | 70° | 80° | 90° |

Egede · Aristoteles · Mitchell · Baily · Atlas · Mercurius · Zeno · Boss
Eudoxus · LACUS · Bürg · Hercules · Chevallier · Carrington · Vashakidze
MORTIS · Mason · Oersted · Shuckburgh · Schumacher · Gauss
Plana · Williams · Hooke · Messala · Rynin
LACUS · Grove · Cepheus · Franklin · Berosus · Vestine
SOMNIORUM · Maury · Berzelius · Bernoulli · Hahn
Daniell · Geminus · Burckhardt
Hall · G. Bond · Cleomedes · Delmotte · Seneca · Liapunov · Joliot
Posidonius · Kirchhoff · Newcomb · Plutarch · Hubble
MARE · Chacornac · MONTES · Eimmart · Cannon · Al Biruni
SERENITATIS · TAURUS · Römer · MARE · Alhazen · Goddard
Linné · Macrobius · Tisserand · CRISIUM · Hansen · Ibn Yunis · Ginzel
Bessel · Littrow · Peirce · MARE · Dreyer
Sulpicius Gallus · Maraldi · Vitruvius · PALUS · Proclus · Yerkes · Picard · Condorcet · MARGINIS · Neper · Jansky
Menelaus · Dawes · Franz · SOMNII · Auzout · Babcock
Plinius · Lyell · Lick · Firmicus · Schubert
Ross · Jansen · Cauchy · Da Vinci · Apollonius · MARE · Dubiago · MARE
Maclear · Taruntius · SPUMANS · SMYTHII
MARE · Sosigenes · Secchi · MARE · Maclaurin · Gilbert · Hirayama
Arago · Censorinus · Webb · Kästner
TRANQUILLITATIS · Messier · FECUNDITATIS · Purkyne
Maskelyne · Torricelli · Gutenberg · Langrenus · La Pérouse · Brunner
Isidorus · Capella · Goclenius · Ansgarius
Theophilus · Mädler · Lohse · Lamé · Behaim · Ritz
Cyrillus · Colombo · Vendelinus · Gibbs · Schorr
MARE · Cook · Holden · Hecataeus
NECTARIS · Monge · Balmer
Catharina · Santbech · Wrottesley · Petavius · Phillips · Humboldt · Barnard
Fracastorius · Borda · Hase · Legendre · Curie
Weinek · Snellius · Adams · Abel
Piccolomini · Reichenbach · Stevinus · Oken · Marinus
Neander · Furnerius · Gum
Rheita · Fraunhofer · Hamilton · Jenner
Metius · Young · Vega · Lyot
Fabricius · Peirescius · Brisbane
Janssen · Steinheil

NORTH

SOUTH

28. What is the origin of the lunar maria?

_____

_____

29. By examining Figure 20.7 (also see Figure 20.1), approximately (20, 40, 70) percent of the near-side of the Moon consists of lunar maria. Circle your answer.

30. What is the name and approximate width of the mare located at 16°N latitude and 59°E longitude?

    Mare _____ , _____ km wide.

31. What is the lunar latitude and longitude of the crater named Copernicus, located near the center of the map?

    Latitude: _____ , longitude: _____

32. Locate the following lunar features on Figure 20.7 and give the lunar latitude and longitude of each. Also, use Figure 20.6 as a guide to indicate the type of feature represented by each.

	LOCATION	TYPE OF FEATURE
Sinus Iridum:	_____ ,	_____
Humboldt:	_____ ,	_____
Mare Orientale:	_____ ,	_____

Rupes Altai: _____ , _____

Kepler: _____ , _____

The number of impact craters within an area can be used to determine the age of a surface. In general, the more craters that are present, the longer the surface has been in existence.

33. Examine the frequency of craters on the lunar highlands compared to those on the maria. Lunar (terrae, maria) have formed most recently. Circle your answer.

34. Rocks brought back from the lunar maria during the manned Apollo landings are about 3.2–3.8 billion years old. Therefore, lunar highlands are (older, younger) than 3.2–3.8 billion years. Circle your answer.

35. Compare the crater frequency on the floor of Mare Smythii (2°S, 87°E) with that of Sinus Iridum (45°N, 31°W). Sinus Iridum appears to be (older, younger) than Mare Smythii.

36. Mare Smythii appears to be (older, younger) than Mare Crisium (16°N, 59°E).

When craters overlap, the rim of the most recent crater will cut through the rim of the older.

37. Using a stereoscope, examine the stereogram of the overlapping lunar craters shown in Figure 20.8

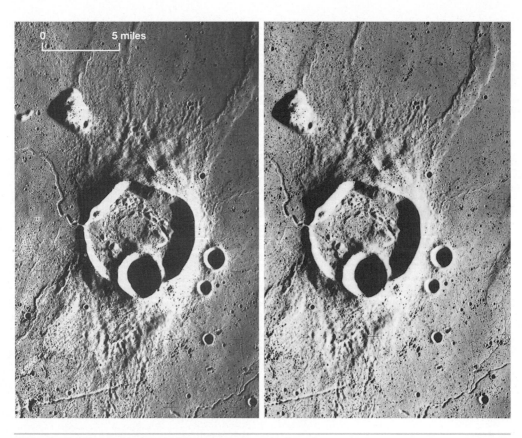

Figure 20.8 Stereogram of lunar craters. (Photo courtesy of NASA)

and label the most recent crater with the word "youngest."

38. Observe the crater Gasserdi (17°S, 39°W) and its relation to Mare Humorum in Figure 20.7. Crater Gasserdi is (older, younger) than Mare Humorum.

39. Locate craters Mee, Hainzel, and the unnamed crater northwest of Hainzel at approximately latitude 42°S and longitude 35°W in Figure 20.7. List the craters in order, from youngest to oldest.

Youngest: _____

_____

Oldest: _____

   Most crater rims become rounded after long periods of bombardment by sand-size particles (micrometeorites). Therefore, the "sharpness" of a crater is an indication of its age.

40. Examine the crater Copernicus on the lunar map, Figure 20.7, and compare the "sharpness" of its rim to other lunar craters. What conclusion can you make about the age of Copernicus?

_____

41. What conclusion can be made about the age of crater Mee compared to crater Tycho, directly east of Mee?

_____

   Examine the crater Copernicus and the area around it closely in Figure 20.7.

42. What type of crater is Copernicus? You may find Figure 20.6 useful.

_____

43. What is the origin of the bright rays that radiate outward from Copernicus?

_____

_____

44. It is likely that Earth was bombarded with meteoroids early in its history at least as frequently as the Moon. If so, why are there so few craters visible on Earth's surface today?

_____

_____

Impact Cratering

Impact cratering is one of the most common processes responsible for altering the surface of many planets and moons. To assist you in understanding the process and how the size and shape of an impact crater is related to the properties of the object that produce it, observe the equipment in the laboratory (Figure 20.9) and then conduct the following experiment by completing each of the indicated steps.

Step 1. Gather the equipment necessary to conduct the impact cratering experiment.

Step 2. Add sand to the sandbox. Flatten the surface of the sand with a wooden ruler.

Step 3. Write a hypothesis describing the suspected relation between an impact crater's diameter and the mass and velocity of the object that produces it.

_____

_____

_____

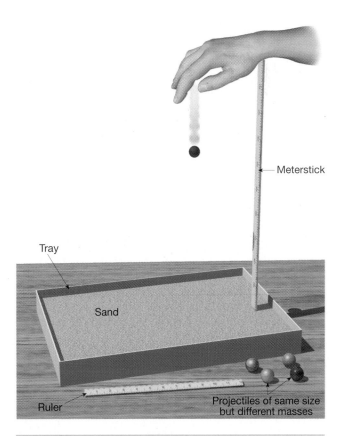

Figure 20.9   Impact cratering lab-equipment setup.

Table 20.1   Impact crater data sheet

Ball Type		Crater Diameter		
Description	Mass	0.5 m Height Crater Diameter (mm)	1.0 m Height Crater Diameter (mm)	1.5 m Height Crater Diameter (mm)

Step 4. One at a time, drop each of the balls from heights of 0.5 m, 1.0 m, and 1.5 m on the sand in the box and measure the diameter in millimeters of the crater produced in each drop. Make sure to flatten the surface of the sand between each drop. Repeat each drop several times, keeping all the variables constant. Record the average for each of the drops on the data sheet, Table 20.1.

45. Which of the variables is directly related to the velocity of the falling object(s)?

_____

_____

46. Examine your data closely and state your conclusions concerning the general relationships between crater size and the 1) mass and 2) velocity of the object that produced the crater.

_____

_____

_____

_____

_____

_____

47. Write a general statement that evaluates your impact-cratering hypothesis with reference to your conclusions.

_____

_____

_____

The Sun

At an average distance of 150 million kilometers (93 million miles), the Sun is the nearest star to Earth. When compared to the other billions of stars in our galaxy, our Sun is considered only "average." However, it is not only important to us as our primary source of energy, but also to astronomers, since it is the only star whose surface can be observed in detail.

Solar Features

As the Sun rotates, it does so differentially, taking fewer days to complete one rotation on its equator than near the poles. The unequal period of rotation causes variations in the Sun's magnetic field, which in turn influence many of its surface features.

Figure 20.10 illustrates several features of the active Sun when the solar disk is photographed in hydrogen

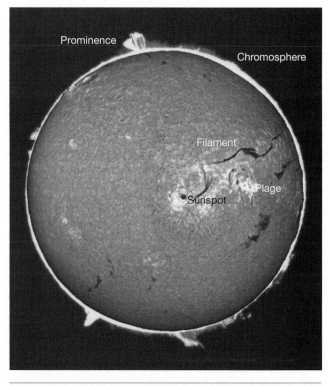

Figure 20.10   The solar disk photographed in hydrogen alpha light. (This composite courtesy of Hale Observatories and National Solar Observatory, Sacramento Peak)

alpha light. The actual diameter of the solar image in the figure is approximately 870,000 miles. Use Figure 20.10 to answer questions 48–54.

48. Using a metric ruler, measure the diameter of the solar image in millimeters. Then determine the scale of the photograph in miles per millimeter and write the scale on the photograph.

$$\text{Scale} = \frac{870,000 \text{ miles}}{\underline{\hspace{2cm}} \text{ mm}} = \underline{\hspace{2cm}} \text{ mi/mm}$$

49. The diameter of Earth is approximately 8,000 miles. Using the scale you calculated in question 48, draw a scale mile Earth on the surface of the Sun in Figure 20.10. Label your drawing "Earth." Approximately how many times larger than the Earth is the diameter of the Sun?

_____

Notice the "granulated" appearance of the solar image on the photograph.

50. What is the cause of the irregular, grainy appearance of the solar surface?

_____

_____

Locate the large solar **prominence** near the top of the photograph.

51. Using the scale you prepared in question 48, approximately how many miles does the prominence extend above the surface of the Sun?

_____ miles

52. Describe the appearance and apparent cause of prominences.

_____

_____

Examine the large **sunspot** (dark "blemish") near the center of the solar disk.

53. What is the approximate diameter of the sunspot?

_____ miles

54. Suggest a reason why sunspots appear as dark areas.

_____

_____

The Moon and Sun on the Internet

Continue your analyses of the topics presented in this exercise by completing the corresponding online activity on the *Applications & Investigations in Earth Science* website at http://prenhall.com/earthsciencelab

Notes and calculations.

The Moon and Sun

Date Due: _____

Name: _____

Date: _____

Class: _____

After you have finished Exercise 20, complete the following questions. You may have to refer to the exercise for assistance or to locate specific answers. Be prepared to submit this summary/report to your instructor at the designated time.

1. Write a brief paragraph explaining the reasons for the changes that occur in the phases of the Moon as observed from Earth during a full lunar cycle of 29.5 days.

_____

_____

_____

2. Each of the four photographs in Figure 20.11 was taken from Earth when the Moon was at its highest position in the sky. In the space provided below each photo, write the name of the phase represented and the time of day when the picture was taken.

3. Figure 20.12 illustrates the Earth-Moon system viewed from above the Northern Hemisphere. If you have completed Exercise 17, "Astronomical

Observations," draw a circle representing the Moon in the proper position in its orbit for each of the lunar phases you observed and recorded in the exercise. Indicate the dark half of the Moon by shading each circle. Label each position with the date and time of your observation. (If you

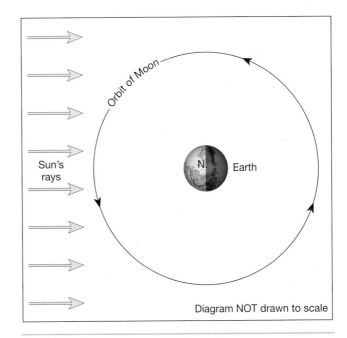

Figure 20.12 Lunar phases diagram.

Figure 20.11 Lunar photos. (Image © UC Regents/Lick Observatory)

have not completed Exercise 17, complete the diagram by illustrating the Moon at its proper position during full-Moon, new-Moon, first quarter, and third quarter. Label each position with the name of the phase.)

4. Explain the difference between a sidereal and synodic month.

_____

_____

_____

5. What are the most obvious differences in appearance between terrae and maria on a lunar photograph or map?

_____

_____

6. Questions 6a–6d refer to Figure 20.13, a photograph of a 20-kilometer-wide lunar crater.

   a. The crater is located in a (terra, mare) region of the Moon. Circle your answer.

Figure 20.13 A 20-km-wide lunar crater. (Photo courtesy of NASA)

b. What evidence suggests that the crater is of comparatively recent origin?

_____

c. The large crater in the photograph is of what type?

_____

d. What is the origin of the rays that extend from the crater rim?

_____

_____

7. When two craters overlap, how can you determine which is the most recent?

_____

_____

8. Define each of the following terms.

New-Moon: _____

_____

Solar eclipse: _____

_____

Sunspot: _____

_____

Lunar terrae: _____

_____

9. Briefly summarize the results of your impact-crater experiment.

_____

_____

_____

_____

Earth Science Skills

Location and Distance on Earth

The ability to find places and features on Earth's surface using maps and globes is an essential skill required of all Earth scientists. This exercise introduces the most commonly used system for determining location on Earth. Using the system as a foundation, you will examine ways to measure distance on Earth's surface.

Objectives

After you have completed this exercise, you should be able to:

1. Explain the most common system used for locating places and features on Earth.
2. Use Earth's grid system to accurately locate a place or feature.
3. Explain the relation between latitude and the angle of the North Star (Polaris) above the horizon.
4. Explain the relation between longitude and solar time.
5. Determine the shortest route and distance between any two places on Earth's surface.

Materials

ruler   calculator
protractor

Materials Supplied by Your Instructor

globe   50–80 cm length of
world wall map   string
atlas

Terms

Earth's grid	South Pole	solar time
latitude	longitude	standard time
parallel of	meridian of	great circle
latitude	longitude	small circle
equator	prime meridian	
North Pole	hemisphere	

Introduction

Globes and maps each have a system of north-south and east-west lines, called the **Earth's grid**, that forms the basis for locating points on Earth (Figure 21.1). The grid is, in effect, much like a large sheet of graph paper that has been laid over the surface of Earth. Using the system is very similar to using a graph; that is, the position of a point is determined by the intersection of two lines.

Latitude is north-south distance on Earth (Figure 21.1). The lines (circles) of the grid that extend around Earth in an east-west direction are called **parallels of latitude**. *Parallels of latitude mark north and south distance from the **equator** on Earth's surface.* As their name implies, these circles are parallel to one another. Two places on Earth, the **North Pole** and **South Pole**, are exceptions; they are points of latitude rather than circles.

Longitude is east-west distance on Earth (Figure 21.1). **Meridians of longitude** are each halves of circles that extend from the North Pole to the South Pole on

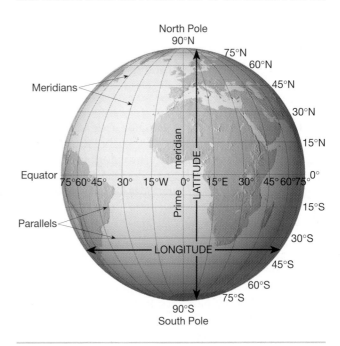

Figure 21.1 Earth's grid system.

one side of Earth. *Meridians of longitude mark east and west distance from the **prime meridian** on Earth's surface.* Adjacent meridians are farthest apart on the equator and converge (come together) toward the poles.

> **The intersection of a parallel of latitude with a meridian of longitude determines the location of a point on Earth's surface.**

Earth's shape is nearly spherical. Since parallels and meridians mark distances on a sphere, their designation, like distance around a circle, is given in *degrees* (°). For more precise location, a degree can be subdivided into sixty equal parts, called *minutes* ('), and a minute of angle can be divided into sixty parts, called *seconds* ("). Thus, 31°10'20" means 31 degrees, 10 minutes, and 20 seconds.

The type of map or globe used determines the accuracy to which a place may be located. On detailed maps it is often possible to estimate latitude and longitude to the nearest degree, minute, and second. On the other hand, when using a world map or globe, it may only be possible to estimate latitude and longitude to the nearest whole degree or two.

In addition to showing location on Earth, latitude and longitude can be used to determine distance. Knowing the shape and size of Earth, the distance in miles and kilometers covered by a degree of latitude or longitude has been calculated. These measurements provide the foundation for navigation.

Determining Latitude

The equator is a circle drawn on a globe that is equally distant from both the North Pole and South Pole. It divides the globe into two equal halves, called **hemispheres**. The equator serves as the beginning point for determining latitude and is assigned the value 0°00'00" latitude.

> **Latitude is distance north and south of the equator, measured as an angle in degrees from the center of Earth (Figure 21.2).**

Latitude begins at the equator, extends north to the North Pole, designated 90°00'00"N latitude (a 90° angle measured north from the equator), and also extends south to the South Pole, designated 90°00'00"S latitude. *The poles and all parallels of latitude, with the exception of the equator, are designated either N (if they are north of the equator) or S (if they are south of the equator).*

1. Locate the equator on a globe. Figure 21.3 represents Earth, with point B its center. Sketch and label the equator on the diagram in Figure 21.3. Also label the Northern Hemisphere and Southern Hemisphere on the diagram.
2. On Figure 21.3, make an angle by drawing a line from point A on the equator to point B (the center

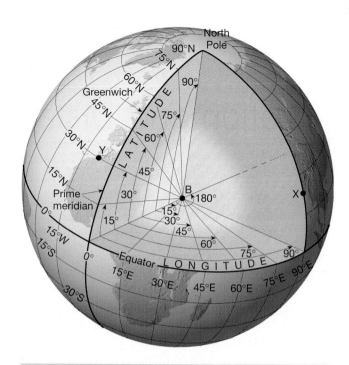

Figure 21.2 Measuring latitude and longitude. The angle measured from the equator to the center of Earth (B) and then northward to the parallel where point Y is located is 30°. Therefore, the latitude of point Y is 30°N. All points on the same parallel as Y are designated 30°N latitude.

The angle measured from the prime meridian where it crosses the equator to the center of Earth (B) and then eastward to the meridian where point X is located, is 90°. Therefore, the longitude of point X is 90°E. All points on the same meridian as X are designated 90°E longitude.

of Earth). Then extend the line from point B to point C in the Northern Hemisphere. The angle you have drawn (∠ABC) is 45°. Therefore, by definition of latitude, point C is at 45°N latitude.

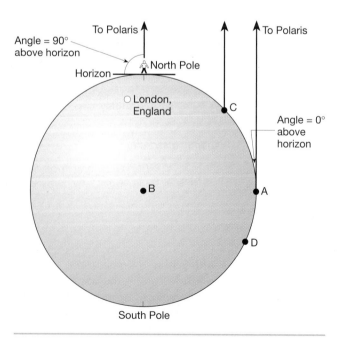

Figure 21.3 Hypothetical Earth.

3. Draw a line on Figure 21.3 parallel to the equator that also goes through point C. All points on this line are 45°N latitude.

4. Using a protractor, measure ∠ABD on Figure 21.3. Then draw a line parallel to the equator that also goes through point D. Label the line with its proper latitude.

On a map or globe, parallels may be drawn at any interval.

5. How many degrees of latitude separate the parallels on the globe you are using?

_____ degrees of latitude between each parallel

6. Keep in mind that the lines (circles) of latitude are parallel to the equator and to each other. Locate some other parallels on the globe. Sketch and label a few of these on Figure 21.3.

7. Use the diagram that illustrates parallels of latitude, Figure 21.4, to answer questions 7a and 7b.

   a. Accurately draw and label the following additional parallels of latitude on the figure.

   5°N latitude

   10°S latitude

   25°N latitude

   b. Refer to Figure 21.4. Write out the latitude for each designated point as was done for points A and B. Remember to indicate whether the point is north or south of the equator by writing an N or S and include the word "latitude."

   Point A: (30°N latitude)   Point D: _____

Point B: (5°S latitude)   Point E: _____

Point C: _____   Point F: _____

8. Use a globe or atlas to locate the cities listed below and give their latitude to the nearest degree. Indicate N or S and include the word "latitude."

   Moscow, Russia: _____

   Durban, South Africa: _____

   Your home city: _____

   Your college campus city: _____

9. By using a globe or atlas, give the name of a city or feature that is equally as far south of the equator as your home city is north.

   _____

10. The farthest one can be from the equator is (45, 90, 180) degrees of latitude. Circle your answer.

11. The two places on Earth that are farthest from the equator to the north and to the south are called the

   _____ and _____ .

There are five special parallels of latitude marked and named on most globes.

12. Use a globe or atlas to locate the following special parallels and indicate the name given to each.

                                        NAME OF PARALLEL

   66°30′00″N latitude: _____

   23°30′00″N latitude: _____

   0°00′00″ latitude: _____

   23°30′00″S latitude: _____

   66°30′00″S latitude: _____

Latitude and the North Star

Today most ships use GPS navigational satellites to determine their location. (For information about the Global Positioning System, visit the website listed at the end of this exercise.) However, early explorers were well aware of the concept of latitude and could use the angle of the North Star (a star named Polaris) above the horizon to determine their north-south position in the Northern Hemisphere. As shown on Figure 21.3, someone standing at the North Pole would look overhead (90° angle above the horizon) to see Polaris. Their latitude is 90°00′00″N. On the other hand, someone standing on the equator, 0°00′00″ latitude, would observe Polaris on the horizon (0° angle above the horizon).

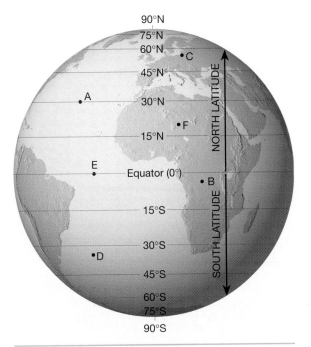

Figure 21.4 Parallels of latitude.

Use Figure 21.3 to answer questions 13–14.

13. The angle of Polaris above the horizon for some-one standing at point C would be (45°, 90°, 180°). Circle your answer.

14. What is the relation between a particular latitude and the angle of Polaris above the horizon at that latitude?

_____

15. What is the angle of Polaris above the horizon at the following cities?

ANGLE OF POLARIS ABOVE THE HORIZON

Fairbanks, AK: _____ degrees

St. Paul, MN: _____ degrees

New Orleans, LA: _____ degrees

Your home city: _____ degrees

Your college campus city: _____ degrees

Determining Longitude

Meridians are the north-south lines (half circles) on the globe that converge at the poles and are farthest apart along the equator. They are used to determine longitude, which is distance east and west on Earth (Figure 21.1). Each meridian extends from pole to pole on one side of the globe.

Notice on the globe that all meridians are alike. The choice of a zero, or beginning, meridian was arbitrary. The meridian that was chosen by international agreement in 1884 to be 0°00'00" longitude passes through the Royal Astronomical Observatory at Greenwich, England, located near London. This internationally accepted reference for longitude is named the *prime meridian*.

Longitude is distance, measured as an angle in degrees **east and west of the prime meridian** (Figure 21.2).

Longitude begins at the prime meridian (0°00'00" longitude) and extends to the east and to the west, halfway around Earth to the 180°00'00" meridian, which is directly opposite the prime meridian. *All meridians, with the exception of the prime meridian and the 180° meridian, are designated either E (if they are east of the prime meridian) or W (if they are west of the prime meridian).*

16. Locate the prime meridian on a globe. Sketch and label it on the diagram of Earth, Figure 21.3.

17. Label the Eastern Hemisphere, that half of the globe with longitudes east of the prime meridian, and the Western Hemisphere on Figure 21.3.

On a map or globe, meridians can be drawn at any interval.

18. How many degrees of longitude separate each of the meridians on your globe?

_____ degrees of longitude between each meridian

19. Keep in mind that meridians are farthest apart at the equator and converge at the poles. Sketch and label several meridians on Figure 21.3.

20. Use the diagram that illustrates meridians of longitude, Figure 21.5, to answer questions 20a and 20b.

a. Accurately draw and label the following additional meridians of longitude on the figure.

35°W longitude

70°E longitude

10°W longitude

b. Refer to Figure 21.5. Write out the longitude for each designated point as was done for points A and B. Remember to indicate whether the point is east or west of the prime meridian by writing an E or W and include the word "longitude."

Point A: <u>(30°E longitude)</u> Point D: _____

Point B: <u>(20°W longitude)</u> Point E: _____

Point C: _____ Point F: _____

21. Use a globe or atlas to locate the cities listed below and give their longitude to the nearest degree. Indicate either E or W and include the word "longitude."

Wellington, New Zealand: _____

Honolulu, Hawaii: _____

Your home city: _____

Your college campus city: _____

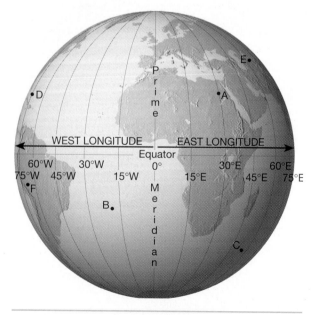

Figure 21.5 Meridians of longitude.

22. Using a globe or atlas, give the name of a city, feature, or country that is at the same latitude as your home city but equally distant from the prime meridian in the opposite hemisphere.

_____

23. The farthest a place can be directly east or west of the prime meridian is (45, 90, 180) degrees of longitude. Circle your answer.

Longitude and Time

Time, while independent of latitude, is very much related to longitude. This fact allows for time to be used in navigation to accurately determine one's location. By knowing the difference in time between two places, one with known longitude, the longitude of the second place can be determined.

Time on Earth can be kept in two ways. **Solar**, or **Sun, time** uses the position of the Sun in the sky to determine time. **Standard time**, the system used throughout most of the world, divides the globe into 24 standard time zones. Everyone living within the same standard time zone keeps the clock set the same. Of the two, solar time is used to determine longitude.

The following basic facts are important to understanding time.

- Earth rotates on its axis from west to east (eastward) or counterclockwise when viewed from above the North Pole (Figure 21.6).

- It is noon, Sun time, on the meridian that is directly facing the Sun (the Sun has reached its highest position in the sky, called the _zenith_) and midnight on the meridian on the opposite side of Earth.

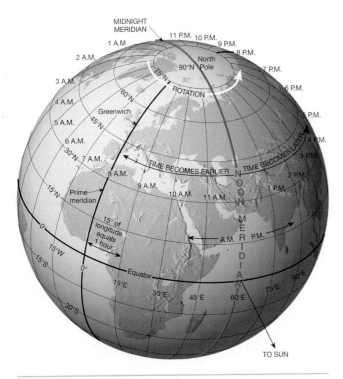

Figure 21.6 The noon meridian and solar time.

- The time interval from one noon by the Sun to the next noon averages 24 hours and is known as the _mean solar day._

- Earth turns through 360° of longitude in one mean solar day, which is equivalent to 15° of longitude per hour or 1° of longitude every 4 minutes of time.

- Places that are east or west of each other, regardless of the distance, have different solar times. For example, people located to the east of the noon meridian have already experienced noon; their time is afternoon [P.M.—_post_ (after) _meridiem_ (the noon meridian)]. People living west of the noon meridian have yet to reach noon; their time is before noon [A.M.—_ante_ (before) _meridiem_ (the noon meridian)]. _Time becomes later going eastward and earlier going westward._

Use the basic facts of time to answer questions 24–26.

24. What would be the solar time of a person living 1° of longitude west of the noon meridian? Be sure to indicate A.M. or P.M. with your answer.

    Solar time: _____ (A.M., P.M.)

25. What would be the solar time of a person located 4° of longitude east of the noon meridian?

    Solar time: _____ (A.M., P.M.)

26. If it is noon, solar time, at 70°W longitude, what is the solar time at each of the following locations?

    SOLAR TIME

    72°W longitude: _____

    65°W longitude: _____

    90°W longitude: _____

    110°E longitude: _____

Early navigators had to wait for the invention of accurate clocks, called _chronometers_, before they could determine longitude. Today most navigation is done using satellites, but ships still carry chronometers as a backup system.

The shipboard chronometer is set to keep the time at a known place on Earth, for example, the prime meridian. If it is noon by the Sun where the ship is located, and at that same instant the chronometer indicates that it is 8 A.M. on the prime meridian, the ship must be 60° of longitude (4 hours difference × 15° per hour) east (the ship's time is later) of the prime meridian (Figure 21.6). The difference in time need not be in whole hours. Thirty minutes difference in time between two places would be equivalent to 7.5° of longitude, twenty minutes would equal 5°, and so forth.

27. It is exactly noon by the Sun at a ship's location. What is the ship's longitude if, at that instant, the time on the prime meridian is the following? (_Note:_ Drawing a diagram showing the prime meridian, the ship's location east or west of the prime meridian, and the difference in hours may be helpful.)

6:00 P.M.: _____

1:00 A.M.: _____

2:30 P.M.: _____

Using Earth's Grid System

Using both parallels of latitude and meridians of longitude, you can accurately locate any point on the surface of Earth.

28. Using Figure 21.7, determine the latitude and longitude of each of the lettered points and write your answers in the following spaces. As a guide, Point A has already been done. Remember to indicate whether the point is N or S latitude and E or W longitude. The only exceptions are the equator, prime meridian, and 180° meridian. They are given no direction because each is a single line and cannot be confused with any other line. Convention dictates that latitude is always listed first.

Point A: (30°N) latitude,   (60°E) longitude

Point B: _____ latitude,   _____ longitude

Point C: _____ latitude,   _____ longitude

Point D: _____ latitude,   _____ longitude

Point E: _____ latitude,   _____ longitude

29. Locate the following points on Figure 21.7. Place a dot on the figure at the proper location and label each point with the designated letter.

Point F: 15°S latitude, 75°W longitude

Point G: 45°N latitude, 0° longitude

Point H: 30°S latitude, 60°E longitude

Point I: 0° latitude, 30°E longitude

30. Use a globe, map, or atlas to determine the latitude and longitude of the following cities.

Kansas City, MO: _____

Miami, FL: _____

Oslo, Norway: _____

Auckland, New Zealand: _____

Quito, Ecuador: _____

Baghdad, Iraq: _____

31. Beginning with a globe or world wall map, and then proceeding to an atlas, determine the city or feature at the following locations.

19°28′N latitude, 99°09′W longitude:

_____

41°52′N latitude, 12°37′E longitude:

_____

1°30′S latitude, 33°00′E longitude:

_____

When you study the Earth sciences, it is important to be familiar with the major physical features of Earth's surface. Identifying the features on a map will help acquaint you with their location for future reference.

32. Use a wall map of the world or world map in an atlas to find the following water bodies, rivers, and mountains. Examine their latitudes and longitudes, and then label each on the world map, Figure 21.8. To conserve space, mark only the

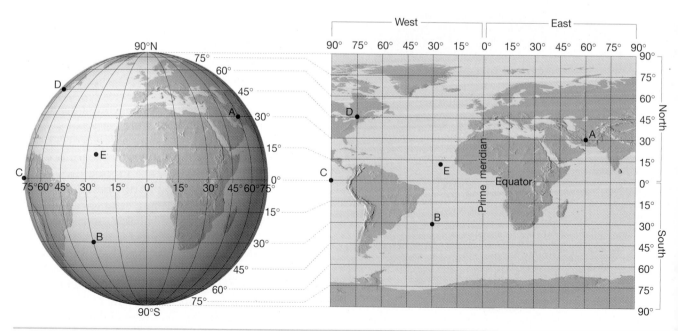

Figure 21.7 Locating places using Earth's grid system.

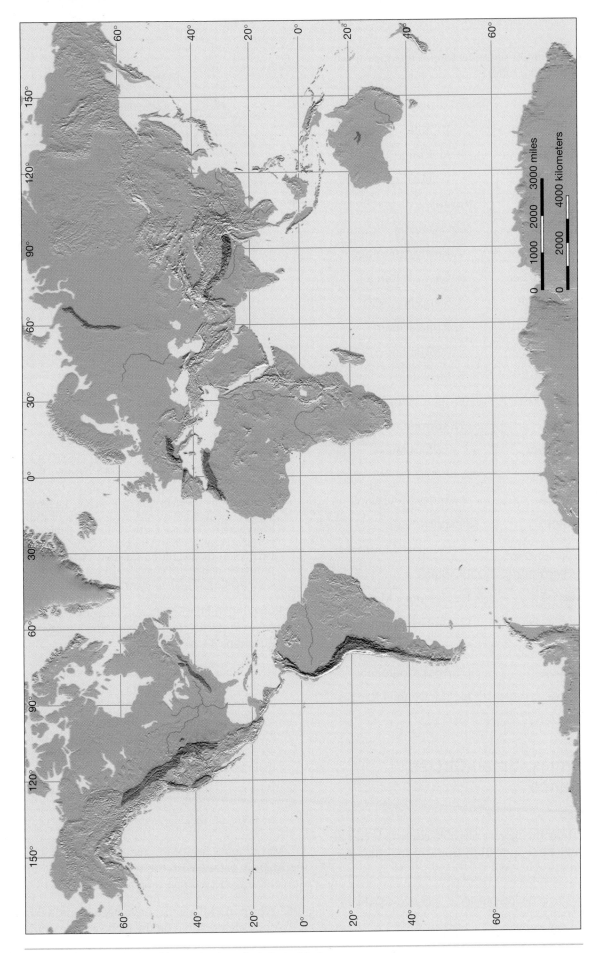

Figure 21.8 Generalized world map showing select physical features.

number or letter of the feature at the appropriate location on the map.

Water Bodies

A. Pacific Ocean

B. Atlantic Ocean

C. Indian Ocean

D. Arctic Ocean

E. Gulf of Mexico

F. Mediterranean Sea

G. Caribbean Sea

H. Persian Gulf

I. Red Sea

J. Sea of Japan

K. Black Sea

L. Caspian Sea

Rivers

North America

a. Mississippi

b. Colorado

c. Missouri

d. Ohio

South America

e. Amazon

Europe and Asia

f. Volga

g. Mekong

h. Ganges

i. Yangtze

Africa and Australia

j. Nile

k. Congo

l. Darling

Mountains

North America

1. Rocky Mountains

2. Cascade Range

3. Sierra Nevada

4. Appalachian Mountains

5. Black Hills

6. Teton Range

7. Adirondack Mountains

South America

8. Andes Mountains

Europe and Asia

9. Pyrenees Mountains

10. Alps

11. Himalaya Mountains

12. Ural Mountains

Africa and Australia

13. Atlas Mountains

14. MacDonnell Ranges

Great Circles, Small Circles, and Distance

Great Circles

A **great circle** is the largest possible circle that can be drawn on a globe (Figure 21.9). Some of the characteristics of a great circle are

- A great circle divides the globe into two equal parts, called *hemispheres*.

- An infinite number of great circles can be drawn on a globe. Therefore, a great circle can be drawn that passes through any two places on Earth's surface.

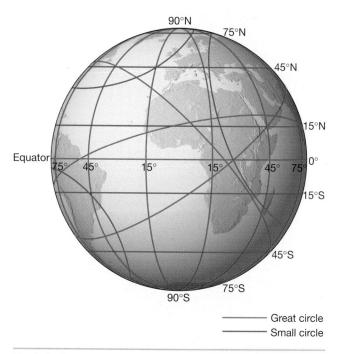

Figure 21.9 Illustrated are a few of the infinite number of great circles and small circles that can be drawn on the globe.

- The shortest distance between two places on Earth is along the great circle that passes through those two places.

- If Earth were a perfect sphere, then one degree of angle along a great circle would cover an identical distance everywhere. Because Earth is slightly flattened at the *poles and bulges slightly* at the equator, there are small differences in the length of a degree. However, for most purposes, *one degree of angle along a great circle equals approximately 111 kilometers or 69 miles.*

Referring to Figure 21.9 and keeping the characteristics of great circles in mind, examine a globe and answer questions 33–35.

33. Remember that great circles do not necessarily have to follow parallels or meridians. Estimate several great circles on the globe. Do this by wrapping a piece of string around the globe that divides the globe into two equal halves. You should be able to see that there are an infinite number of great circles that can be marked on the globe.

34. Which parallel(s) of latitude is/are a great circle(s)?

_____

Meridians of longitude are each half circles. If each meridian is paired with the meridian on the opposite side of the globe, a circle is formed.

35. Which meridians that have been paired with their opposite meridian on the globe are great circles?

_____

Small Circles

Any circle on the globe that does not meet the characteristics of a great circle is considered a **small circle** (Figure 21.9). Therefore, a small circle *does not* divide the globe into two equal parts and *is not* the shortest distance between two places on Earth. Referring to Figure 21.9 and keeping the characteristics of small circles in mind, examine a globe and answer questions 36–38.

36. In general, which parallels of latitude are small circles?

    _____

37. Which two latitudes are actually points, rather than circles?

    _____

38. In general, which meridians that have been paired with their opposite meridian on the globe are small circles?

    _____

39. Indicate, by placing an "X" in the appropriate column, which of the following pairs of points illustrated on the Earth's grid in Figure 21.7 are on a great circle and which are on a small circle.

	GREAT CIRCLE	SMALL CIRCLE
Points A–H	_____	_____
Points D–G	_____	_____
Points C–I	_____	_____
Points B–H	_____	_____

40. Now that you know the characteristics of great and small circles, complete the following statements by circling the correct response.

    a. All meridians are halves of (great, small) circles.

    b. With the exception of the equator, all parallels are (great, small) circles.

    c. The equator is a (great, small) circle.

    d. The poles are (points, lines) of latitude, rather than circles.

Determining Distance Along a Great Circle

Determining the distance between two places on Earth when both are on the equator or the same great circle meridian requires two steps:

Step 1. Determine the number of degrees along the great circle between the two places (degrees of longitude on the equator or degrees of latitude on a meridian).

Step 2. Multiply the number of degrees by 111 kilometers or 69 miles (the approximate number of kilometers or miles per degree for any great circle).

Use a globe and these steps to answer questions 41 and 42.

41. Approximately how many miles would you journey if you traveled from 10°W longitude to 40°E longitude at the equator by way of the shortest route?

    _____ miles

42. Approximately how many kilometers is London, England, directly north of the equator?

    _____ kilometers

Determining the shortest distance between two places on Earth that are *not* both on the equator or the same great circle meridian requires the four steps (Figure 21.10):

Step 1. On a globe, determine the great circle that intersects both places.

Step 2. Stretch a piece of string along the great circle between the two places on the globe and mark the distance between them on the string with your fingers (Figure 21.10A).

Step 3. While still marking the distance with your fingers, place the string on the equator with one end on the prime meridian. Determine the number of degrees along the great circle between the two places by measuring the marked string's length in degrees of longitude along the equator, which is also a great circle (Figure 21.10B).

Step 4. Multiply the number of degrees along the great circle by 69 miles (111 kilometers) to arrive at the approximate distance. (For example, the great circle distance between X and Y in Figure 21.10 would be approximately 2070 miles, 30° × 69 miles/degree, or 3330 kilometers, 30° × 111 kilometers/degree.)

Use a globe, a piece of string, and the four steps to answer questions 43 and 44.

43. Determine the approximate great circle distance in degrees, miles, and kilometers from Memphis, TN, to Tokyo, Japan.

    Degrees along the great circle between Memphis and Tokyo = _____°

    Distance along the great circle between Memphis and Tokyo = _____ miles (_____ km)

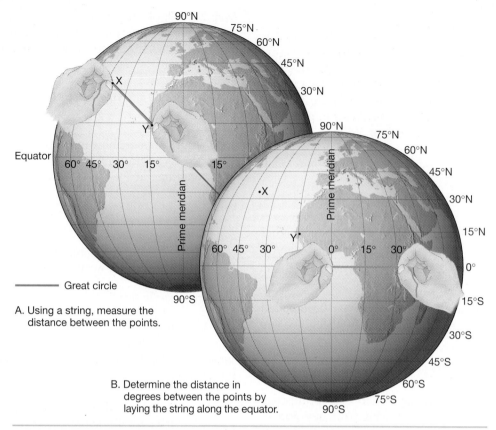

Great circle

A. Using a string, measure the distance between the points.

B. Determine the distance in degrees between the points by laying the string along the equator.

Figure 21.10 Determining the distance between two places on Earth along a great circle other than the equator or great circle meridian. In the example illustrated, the distance between X and Y along the great circle is 30°, which is approximately equivalent to 2070 miles (3330 kilometers).

44. Describe the flight route, by listing states, countries, etc., that a plane would follow as it flew by way of the shortest route between Memphis, TN, and Tokyo, Japan.

_____

_____

Determining Distance Along a Parallel

Since all parallels except the equator are small circles, the length of one degree of longitude along a parallel, other than the equator, will always be less than 69 miles or 111 kilometers. Table 21.1 shows the length of a degree of longitude at various latitudes on Earth.

45. Examine a globe. What do you observe about the distance around Earth along each parallel as you get farther away from the equator?

_____

46. Use Table 21.1, "Longitude as distance," to determine the length of one degree of longitude at each of the following parallels.

LENGTH OF 1° OF LONGITUDE

15° latitude:	_____ km,	_____ miles
30° latitude:	_____ km,	_____ miles
45° latitude:	_____ km,	_____ miles
80° latitude:	_____ km,	_____ miles

47. Use the Earth's grid illustrated in Figure 21.7 to determine the distances between the following points.

Distance between points D and G:

_____ degrees × _____ miles/degree

= _____ miles

Distance between points B and H:

_____ degrees × _____ km/degree

= _____ km

Memphis, TN, and Tokyo, Japan, are both located at about 35°N latitude.

48. Use a globe or world map to determine how many degrees of longitude separate Memphis, TN, from Tokyo, Japan.

Table 21.1   Longitude as distance

°Lat.	Length of 1° Long. km	Length of 1° Long. miles	°Lat.	Length of 1° Long. km	Length of 1° Long. miles	°Lat.	Length of 1° Long. km	Length of 1° Long. miles
0	111.367	69.172	30	96.528	59.955	60	55.825	34.674
1	111.349	69.161	31	95.545	59.345	61	54.131	33.622
2	111.298	69.129	32	94.533	58.716	62	52.422	32.560
3	111.214	69.077	33	93.493	58.070	63	50.696	31.488
4	111.096	69.004	34	92.425	57.407	64	48.954	30.406
5	110.945	68.910	35	91.327	56.725	65	47.196	29.314
6	110.760	68.795	36	90.203	56.027	66	45.426	28.215
7	110.543	68.660	37	89.051	55.311	67	43.639	27.105
8	110.290	68.503	38	87.871	54.578	68	41.841	25.988
9	110.003	68.325	39	86.665	53.829	69	40.028	24.862
10	109.686	68.128	40	85.431	53.063	70	38.204	23.729
11	109.333	67.909	41	84.171	52.280	71	36.368	22.589
12	108.949	67.670	42	82.886	51.482	72	34.520	21.441
13	108.530	67.410	43	81.575	50.668	73	32.662	20.287
14	108.079	67.130	44	80.241	49.839	74	30.793	19.126
15	107.596	66.830	45	78.880	48.994	75	28.914	17.959
16	107.079	66.509	46	77.497	48.135	76	27.029	16.788
17	106.530	66.168	47	76.089	47.260	77	25.134	15.611
18	105.949	65.807	48	74.659	46.372	78	23.229	14.428
19	105.337	65.427	49	73.203	45.468	79	21.320	13.242
20	104.692	65.026	50	71.727	44.551	80	19.402	12.051
21	104.014	64.605	51	70.228	43.620	81	17.480	10.857
22	103.306	64.165	52	68.708	42.676	82	15.551	9.659
23	102.565	63.705	53	67.168	41.719	83	13.617	8.458
24	101.795	63.227	54	65.604	40.748	84	11.681	7.255
25	100.994	62.729	55	64.022	39.765	85	9.739	6.049
26	100.160	62.211	56	62.420	38.770	86	7.796	4.842
27	99.297	61.675	57	60.798	37.763	87	5.849	3.633
28	98.405	61.121	58	59.159	36.745	88	3.899	2.422
29	97.481	60.547	59	57.501	35.715	89	1.950	1.211
30	96.528	59.955	60	55.825	34.674	90	0.000	0.000

_____ degrees of longitude separate Memphis, TN, and Tokyo, Japan.

49. From the longitude as distance table, Table 21.1, the length of one degree of longitude at latitude 35°N is

    _____ km (_____ miles).

50. How many miles is Tokyo, Japan, *directly* west of Memphis, TN? Show your calculation below.

    _____ miles

51. In question 43 you determined the great circle distance between Memphis, TN, and Tokyo,

Japan. How many miles shorter is the great circle route between these cities than the east-west distance along a parallel (question 50)?

The great circle route is _____ miles shorter.

Location and Distance on Earth on the Internet

Continue your analyses of the topics presented in this exercise by completing the corresponding online activity on the *Applications & Investigations in Earth Science* website at http://prenhall.com/earthsciencelab

Notes and calculations.

Location and Distance on Earth

Date Due: _____

Name: _____

Date: _____

Class: _____

After you have finished Exercise 21, complete the following questions. You may have to refer to the exercise for assistance or to locate specific answers. Be prepared to submit this summary/report to your instructor at the designated time.

1. In Figure 21.11, prepare a diagram illustrating Earth's grid system. Include and label the equator and prime meridian. Refer to the diagram to explain the system used for locating points on the surface of Earth.

Figure 21.11 Diagram of Earth's grid system.

Explanation: _____

_____

_____

2. Define the following terms.

Parallel of latitude: _____

_____

Meridian of longitude: _____

_____

Great circle: _____

_____

3. Determine whether or not the following statements are true or false. If the statement is false, correct the word(s) so that it reads as a true statement.

T F a. The distance measured north or south of the prime meridian is called latitude.

T F b. All meridians, when paired with their opposite meridian on Earth, form great circles.

T F c. The equator is the only meridian that is a great circle.

4. What is the relation between the latitude of a place in the Northern Hemisphere and the angle of Polaris above the horizon at that place?

_____

5. Approximately how many miles does one degree equal along a great circle?

One degree along a great circle equals _____ miles.

6. What is the latitude and longitude of your home city?

_____ latitude, _____ longitude

7. Use a globe or map to determine, as accurately as possible, the latitude and longitude of Athens, Greece.

_____ latitude, _____ longitude

8. Write a brief paragraph describing how to determine the shortest distance between two places on Earth's surface.

_____

_____

_____

9. From question 51 of the exercise, how many miles shorter is the great circle route between Memphis, TN, and Tokyo, Japan, than the straight east-west distance along a parallel?

_____ miles shorter

10. Approximately how many miles is it from London, England, to the South Pole? (Show your calculation.)

_____ miles

11. Using Figure 21.12, determine the latitude and longitude of each of the lettered points and write your answers in the following spaces.

Point A: _____

Point B: _____

Point C: _____

Point D: _____

Point E: _____

12. You are shipwrecked and floating in the Atlantic Ocean somewhere between London, England, and New York, NY. Fortunately, you managed to save your globe. You have been in London so your watch is still set for London time. It is noon, by the Sun, at your location. Your watch indicates that it is 4 P.M. in London. Are you closer to the

United States or to England? Explain how you arrived at your answer.

_____

_____

_____

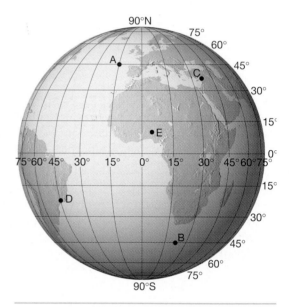

Figure 21.12 Locating places using Earth's grid.

The Metric System, Measurements, and Scientific Inquiry

Earth science, the study of Earth and its neighbors in space, involves investigations of natural objects that range in size from the very smallest divisions of atoms to the largest of galaxies (Figure 22.1). From atoms to galaxies, objects are each unique in their size, mass, and volume; and yet all are related when it comes to understanding the nature of Earth and its place in the universe.

Almost every scientific investigation requires accurate measurements. One important purpose of this exercise is to examine the metric system as a method of scientific measurement used in the Earth sciences. In addition, a few special units of measurement are also examined. The exercise concludes with an activity that focuses on the nature of scientific inquiry.

Objectives

After you have completed this exercise, you should be able to:

1. List the units for length, mass, and volume that are used in the metric system.
2. Use the metric system for measurements.
3. Convert units within the metric system.
4. Understand and use the micrometer and nanometer for measuring very small distances as well as the astronomical unit and light-year for measuring large distances.
5. Determine the approximate density and specific gravity of a solid substance.
6. Conduct a scientific experiment using accepted methods of scientific inquiry.

Materials

metric ruler   calculator

Materials Supplied by Your Instructor

metric tape measure   paper clip   nickel coin
   or meterstick   paper cup   small rock
metric balance   thread
"bathroom" scale   large graduated cylinder
   (metric)   (marked in milliliters)

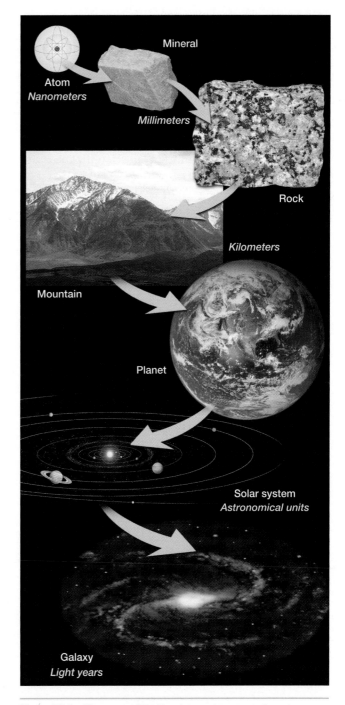

Figure 22.1 The range of Earth science measurements.

Terms

Introduction

To describe objects, Earth scientists use units of measurement that are relative to the particular feature being studied. For example, centimeters or inches, instead of kilometers or miles, would be used to measure the width of this page; and kilometers or miles, rather than centimeters or inches, to measure the distance from New York to London, England.

Most areas of science have developed units of measurement that meet their particular needs. However, regardless of the unit used, all scientific measurements are defined within a broader system so that they may be understood and compared. In science, the fundamental units have been established by the *International System of Units* (SI, Système International d'Unités) (Table 22.1).

The Metric System

The **metric system** is a decimal system (based on fractions or multiples of ten) that uses only one basic unit for each type of measurement: the **meter** (m) as the unit of length (Figure 22.2), the **liter** (l) as the unit of volume (One liter is equal to the volume of one kilogram of pure water at 4°C (39.2°F), about 1.06 quarts.), and the **gram** (g) as the unit of mass (Figure 22.3). In the English system, the units used to express the same relations are feet, quarts, and ounces.

Table 22.1 Base units of the SI. From these base units, other units are derived to express quantities such as power (watt, W), force (newton, N), energy (joule, J), and pressure (pascal, Pa).

Unit	Quantity measured	Symbol
meter	length	m
kilogram	mass	kg
second	time	s
kelvin	thermodynamic temperature	K
ampere	electric current	A
mole	quantity of a substance	mol
candela	luminous intensity	cd

METER (m)

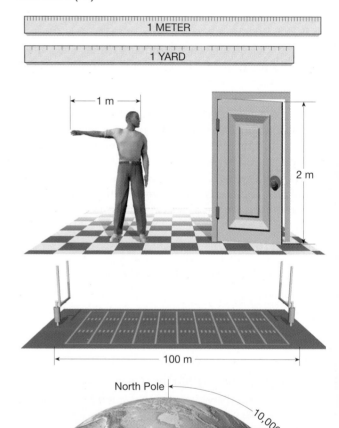

Figure 22.2 The SI unit of length is the meter (m), which is slightly longer than a yard. Originally described as one ten-millionth of the distance from the equator to the North Pole, it is currently defined as the distance traveled by light in a vacuum in 0.0000000033 of a second.

GRAM (g)

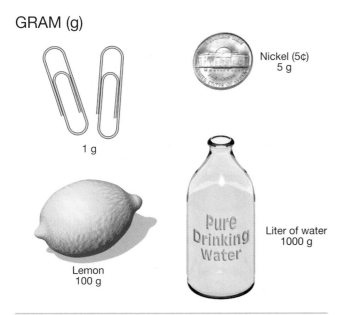

Nickel (5¢)
5 g

Lemon
100 g

Liter of water
1000 g

Figure 22.3 The basic unit of mass in the metric system is the gram (g), approximately equal to the mass of one cubic centimeter of pure water at 4°C (39.2°F). A gram is about the weight of two paper clips, while an ounce is about the weight of 40 paper clips.

Working with the Metric System

In the metric system the basic units of weights and measures are in "tens" relations to each other. It is similar to our monetary system where 10 pennies equal one dime and ten dimes equal one dollar. However, in the English system of weights and measures no such regularity exists; for example, 12 inches equal a foot and 5280 feet equal a statute mile. Thus, the advantage of the metric system is *consistency.*

Table 22.2 illustrates the prefixes that are used in the metric system to indicate how many times more (in multiples of 10) or what fraction (in fractions of ten) of the basic unit you have. Therefore, from the information in the table, a *kilo*gram (kg) means one thousand grams, while a *milli*gram (mg) is one one-thousandth of a gram.

To familiarize yourself with metric units, determine the following measurements using the equipment provided in the laboratory.

Measuring length:

1. Use a metric measuring tape (or meterstick) to measure your height as accurately as possible to the nearest hundredth of a meter (called a centimeter).

   _____ . _____ meters (m)

2. Use a metric ruler to measure the length of this page as accurately as possible to the nearest tenth of a centimeter (called a millimeter).

   _____ . _____ centimeters (cm)

3. Accurately measure the length of your shoe to the nearest millimeter.

   _____ millimeters (mm)

Measuring volume:

4. Use a graduated measuring cylinder to measure the volume of the paper cup to the nearest milliliter.

   _____ milliliters (ml)

Table 22.2 Metric prefixes and symbols

Prefix[1]	Symbol[2]	Meaning
giga-	G	one billion times base unit (1,000,000,000 × base)
mega-	M	one million times base unit (1,000,000 × base)
kilo-	k	one thousand times base unit (1000 × base)
hecto-	h	one hundred times base unit (100 × base)
deka-	da	ten times base unit (10 × base)
BASE UNIT	m (meter)–base unit of length l (liter)–base unit of volume g (gram)–base unit of mass	
deci-	d	one-tenth times base unit (.1 × base)
centi-	c	one-hundredth times base unit (.01 × base)
milli-	m	one-thousandth times base unit (.001 × base)
micro-	μ	one-millionth times base unit (.000001 × base)
nano-	n	one-billionth times base unit (.000000001 × base)

[1] A prefix is added to the base unit to indicate how many times more, or what fraction of, the base unit is present. For example, a kilometer (km) means one thousand meters and a millimeter (mm) means one thousandth of a meter.
[2] When writing in the SI system, periods are not used after the unit symbols and symbols are not made plural. For example, if the length of a stick is 50 centimeters, it would be written as "50 cm" (not "50 cm." or "50 cms").

Measuring mass:

5. Weigh the following and record your results. (Follow the directions of your instructor for using a metric balance.)

   Sample of rock: _____ grams (g)

   Paper clip: _____ grams (g)

   Nickel coin: _____ grams (g)

(*Note:* Two terms that are often confused are *mass* and *weight*. Mass is a measure of the amount of matter an object contains. Weight is a measure of the force of gravity on an object. For example, the mass of an object would be the same on both Earth and the Moon. However, because the gravitational force of the Moon is less than that of Earth, the object would weigh less on the Moon. On Earth, mass and weight are directly related, and often the same units are used to express each.)

6. Use the metric "bathroom" scale. Weigh yourself as accurately as possible to the nearest tenth of a kilogram. (*Note:* If a metric scale is not available, convert your weight in pounds to kilograms by multiplying your weight (in pounds) by 0.45.)

   _____ . _____ kilograms (kg)

Metric Conversions

As stated earlier, one important advantage of the metric system is that it is based on "tens." As shown on the metric conversion diagram, Figure 22.4, conversion from one unit to another can be accomplished simply by *moving the decimal point* to the left if going to larger units, or by moving the decimal point to the right if going to smaller units.

For example, if you measure the length of a piece of string and it is 1.43 decimeters long, in order to convert its length to millimeters, start with 1.43 on the "deci-" step of the diagram. Then move the decimal two places (steps) to the right (the "milli-" step). The length, in millimeters, becomes 143.0 millimeters.

7. Use the metric conversion diagram, Figure 22.4, to convert the following:

   a. 2.05 meters (m) = _____ centimeters (cm)

   b. 1.50 meters (m) = _____ millimeters (mm)

   c. 9.81 liters (l) = _____ deciliters (dl)

   d. 5.4 grams (g) = _____ milligrams (mg)

   e. 6.8 meters (m) = _____ kilometer (km)

   f. 4214.6 centimeters = _____ meters (m)

   g. 321.50 grams = _____ kilogram (kg)

   h. 70.73 hectoliters = _____ dekaliters (dal)

8. Use a metric tape measure (or meterstick) to determine the length of your laboratory table as accurately as possible to the nearest hundredth of a meter. Then convert the length to each of the units in question 8b.

   a. Length of table: _____ . _____ meters

   b. Length of table equals:

      _____ millimeters (mm)

      _____ centimeters (cm)

      _____ km

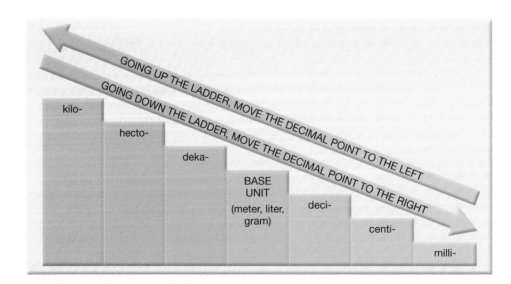

Figure 22.4 Metric conversion diagram. Beginning at the appropriate step, if going to larger units (left), move the decimal to the left for each step crossed. When going to smaller units (right), move the decimal to the right for each step crossed. For example, 1.253 meters (base unit step) would be equivalent to 1253.0 millimeters (decimal moved three steps to the right, the milli- step).

Metric-English Conversions

Because the change to the metric system will occur gradually over the next few decades, we will be forced to use both systems simultaneously. If we are not able to convert from one system to the other, we will occasionally be inconvenienced or mildly frustrated.

By using the conversion tables on the inside back cover of this manual, show the metric equivalent for each of the following units.

Length conversion:

9. 1 inch = _____ centimeters
10. 1 meter = _____ feet
11. 1 mile = _____ kilometers

Volume conversion:

12. 1 gallon = _____ liters
13. 1 cubic centimeter = _____ cubic inch

Mass conversion:

14. 1 gram = _____ ounce
15. 1 pound = _____ kilogram

Temperature

Temperature represents one relatively common example of using different systems of measurement. On the Fahrenheit temperature scale, 32°F is the melting point of ice and 212°F marks the boiling point of water (at standard atmospheric pressure). On the **Celsius scale**, ice melts at 0°C and water boils at 100°C. On the **Kelvin scale**, ice melts at 273 K.

Conversion from one temperature scale to the other can be accomplished using either an equation or graphic comparison scale. To convert Celsius degrees to Fahrenheit degrees, the equation is °F = (1.8)°C + 32°. To convert Fahrenheit degrees to Celsius degrees, the equation is °C = (°F − 32°)/1.8. To convert Kelvins (K) to Celsius degrees, subtract 273 and add the degree symbol.

16. Convert the following temperatures to their equivalents. Do the first four conversions using the appropriate equation, and the others using the temperature comparison scale on the inside-back cover of this manual.

    a. On a cold day it was 8°F = _____ °C

    b. Ice melts at 0°C = _____ °F

    c. Room temperature is 72°F = _____ °C

    d. A hot summer day was 35°C = _____ °F

    e. Normal body temperature is 98.6°F = _____ °C

    f. A warm shower is 27°C = _____ °F

    g. Hot soup is 72°C = _____ °F

    h. Water boils at 212°F = _____ K

17. Using the temperature comparison scale, answer the following:

    a. The thermometer reads 28°C. Will you need your winter coat? _____

    b. The thermometer reads 10°C. Will the outdoor swimming pool be open today? _____

    c. If your body temperature is 40°C, do you have a fever? _____

    d. The temperature of a cup of cocoa is 90°C. Will it burn your tongue? _____

    e. Your bath water is 15°C. Will you have a scalding, warm, or chilly bath? _____

    f. "Who's been monkeying with the thermostat? It's 37°C in this room." Are you shivering or perspiring? _____

Metric Review

Use what you have learned about the metric system to determine whether or not the following statements are *reasonable*. Write "yes" or "no" in the blanks. *Do not* convert these units to English equivalents, only *estimate* their value.

18. A man weighs 90 kilograms. _____
19. A fire hydrant is a meter tall. _____
20. A college student drank 3 kiloliters of coffee last night. _____
21. The room temperature is 295 K. _____
22. A dime is 1 millimeter thick. _____
23. Sugar will be sold by the milligram. _____
24. The temperature in Paris today is 80°C. _____
25. The bathtub has 80 liters of water in it. _____
26. You will need a coat if the outside temperature is 30°C. _____
27. A pork roast weighs 18 grams. _____

Special Units of Measurement

Scientists often use special units to measure various phenomena. Most of them are defined using the units of the International System of Units. Throughout this course you will encounter several of these units in your reading and laboratory studies. Only a few are introduced here.

Very Small Distances

Two units commonly used to measure very small distances are the **micrometer** (symbol, μm), also known as the **micron**, and the **nanometer** (symbol, nm).

By definition, one micrometer equals .000001 m (one millionth of a meter). There are one million micrometers in one meter and 10,000 micrometers in a centimeter. A nanometer equals .000000001 m (one billionth of a meter).

28. There are (10, 100, 1,000) nanometers in a micrometer. Circle your answer.

29. What would be the length of a 2.5 centimeter line expressed in micrometers and nanometers?

_____ micrometers in a 2.5 cm line

_____ nanometers in a 2.5 cm line

30. Some forms of *radiation* (e.g. light) travel in very small waves with distances from crest to crest of about 500 nanometers (0.5 μm). How many of these waves would it take to equal one centimeter?

_____ waves in one centimeter

Very Large Distances

On the other extreme of size, astronomers must measure very large distances, such as the distances between planets or to the stars and beyond. To simplify their measurements, they have developed special units including the **astronomical unit** (symbol, AU) and the **light-year** (symbol, LY).

The astronomical unit is a unit for measuring distance within the solar system. One astronomical unit is equal to the average distance of Earth from the Sun. This average distance is 150 million kilometers, which is approximately equal to 93 million miles.

31. The planet Saturn is 1,427 million kilometers from the Sun. How many AUs is Saturn from the Sun?

_____ AUs from the Sun

The light-year is one unit for measuring distances to the stars and beyond. One light-year is defined as the distance that light travels in a vacuum in one year. This distance is about 6 trillion miles (6,000,000,000,000 miles).

32. Approximately how many kilometers will light travel in one year?

_____ kilometers per year

33. The nearest star to Earth, excluding our Sun, is named Proxima Centauri. It is about 4.27 light-years away. What is the distance of Proxima Centauri from Earth in both miles and kilometers?

_____ miles

_____ kilometers

Density and Specific Gravity

Two important properties of a material are its **density** and **specific gravity**. Density is the mass of a substance per unit volume, usually expressed in grams per cubic centimeter (g/cm^3) in the metric system. The specific gravity of a solid is the *ratio* of the mass of a given volume of the substance to the mass of an equal volume of some other substance taken as a standard (usually water at 4°C). Because specific gravity is a ratio, it is expressed as a pure number and has no units. For example, a specific gravity of 6 means that the substance has six times more mass than an equal volume of water. Because the density of pure water at 4°C is 1 g/cm^3, the specific gravity of a substance will be numerically equal to its density.

The approximate density and specific gravity of a rock, or other solid, can be arrived at using the following steps:

Step 1. Determine the mass of the rock using a metric balance.

Step 2. Fill a graduated cylinder that has its divisions marked in milliliters approximately two-thirds full with water. Note the level of the water in the cylinder in milliliters.

Step 3. Tie a thread to the rock and immerse the rock into the water in the graduated cylinder. Note the new level of the water in the cylinder.

Step 4. Determine the difference between the beginning level and after-immersion level of the water in the cylinder.

Step 5. Calculate the density and specific gravity using the following information and appropriate equations.

A milliliter of water has a volume approximately equal to a cubic centimeter (cm^3). Therefore, the difference between the beginning water level and the after-immersion water level in the cylinder equals the volume of the rock in cubic centimeters. Furthermore, *a cubic centimeter (one milliliter) of water has a mass of approximately one gram.* (*Note:* Using the equipment already present in the lab, you may want to devise a simple experiment to confirm this fact.) Therefore, the difference between the beginning water level and the after-immersion water level in the cylinder is the mass of a volume of water equal to the volume of the rock.

Using the steps listed above for determining density and specific gravity, complete questions 34 and 35.

34. Determine the density and specific gravity of a small rock sample by completing questions 34a–34f.

a. Mass of rock sample: _____ grams

b. After-immersion level of water: _____ ml

Beginning level of water in cylinder: _____ ml
_____

Difference: _____ ml

c. Volume of rock sample: _____ cm^3

d. Mass of a volume of water equal to the volume of the rock: _____ g

e. Density of rock:

$$\text{Density} = \frac{\text{mass of rock (g)}}{\text{volume of rock (cm}^3)}$$

$$= \text{_____} \text{ g/cm}^3$$

f. Specific gravity of rock:

Specific gravity

$$= \frac{\text{mass of rock (g)}}{\text{mass of an equal volume of water (g)}}$$

$$= \text{_____}$$

35. As a means of comparison, your instructor may require that you determine the density and/or specific gravity of other objects. If so, record your results in the following spaces.

a. Object: _____

   Density: _____ g/cm^3

   Specific gravity: _____

b. Object: _____

   Density: _____ g/cm^3

   Specific gravity: _____

36. If you have investigated the densities and/or specific gravities of several objects, write a brief paragraph comparing the objects.

_____

_____

_____

Methods of Scientific Inquiry

Scientists use many methods in an attempt to understand natural phenomena. Some scientific discoveries represent purely theoretical ideas, while others may occasionally occur by chance. However, scientific knowledge is often gained by following a sequence of steps which involve

Step 1. Establishing a **hypothesis**—a tentative, or untested, explanation.

Step 2. Gathering data and conducting experiments to validate the hypothesis.

Step 3. Accepting, modifying, or rejecting the hypothesis on the basis of extensive data gathering or experimentation.

The following simple inquiry should help you understand the process.

Step 1—Establishing a Hypothesis

Observe all the people in the laboratory and pay particular attention to each individual's height and shoe length.

37. Based on your observations, write a hypothesis that relates a person's height to their shoe length.

   Hypothesis: _____

   _____

   _____

Step 2—Gathering Data

Previously, in questions 1 and 3 of the exercise, each person in the laboratory measured his or her height using a metric tape measure (or meterstick) and shoe length.

38. Gather your data by asking ten or fifteen people in the lab for their height and shoe length measurements. Enter your data in Table 22.3 by recording height to the nearest hundredth of a meter and shoe length to the nearest millimeter.

Step 3—Evaluating the Hypothesis Based Upon the Data

Plot all your data from Table 22.3 on the Height vs. Shoe Length graph, Figure 22.5, by locating a person's height on the vertical axis and his or her shoe length on the horizontal axis. Then place a dot on the graph where the two intersect.

39. Describe the pattern of the data points (dots) on the Height vs. Shoe Length graph, Figure 22.5. For example, are the points scattered all over the graph, or do they appear to follow a line or curve?

   _____

   _____

40. Draw a single line on the graph that appears to average, or best fit, the pattern of the data points.

41. Describe the relation of height to shoe length that is illustrated by the line on your graph.

   _____

   _____

   _____

42. Ask several people, whose height and shoe length you have not used to prepare the graph, for their height. Then see how accurately your line predicts what their shoe length should be. Do this by marking each person's height on the vertical axis and then follow a line straight across to the right until you intersect the line on the graph. Read the

Table 22.3   Data table for recording height and shoe length measurements of people in the lab

Person	Height (nearest hundredth of a meter)	Shoe Length (nearest millimeter)
1	_____._____ m	_____ mm
2	_____._____ m	_____ mm
3	_____._____ m	_____ mm
4	_____._____ m	_____ mm
5	_____._____ m	_____ mm
6	_____._____ m	_____ mm
7	_____._____ m	_____ mm
8	_____._____ m	_____ mm
9	_____._____ m	_____ mm
10	_____._____ m	_____ mm
11	_____._____ m	_____ mm
12	_____._____ m	_____ mm
13	_____._____ m	_____ mm
14	_____._____ m	_____ mm
15	_____._____ m	_____ mm

predicted shoe length from the axis directly below the point of intersection.

43. Summarize how accurately your graph predicts a person's shoe length, knowing only his or her height.

44. Using your graph's ability to make predictions as a guide, do you think you should accept, reject, or modify your original hypothesis? Give the reason(s) for your choice.

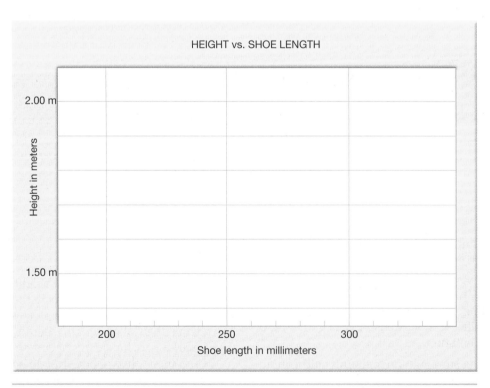

Figure 22.5   Height vs. Shoe Length graph.

Your study has been restricted to people in your laboratory.

45. Why would your ability to make predictions have been more accurate if you had used the heights and shoe lengths of ten thousand people to construct your graph?

_____

_____

_____

Drawing hasty conclusions with limited data can often cause problems. In science you can never have too much data. Experiments are repeated many times by many different people before the results are accepted by the scientific community.

The Metric System on the Internet

Continue your analyses of the topics presented in this exercise by completing the corresponding online activity on the *Applications & Investigations in Earth Science* website at http://prenhall.com/larthsciencelab

Notes and calculations.

The Metric System, Measurements, and Scientific Inquiry

Date Due: _____

Name: _____

Date: _____

Class: _____

After you have finished Exercise 22, complete the following questions. You may have to refer to the exercise for assistance or to locate specific answers. Be prepared to submit this summary/report to your instructor at the designated time.

1. List the basic metric unit and symbol used for these measurements:

   Length: _____

   Mass: _____

   Volume: _____

2. Convert the following units:

   a. 2 liters = _____ deciliters

   b. 600 millimeters = _____ meter

   c. 72°F = _____ °C

   d. 0.32 kilograms = _____ grams

   e. 12 grams = _____ milligrams

3. Indicate by answering "yes" or "no" whether or not the following statements are reasonable:

   a. A person is 600 centimeters tall. _____

   b. A bag of groceries weighs 5 kilograms. _____

   c. It took 52 liters of gasoline to fill the car's empty gasoline tank. _____

4. List your height and shoe length using the metric system.

   a. Height: _____._____ meters

   b. Shoe length: _____ millimeters

5. How many micrometers are there in 3.0 centimeters?

   _____ micrometers in 3.0 centimeters

6. How many waves, each 500 nanometers wide, would fit along a two centimeter line?

   _____ waves along a two centimeter line

7. What would be the distance in kilometers of a star that is 6.5 light-years from Earth?

   _____ kilometers from Earth

8. Uranus, one of the most distant planets, is 2,870 million kilometers from the Sun. What is its distance from the Sun in astronomical units?

   _____ astronomical units from the Sun

9. Explain the difference between the two terms, *density* and *specific gravity*.

   _____

   _____

   _____

10. At the conclusion of your height–shoe length experiment, in question 44 you (accepted, rejected, modified) your original hypothesis. Circle your answer and give the reason for this decision.

    _____

    _____

    _____

Appendix

A. World

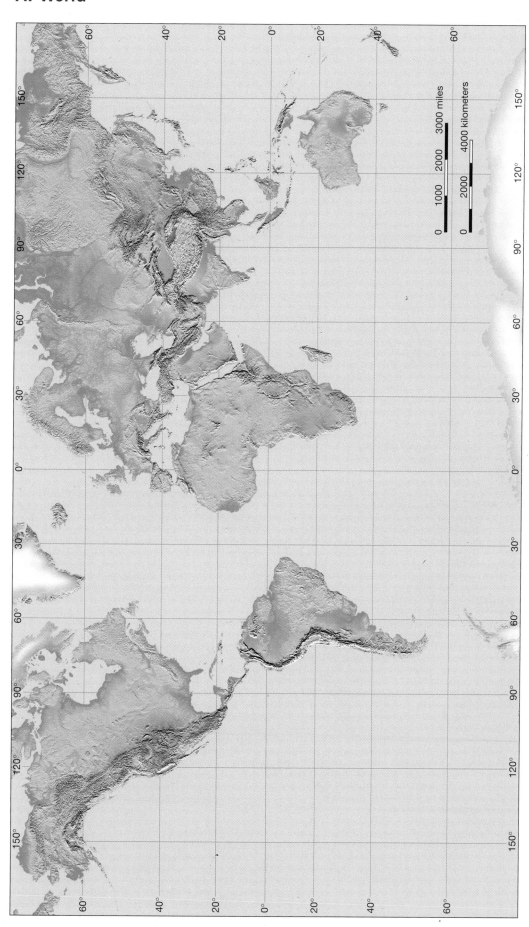

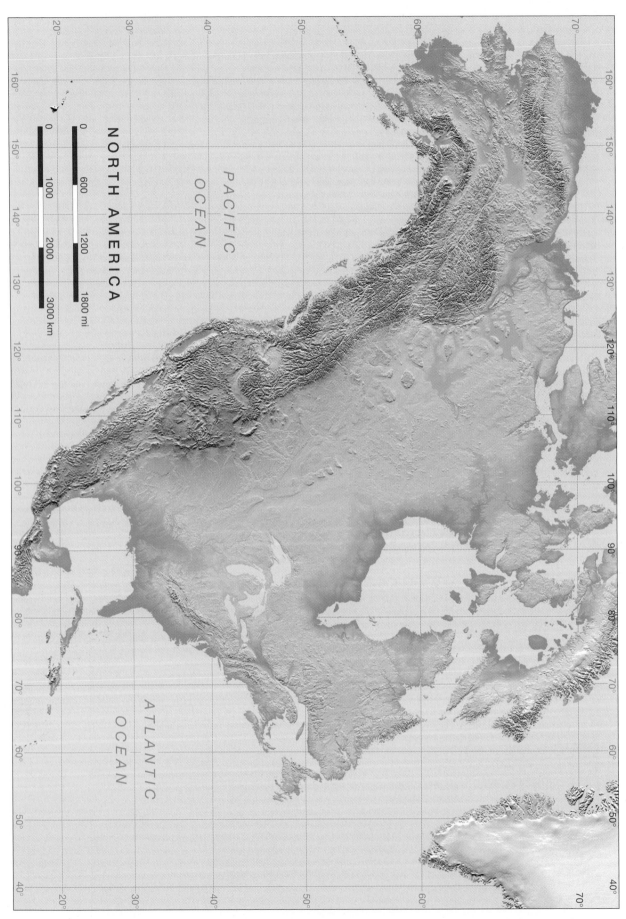

NORTH AMERICA

PACIFIC
OCEAN

ATLANTIC
OCEAN

| 0 | 1000 | 2000 | 3000 km |
| 0 | 600 | 1200 | 1800 mi |

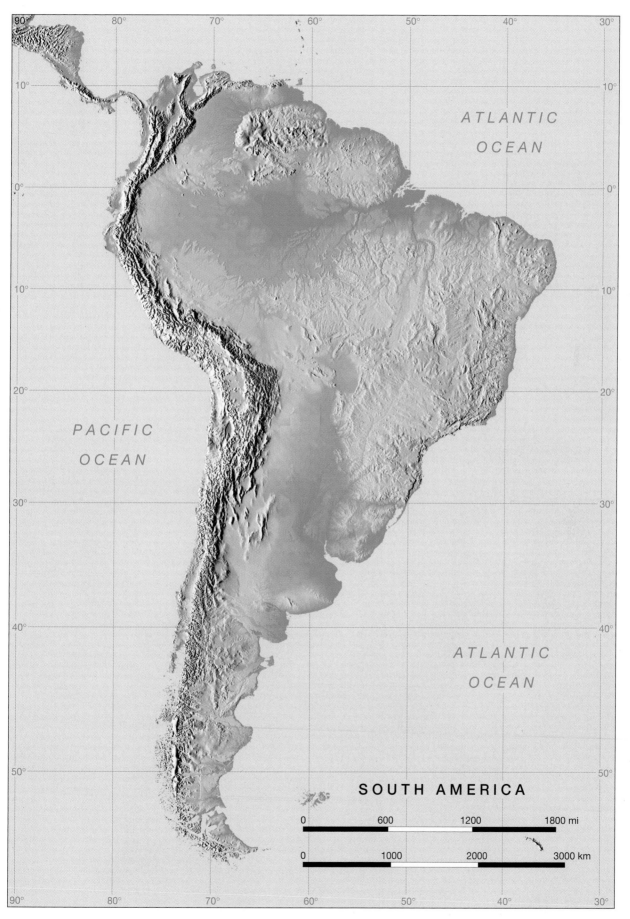

ATLANTIC
OCEAN

PACIFIC
OCEAN

ATLANTIC
OCEAN

SOUTH AMERICA

| 0 | 600 | 1200 | 1800 mi |

| 0 | 1000 | 2000 | 3000 km |

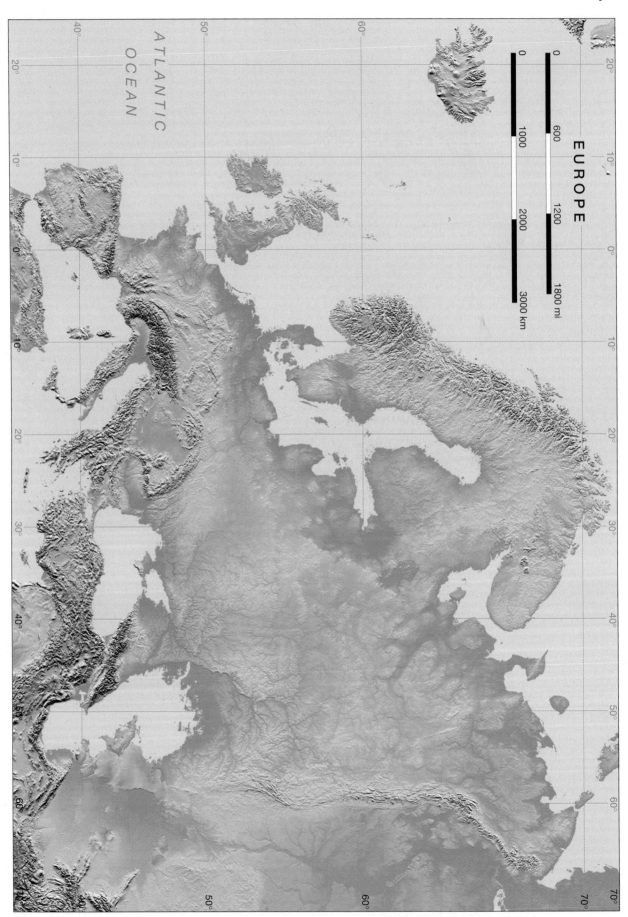

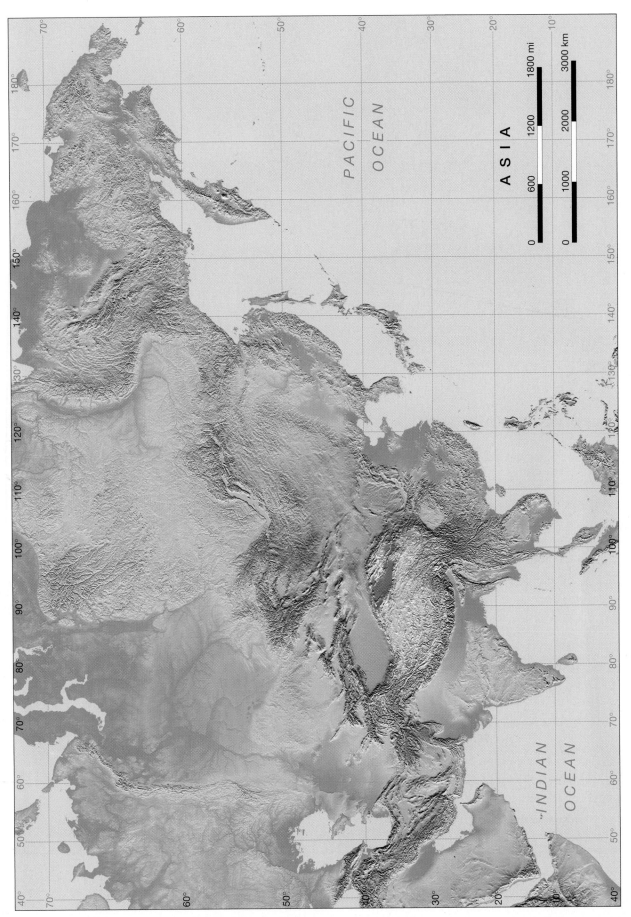

PACIFIC OCEAN

INDIAN OCEAN

A S I A

ASIA

0   600   1200   1800 mi

0   1000   2000   3000 km

ATLANTIC

OCEAN

INDIAN

OCEAN

AFRICA

| 0 | 600 | 1200 | 1800 mi |

| 0 | 1000 | 2000 | 3000 km |